Lecture Notes in Artificial Intelligence 12294

Subseries of Lecture Notes in Computer Science

Series Editors

Randy Goebel
University of Alberta, Edmonton, Canada
Yuzuru Tanaka
Hokkaido University, Sapporo, Japan
Wolfgang Wahlster
DFKI and Saarland University, Saarbrücken, Germany

Founding Editor

Jörg Siekmann
DFKI and Saarland University, Saarbrücken, Germany

More information about this series at http://www.springer.com/series/1244

Frank-Peter Schilling · Thilo Stadelmann (Eds.)

Artificial Neural Networks in Pattern Recognition

9th IAPR TC3 Workshop, ANNPR 2020
Winterthur, Switzerland, September 2–4, 2020
Proceedings

 Springer

Editors
Frank-Peter Schilling 🆔
Zurich University of Applied Sciences
ZHAW
Winterthur, Switzerland

Thilo Stadelmann 🆔
Zurich University of Applied Sciences
ZHAW
Winterthur, Switzerland

ISSN 0302-9743   ISSN 1611-3349 (electronic)
Lecture Notes in Artificial Intelligence
ISBN 978-3-030-58308-8   ISBN 978-3-030-58309-5 (eBook)
https://doi.org/10.1007/978-3-030-58309-5

LNCS Sublibrary: SL7 – Artificial Intelligence

This Springer imprint is published by the registered company Springer Nature Switzerland AG
The registered company address is: Gewerbestrasse 11, 6330 Cham, Switzerland

Preface

This volume contains the papers presented at the 9th IAPR TC3 Workshop on Artificial Neural Networks for Pattern Recognition (ANNPR 2020), held in virtual format and organized by Zurich University of Applied Sciences ZHAW, Switzerland, during September 2–4, 2020. ANNPR 2020 follows the success of the ANNPR workshops of 2003 (Florence), 2006 (Ulm), 2008 (Paris), 2010 (Cairo), 2012 (Trento), 2014 (Montreal), 2016 (Ulm), and 2018 (Siena). The series of ANNPR workshops have served as a major forum for international researchers and practitioners from the communities of pattern recognition and machine learning based on artificial neural networks.

From the 34 manuscripts submitted, the Program Committee selected 22 papers for the scientific program, organized in several sessions on foundations as well as practical applications of neural networks for pattern recognition. The workshop was enriched by four keynote presentations, given by Bernd Freisleben (University of Marburg, Germany), Pascal Paysan (Varian Medical Systems Imaging Laboratory, Switzerland), Jürgen Schmidhuber (Dalle Molle Institute for Artificial Intelligence, USI-SUPSI, Switzerland), and Naftali Tishby (Hebrew University of Jerusalem, Israel).

A broad range of topics were discussed at the workshop. In the area of foundations of machine learning and neural networks for pattern recognition, papers discussing novel developments using predictive autoencoders, support vector machines, as well as recurrent and convolutional neural networks were presented. On the other hand, applications and solutions to real-world problems were discussed for topics ranging from medical applications (pain recognition, dermatologic imaging, cancer histology, mammography), finance, industrial applications (QR-Code detection, wear detection of industrial tools, geotechnical engineering, 3D point clouds processing), text-to-speech synthesis, and face recognition to weather forecasting and crop species classification.

The planning of ANNPR 2020 was significantly impacted by the consequences of the COVID-19 pandemic, which forced us in the end to move the workshop to an online format; the first time in its history. At the time of writing this foreword, the workshop was still ahead of us, but our goal was to create an engaging and stimulating online conference.

A new feature for the series of ANNPR workshops, stimulated by the fact that the 2020 edition was hosted by a university of applied sciences, was the applied research session on the afternoon of the second day of the event, which aimed to bridge the gap between purely academic research on the one hand, and applied research as well as practical use-cases in an industry setting on the other hand. To this end, we invited companies, from start-ups to multinationals, to showcase their own research and applications of neural networks for pattern recognition, to solve real-world problems.

The workshop would not have been possible without the help of many people and organizations. We are grateful to the authors who submitted their contributions to the workshop, despite the difficult situation caused by the pandemic. We thank the members of the Program and Extended Organizing Committees for performing the difficult task of selecting the best papers from a large number of high-quality submissions.

ANNPR 2020 was supported by the International Association for Pattern Recognition (IAPR), by the IAPR Technical Committee on Neural Networks and Computational Intelligence (TC3), by the Swiss Alliance for Data-Intensive Services, and by the School of Engineering of Zurich University of Applied Sciences ZHAW. We are grateful to our sponsoring partners, who not only have financially supported us, but also contributed relevant examples for practical applications and use-cases for using neural networks in pattern recognition during the applied research session. Finally, we are indebted to the secretariat of the Institute of Applied Information Technology (InIT) at ZHAW for their administrative support. Finally, we wish to express our gratitude to Springer for publishing these proceedings within their LNCS/LNAI series.

July 2020   Frank-Peter Schilling
                                                                    Thilo Stadelmann

Organization

General Chairs

Frank-Peter Schilling   Zurich University of Applied Sciences ZHAW, Switzerland

Thilo Stadelmann   Zurich University of Applied Sciences ZHAW, Switzerland

Extended Organizing Committee

Oliver Dürr   Konstanz University of Applied Sciences, Germany

Stefan Glüge   Zurich University of Applied Sciences ZHAW, Switzerland

Gundula Heinatz-Bürki   Swiss Alliance for Data-Intensive Services, Switzerland

Martin Jaggi   École Polytechnique Fédérale de Lausanne EPFL, Switzerland

Thomas Ott   Zurich University of Applied Sciences ZHAW, Switzerland

Beate Sick   University of Zurich and ZHAW, Switzerland

Program Committee

Hazem Abbas   Ain Shams University, Egypt

Erwin Bakker   Leiden University, The Netherlands

Marcel Blattner   Tamedia Digital, Switzerland

Ricardo Da Silva Torres   Norwegian University of Science and Technology, Norway

Oliver Dürr   Konstanz University of Applied Sciences, Germany

Andreas Fischer   Haute école spécialisée de Suisse occidentale HES-SO, Switzerland

Markus Hagenbuchner   University of Wollongong, Australia

Martin Jaggi   École Polytechnique Fédérale de Lausanne EPFL, Switzerland

Marco Maggini   University of Siena, Italy

Simone Marinai   University of Florence, Italy

Elena Mugellini   Haute école spécialisée de Suisse occidentale HES-SO, Switzerland

Pascal Paysan   Varian Medical Systems Imaging Laboratory, Switzerland

Marcello Pelillo   University of Venice, Italy

Marc Pouly   Lucerne University of Applied Sciences and Arts
                              HSLU, Switzerland
Friedhelm Schwenker   Ulm University, Germany
Eugene Semenkin   Siberian State Aerospace University, Russia
Beate Sick   University of Zurich and ZHAW, Switzerland
Nicolaj Stache   Heilbronn University of Applied Sciences, Germany
Susanne Suter   Zurich University of Applied Sciences ZHAW,
                              Switzerland
Edmondo Trentin   University of Siena, Italy

Supporting Institutions

International Association for Pattern Recognition (IAPR)
Technical Committee 3 (TC 3) of the IAPR
Swiss Alliance for Data-Intensive Services, Switzerland
Zurich University of Applied Sciences ZHAW, Switzerland

Supporting Companies

Gold

Schweizerische Mobiliar Versicherungsgesellschaft AG, Switzerland

Silver

CSEM SA, Switzerland
Microsoft Schweiz GmbH, Switzerland
Roche Diagnostics International Ltd., Switzerland
turicode Inc., Switzerland

Bronze

leanBI AG, Switzerland
querdenker engineering GmbH, Germany
Schweizerische Bundesbahnen SBB AG, Switzerland
ScorePad AG, Switzerland

Contents

Contents

Invited Papers

Deep Learning Methods for Image Guidance in Radiation Therapy

Pascal Paysan$^{(\boxtimes)}$, Igor Peterlik, Toon Roggen , Liangjia Zhu,
Claas Wessels, Jan Schreier, Martin Buchacek, and Stefan Scheib

Varian Medical Systems Imaging Laboratory GmbH,
Taefernstr. 7, 5405 Daettwil, Switzerland
pascal.paysan@varian.com
http://www.varian.com/

Abstract. Image guidance became one of the most important key technologies in radiation therapy in the last two decades. Nowadays medical images play a key role in virtually every aspect of the common treatment workflows. Advances in imaging hardware and algorithmic processing are enablers for substantial treatment quality improvements like online adaptation of the treatment, accounting for anatomical changes of the day, or intra-fraction motion monitoring and organ position verification during treatment. Going through this rapid development, an important observation is that further improvements of various methods heavily rely on model knowledge. In a classical sense such model knowledge is for example provided by mathematically formulated physical assumptions to ill-posed problems or by expert systems and heuristics. Recently, it became evident that in various applications such classical approaches get outperformed by data driven machine learning methods. Especially worth to mention is that this not only holds true in terms of precision and computational performance but also in terms of complexity reduction and maintainability. In this paper we provide an overview about the different stages in the X-ray based imaging pipeline in radiation therapy where machine learning based algorithms show promising results or are already applied in clinical routine.

Keywords: Radiation therapy · X-ray imaging · Cone Beam Computed Tomography (CBCT) · Image Guided Radiation Therapy (IGRT) · During treatment motion monitoring · Tissue tracking · Automatic segmentation · Artificial intelligence · Deep learning

1 Introduction

A typical radiation therapy treatment starts with a diagnosis based on Computed Tomography (CT), Magnetic Resonance Imaging (MRI), Single-Photon Emission Computed Tomography (SPECT), Positron Emission Tomography - CT (PET/CT), or a combination of medical imaging modalities [48]. A first step in radiation therapy is typically the acquisition of a so-called simulation

© Springer Nature Switzerland AG 2020
F.-P. Schilling and T. Stadelmann (Eds.): ANNPR 2020, LNAI 12294, pp. 3–22, 2020.
https://doi.org/10.1007/978-3-030-58309-5_1

CT scan, where the patient is positioned in its treatment position. The simulation CT serves as a quantitative patient model which is used in the treatment planning process. Typical tasks are anatomy delineation, optimization of the radiation beam entrance angles and dose calculation. This acquired CT scan serves also as the baseline anatomy of the patient. During the different treatment sessions, the patient will be positioned as closely to this simulated position as possible to minimize position uncertainties. For imaging during treatment nowadays, the vast majority of radiation therapy systems rely on x-ray imaging using flat panel detectors. Figure 1 demonstrates this kV image acquisition with a patient in treatment position.

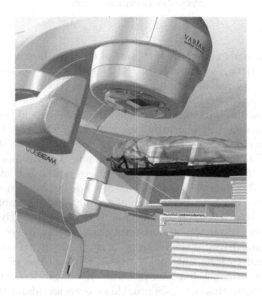

Fig. 1. Varian TrueBeam radiation therapy delivery platform, acquiring kV images of a patient in treatment position.

Reasons for that are the relatively low complexity, versatile applications, and affordability of these x-ray imaging systems. The x-ray system is used mainly for position verification relative to the treatment planning position but also for during treatment motion monitoring and rather lately for plan adaptation. Dependent on the needs for the different tasks in radiation therapy the x-ray based imaging system mounted on the treatment delivery device (on-board imaging) can be used to acquire different types of images such as 2D x-ray projections, fluoroscopic projection sequences, 3D Cone Beam CT (CBCT) images, and motion resolved 4D CBCT images. Future trends are clearly to use the on-board imaging systems for additional tasks throughout the therapy such as direct organ segmentation and direct dose calculation based on the acquired 3D or 4D CBCT data and for soft-tissue motion monitoring during treatment. Finally, the patient being under fractionated treatment (up to 40 fractions) is sometimes in parallel monitored using CT, MRI, PET/CT, or SPECT/CT to assess the treatment response. This should preferably be (at least partially) replaced by on-board imaging-based procedures available at the treatment device. Technically, the imaging pipeline of the on-board imaging system starts with the x-ray imaging hardware consisting of the x-ray tube and the flat panel detector. The detector acquires 2D projections for example as single frames (triggered images) to perform an image match against the planning CT prior each fraction or to verify the internal anatomy at certain control points during radiation beam delivery within a fraction. The projections can be as well acquired as a sequence either from a single viewing direction to verify for example certain internal motion trajectories or during rotation of the treatment delivery device (typically called gantry) around the patient to perform motion management. An elementary part of the motion management is tracking internal structures of the patient on acquired projections. For volumetric imaging the system rotates around the patient and acquires a sequence of projections which allows to reconstruct a 3D CBCT image. This includes reconstructions that are resolved with respect to phases of respiratory motion (4D CBCT) [75] and even respiratory as well as cardiac motion (5D CBCT) [68]. Having the images at hand, deformable image registration and automatic segmentation [26] are recently a topic of growing interest in the context of adaptive radiotherapy. In the following, we want to discuss where it is already state of the art or where we see the potential (or already evidence) to apply learning-based methods to improve X-ray guided radiation therapy and thus clinical outcome.

2   Motion Monitoring During Treatment

Unlike conventional radiation therapy delivery schemes that typically deliver 2 Gy per daily fraction over several weeks, Stereotactic Body Radiation Therapy (SBRT) delivers high doses in a few or even a single fraction (6–24 Gy, 1–8 fractions) [19,63]. This allows for an increased tumor control and reduces toxicity in healthy tissues. To ensure a sub-millimeter accuracy for the high dose deposition in SBRT, during-treatment motion management is indispensable. Correct patient setup on the day of the treatment is accomplished by image registration on a 3D CBCT, followed by a couch adaptation. A position verification during delivery of the fraction could be a 3D-3D match at mid- and post-delivery time with a re-acquired CBCT. Alternatively, a 2D-2D match or a 2D-3D match at certain angles can be applied. In the above cases the treatment beam is interrupted and resumed after the position had been confirmed. The drawback is that during actual treatment delivery the therapist is blind to any motion that occurs. An excellent overview of classical state of the art motion monitoring methods with the beam on is given in [6]. Of interest are the tracking methods that use kV/kV imaging with triangulation for 3D position information (CyberKnife®), Sequential Stereo (Varian Medical Systems), which sequentially acquires 2D images with a single imager, and approaches based on a Probability Density Function (PDF). Proclaimed accuracy of the Euclidean distance in 3D space is 0.33 mm on phantom data and a time delay to acquire sufficient images for the 3D reconstruction is to be considered. Alternatively, updating the raw image stack with the latest 2D acquisition, smears out the position change that occurred on the last 2D image in the 3D volume that also contains all previous images that did not undergo this motion. Methods based on a PDF map based on pre-treatment-day acquired 4DCT data [36] proclaim an accuracy of the Euclidean distance in 3D space of 1.64 ± 0.73 mm. Deep learning methods are being developed to further improve these results and are being discussed below for both bony structures and soft tissue.

2.1 Tracking of Bony Structures

For spine tumors, neighboring vertebrae can serve as anatomical landmarks and periodically acquired 2D kilovoltage (kV) images during treatment allow for a fast detection model to compare the vertebræ positions to the reference positions for a specific gantry angle. A CTV to PTV margin, representing all expected uncertainties including motion during delivery, is recommended to be <3 mm in [49]. Recommendations on patient setup accuracy (positioning of the patient on the couch before delivery of the treatment) are <2 mm for translations, <3°– 4° for roll and pitch and <10° for yaw, according to [5]. A dosimetric study for spine stereotactic treatments recommends a patient setup translational error ≤ 1 mm and a rotational error $\leq 2°$ [79] while the rotational setup error recommendation is reduced to $\leq 1°$ in [15]. Note that in all above situations a 3D setup CBCT or other volumetric verification is available. However, as the patient is not supposed to move anymore after correct setup, the above recommendations can also be projected to in-treatment position monitoring. An intrafraction study of spine SBRT treatments that acquired CBCTs during treatment delivery reports position standard deviations of up to 1.3 mm, and this for each of the three main axes: chest-back, left-right and head-feet [49].

In [66] a Deep Learning (DL) model based on Mask R-CNN (Regional Convolutional Neural Network) is described for vertebra detection of the thoracic spine (T9–T12) and the lumbar spine (L1–L4). It differs from the above methods in two key aspects: First, the model does not rely on temporal imaging information, acquired prior to the delivery time instance where the position is verified. Second: The model generalizes for vertebra in a human corpus, which means no patient-specific information is needed or models need to be trained prior to treatment delivery. The model allows for a fast structure localization (<2 Hz) on 2D kV projection images that are acquired during the VMAT treatment delivery. It allows assessing instant 2D position verification, using segmentation, along the delivery of the VMAT arc as well as sequential (delayed) 3D position verification, when the subsequent projection images are included in a Digital TomoSynthesis (DTS) or CBCT. Alternatively, making use of a stereoscopic dual imager setup, the 2D position pairs can be triangulated to obtain an instant 3D position. Intensity-Modulated Radiation Therapy (IMRT) and 3D Conformal Radiotherapy (CRT) can benefit from the DL model for fast structure localization as well, as long as 2D kV projection images are acquired. Typical model training times vary from 1–3 days on an Intel Xeon W-2102 2.90 GHz CPU with 32 GB RAM and an NVIDIA GeForce GTX 1080 Ti with 12 GB RAM. The model's accuracy to detect and estimate motion is assessed offline using the well-known Mean Average Precision (mAP) metric [57]. Although the mAP metric makes a lot of sense in the computer vision domain, from a clinical perspective there are other more important metrics to consider: In this study the motion of the 2D Centre of Mass (CoM) of the vertebra is assessed for the best model as identified by the mAP. The test data in the first assessment contains actual patient data. In addition, a patient-like full-body phantom with vertebrae (PIXY TPO-1067 [38]) in treatment position is moved in a controlled setup and the motion detection is assessed by the DL landmark detection model for vertebra and compared to its ground truth.

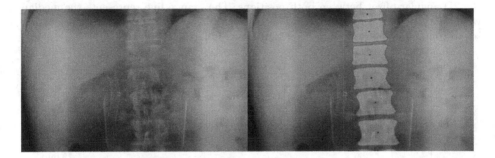

Fig. 2. Left: 2D projection image of a patient not previously seen by the model. Right: Vertebrae detected by the model and their derived CoMs (blue dots). (Color figure online)

An ordinary 2D kV projection image (Fig. 2, left) needs to be provided to the model, which returns a segmentation mask, a bounding box and a classification label (not shown) for each vertebra that is detected (Fig. 2, right). Additionally, the 2D CoM is calculated from the segmentation. Figure 3 summarizes the model performance on CoM motion detection (for isocentric shifts/rotations), based on 50 structures, detected on different projection angles uniformly distributed over the acquired arc. The motion was introduced in the head-feet (vertical) direction and the horizontal direction. Depending on the gantry angle this would correspond to a combination of a chest-back and a lateral motion. To detect a rotational change based on the CoM (a single point), at least 2 vertebrae CoMs are required. This study considers all vertebrae in the field of view. Figure 4 shows the probability of a tracking error in the range of 0–2 mm. The different curves show the probability at different shift amplitudes that were carried out as well as the probability when all shift amplitudes are evaluated together.

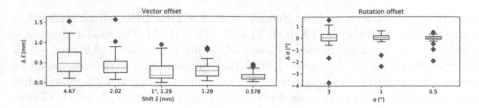

Fig. 3. Detection accuracy of CoM motion for four shifts in chest-back or lateral (equivalent to the horizontal direction on the image) and head-feet (vertical direction on the detector) direction (left graph). A combined shift-rotation was introduced as well, where its equivalent shift of the CoM in 2D was assessed. The right graph shows three rotation offsets for angular directions α (right graph). Detections were performed for 5 vertebrae fully visible on all 10 projection images of a patient previously not seen by the model (blind data set). Δs is the 2D vector offset between both positions. $\Delta \alpha$ is the 2D angular offset between both positions. The orange line represents the median value, and the box comprises all values between quantiles 1 and 3. The whiskers are set to contain all values within 1.5 x IQR (Inter-Quartile Range: Q3–Q1).

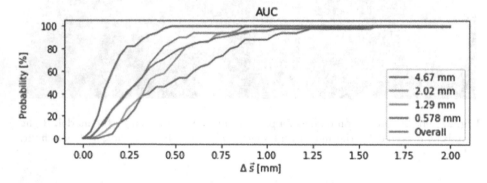

Fig. 4. Probability of a tracking error Δs in the range of 0–2 mm. The different curves show the probability at the different shift amplitudes of Fig. 2, as well as the probability when all shift amplitudes are evaluated together.

The second assessment involves the PIXY patient-like full-body phantom with vertebrae. The results for a position change detection based on a single vertebra are shown in Figure 5. The data for one shift s contains a total of 40 structure shifts, detected on projection images that are orthogonal to the chest-back or lateral axes (Gantry angles: 0°, 90°, 180° and 270°). In total three such shifts s were analyzed: 25.46 mm, 11.31 mm and 4.24 mm. Figure 6 shows the probability of a tracking error in the range of 0–2 mm. The different curves show the probability at the different shift amplitudes that were carried out as well as the probability when all shift amplitudes are evaluated together.

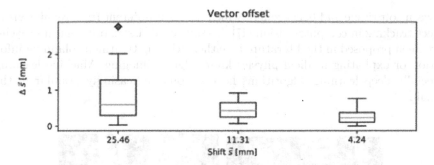

Fig. 5. Detection accuracy of CoM motion of a phantom on the treatment couch for shifts in the chest-back (x) and head-feet (y) direction, where s is the resulting shift vector. A total of 40 vertebrae positions were analyzed for each of the three shifts. Δs is the 2D vector offset between both positions. The orange line represents the median value, and the box comprises all values between quantiles 1 and 3. The whiskers are set to contain all values within 1.5 x IQR (Inter-Quartile Range: Q3–Q1).

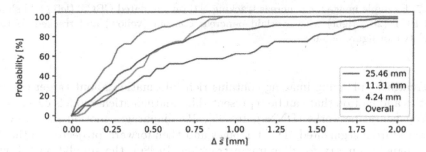

Fig. 6. Probability of a tracking error Δs in the range of 0–2 mm. The different curves show the probability at the different shift amplitudes of Fig. 5, as well as the probability when all shift amplitudes are evaluated together.

The above results show a sensitivity for positional changes in the range of 1.5 mm, with a median below 0.5 mm. Combining positional information of all vertebrae visible on a single projection image yields a sub-millimeter motion detection up to the smallest shift of 1 pixel-equivalent on the detector. The experiments with the phantom confirm these results. Spine rotations above 1° can be identified, at 0.5° detection becomes unstable.

2.2 Soft Tissue Tracking

Soft tissue position tracking is crucial in motion monitoring during radiation therapy to ensure that high dose delivery is confined to the tumor (Fig. 7) and not to the surrounding healthy tissue. Significant efforts have been made to improve the accuracy and robustness of soft tissue tracking in the past [39], where kV x-ray imaging is probably the most commonly used imaging modality for a number of practical reasons. One big challenge is the lack of sufficient contrast between soft tissue and background, which makes it different from regular visual object tracking in computer vision [47]. To tackle this issue, different approaches have been proposed in the literature by either utilizing treatment planning information or exploiting medical physics knowledge in imaging. Machine learning, especially deep learning, algorithms have become increasingly popular in this domain.

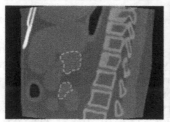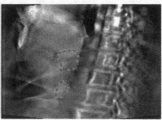

Fig. 7. Example image for pancreas tracking [41] on simulated CBCT (left) and short-arc CBCT (20°, right) with overlaid pancreas contour (yellow) and tracked contour (blue) (Color figure online)

Treatment planning imaging contains rich information about target characteristics and motion that can be represented by mathematical models or encoded in deep neural networks (DNNs). In [35], 3D diaphragm motion models were generated from segmented 4D CT images and then forward projected to the 2D X-ray panel geometry for diaphragm tracking. In [86], the simulation CT was deformed and transformed to generate enough synthetic digitally reconstructed radiographs (DRRs) along with known tumor locations. Then, a DNN was used to model the relation between DRRs and their corresponding bounding boxes of tumors. This model was applied to predict tumor locations in real projections acquired during the actual treatment.

Physics-based approaches aim at exploiting hardware advances to improve soft tissue contrast. Dual energy (DE) imaging and multi-layer detectors are among these promising technologies. For example, a fast-kV switching DE fluoroscopy was implemented on a bench top system by alternating between high and low x-ray energies. Bony anatomy was suppressed using the classical weighted logarithm subtraction (WLS) method [34]. A deep learning model was used to improve the accuracy of WLS [33]. In X-ray imaging, a stacked flat panel detector design allows to get a plurality of images with low and high signal to noise ratio (SNR) and high and low spatial resolution, respectively. Image fusion schemes are available to take advantage of such "low - high" and "signal - resolution" information to combine images together with the aim to maximize SNR of the fused image and prevent loss of spatial resolution [88].

3   CBCT Image Reconstruction

Nowadays, volumetric imaging is arguably an integral part of the workflow in radiation therapy. While initially it was mainly intended for bone-based 3D positioning of the patient, it has progressively become an important tool for soft tissue matching thanks to improvements in image quality of the reconstructed volume. Recently, these improvements allowed for adaptive radiation therapy where the treatment plan is being adapted to the anatomical changes directly during a fraction prior the actual treatment. Clearly, improving image quality is the key aspect of the successful deployment of volume reconstruction methods and the actual progress in machine learning brings new opportunities in this area.

A typical image reconstruction pipeline consists of pre-processing performed in the projection space, analytical or iterative reconstruction, and volume post-processing.

3.1   X-ray Projection Pre-processing

The pre-processing phase already provides several opportunities for a successful application of deep learning (DL) methods. A good example is the correction of signal degradation caused by X-ray photons that are scattered within the patient body [22,54]. The approach is to use Monte Carlo Methods as forward simulation of primary and scatter signal and train a U-net type regression to predict the scatter component out of the combined signal acquired by the flat panel detector.

Metal artifacts in CBCT are another prominent example where projection-based corrections with the help of DL show advances compared to classical approaches [50,52,53,60,83]. The speciality here is that the X-ray beams penetrating metal objects are affected by various physical effects, namely beam hardening, increased scatter, and high noise because of the strong attenuation. This makes it nearly impossible to use the affected information directly and thus requires to include the prior knowledge into the reconstruction process.

3.2   CBCT Volume Post-processing

An obvious application of DL methods is post-processing of CBCT volumes performed to correct for inaccuracies and artifacts. A beneficial aspect here is that the generation of training data for supervised learning is often quite straightforward: A prominent scenario is to generate training pairs with the complete projection set as ground truth and a projection subset (sparse-view) as a simulation of low dose scans [31,43,85]. All these methods apply classical filtered back-projection (FBP) to perform the first reconstruction affected by typical sparsity streaks and use neural networks, such as U-Nets, to improve the quality of the final image. The reconstruction and subsequent correction of limited-angle acquisitions have been addressed using a similar approach [70,80]. However, it has been pointed out that these approaches cannot guarantee that the output image faithfully represents the anatomy of the patient and does not fabricate fictitious structures due to the prior knowledge trained on a patient population. A possible mitigation is through the comparison of the reconstructed volume against the acquired projections by applying forward projections and enforce minimal differences [46]. A hybrid approach has been proposed in [37] to reduce artifacts related to the incompleteness of input projections due to limited-angle, sparseness and truncated acquisition: In the first phase, an U-net is employed to complete the insufficient input data. The completed set combining both the measured and computed projections is then reconstructed by conventional iterative reconstruction technique. A rather rarely addressed topic is the field of motion artifact reduction. This is presumably because of the general lack of motion-free ground truth training data. In [61] we proposed a framework for CBCT motion artifact simulation and applied it in a proof of principal study [62] to train a U-net based artifact reduction method in the image domain (see Fig. 8).

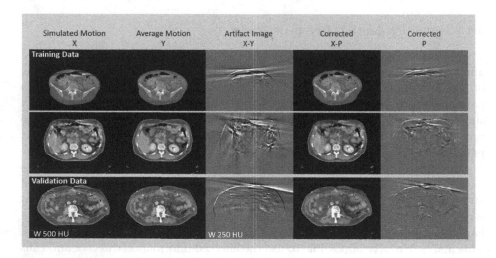

Fig. 8. Examples of training data used for DL-based motion artifact reduction [62] generated by applying 4DCT based motion simulation using recorded breathing curves [61]. The columns show from left to right the simulation with motion artifacts X, the desired output representing the average motion during scan Y, the artifact image $X - Y$, the image after applying the predicted artifact $Y - P$, and the predicted artifact P.

3.3   Iterative CBCT Reconstruction Methods

Apart from FBP, iterative methods represent a common technique for CT image reconstruction. Here the reconstruction corresponds to a step-by-step minimisation of the objective loss function $\ell(f, p, A)$ defined in terms of the reconstructed volume f, acquired projections p and the system matrix A relating the volume voxels to pixels in the projection space. The loss function

$\ell(f,p,A) = \psi(f,p,A) + r(f)$ consists of the data fidelity term $\psi(f,p,A)$ enforcing consistency of the reconstructed volume with acquired projections and the regularization term $r(f)$ encouraging reconstructions to satisfy a priori assumed properties, e.g. piece-wise smoothness. The common choices for the former include the L_2-norm projection error $||Af-p||_2^2$ [28] or the statistical loss function [20,65] taking into account the stochastic nature of signal detected in the projections. The regularization is often represented by variants of the total-variation [23]. During each iteration step, the update of f is calculated by comparing the forward-projected volume Af to the acquired projections; the mathematical formulation depends on the precise form of the objective loss function, the chosen regularizer and the iteration scheme [2,28]. Machine learning techniques can then alter this general scheme in a number of ways.

The first set of methods includes learning prior information from a training dataset containing high-quality reconstruction or alternatively general images. The learned information is then used at each iteration step to enhance the quality of limited-angle or sparse-view reconstructions. The learned information can be as simple as texture content in similar patches [42] while deploying deep neural networks allows for extracting higher-level features and for greater expressivity. In [10], a deep residual convolutional neural network (CNN) was trained for image denoising on the COCO dataset [51] and then used as a filter at each iteration step. In [13], a CNN is trained on a dataset containing high-quality reconstructions to yield ground truth images by refining unfinished iterations. The trained CNN then defines a regularization term enforcing the volume to lie close to the ground truth.

In another set of methods, each iteration step is partially replaced by a deep neural network and the whole unrolled system is trained at once, having the subsampled set of projections as an input and high-quality reconstructions as target. Examples include [77] or [11,17]; in the latter, a DenseNet-inspired network is used in each iteration step to propose an optimal volume update based on the current as well as the previous gradients of the loss function; this is in fact a generalization of the Nesterov momentum [78] used for the speedup of iterative reconstruction.

3.4   End-to-End CBCT Image Reconstruction Learning

The last class of reconstruction algorithms that we want to mention here is applying deep learning in an end-to-end approach where pre- and post-processing (in projection and volume domain) are jointly trained. Here one of the foundation papers by Würfl et al. [81] uses neural networks to learn filter and weighting in projection space while evaluating loss functions in the image domain. Other prominent examples are the previously mentioned methods for metal artifact reduction by Lin and Lyu et al. [52,53] but also for the limited angle scans [30]. Zhu et al. [87] proposed to learn the complete reconstruction process including domain transformation between the projection and volume space. That implies learning the system matrix which is normally well known.

An alternative approach to DL-based end-to-end reconstruction is presented in [24]: a continuum of intermediate representations is employed to break down the original problem, where line integrals are gradually restricted via partial line integrals until the level of image voxels is attained. The resulting hierarchy is mapped onto the network architecture, allowing for significant reduction of the computational complexity.

3.5   4D CBCT Reconstruction

In 4D reconstruction normally 10 volumes with respect to their respiratory phase are reconstructed from acquisitions with approximately the same number of projections as needed for a 3D scans and subsequently utilizes approximately the same dose. This leads to strong under-sampling of the phases and makes it even harder to obtain a certain image quality. Classical approaches try to join information from all phases by using an initial combined reconstruction (MKB) [76], temporal regularization (4DTV) [64], or by applying deformable registration between the phases (MoCo) [7].

In [12] an iterative deep learning approach derived from the 3D AirNet method [11] has been applied to reduce sparseness streaks in 4D CBCT reconstructions. Zhang et al. [84] proposes a motion compensated reconstruction algorithm applying deep learning for patient population based deformation field refinements. A method for the suppression of sparseness artifacts in cardiac CT imaging based on learned data exchange between phases with cyclic loss has been presented by Kang et al. [44].

In motion resolved reconstructions the challenge is to overcome the sparse sampling of the motion resolved images by reusing information from other motion states. This implies that the ongoing motion needs to be resolved up to a certain extent, what makes the problem even more ill-posed and therefore prior information about anatomy and physiological motion needs to be taken into account. Further challenges are to overcome the phase correlated reconstruction and to address motion amplitudes [69].

4   Deep Learning for Organ Segmentation

For the generation of a radiotherapy treatment plan, the position of the tumor as well as surrounding organs need to be known. In a typical workflow, a clinician contours these structures on either a CT or an MRI image.

Previously, the automatic segmentation of anatomical structures was performed by heuristic algorithms, such as thresholding [3] or watershed [74] and joined to be applied for a complete anatomical site [29]. However, these algorithms need to be specifically designed for each organ. The advancements in deep learning now make it feasible to generate segmentation solutions for a multitude of different anatomical structures using the same or similar underlying neural networks. These are then trained on example segmentations to adapt to the particular structure. The underlying neural network is most often a convolutional neural network as its architecture is especially suited for image-related tasks. More specifically the U-Net [67] and derivations from it, such as the Tiramisu [40] and BibNet [71], are commonly employed for anatomical segmentations.

One limitation of the above-mentioned networks is their incapability to learn strong shape priors. This leads to inadequate performance in case of weak image quality. Methods such as anatomically constrained neural networks [59], try to circumvent it by forcing the network to learn a shape representation. With the described methods above, organ segmentation algorithms have been developed for many different anatomical sites such as abdomen [9], female breast [56,71], head and neck [58], female pelvis [32], male pelvis [73] and thorax [18]. In the case of head and neck and male pelvis, performance on par with clinicians have been reported.

A special challenge is to perform automatic segmentation directly on CBCT reconstructions [1] due to its, in some aspects, inferior image quality compared to CT (see Fig. 9). Residual motion is apart from scatter one of the most prominent challenges that especially makes it hard to define the ground truth.

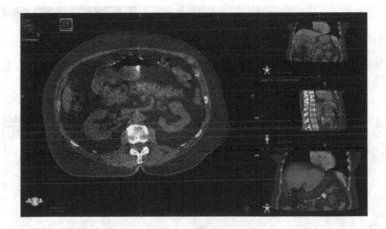

Fig. 9. Exemplary pancreas automatic segmentation result (red) on a CBCT volume [1]. (Color figure online)

The integration of these algorithms in clinical practice faces one additional hurdle: The data used for training the deep neural network may originate from a different hospital or even geography. For a CT segmentation, there is evidence that for most structures the quality remains unimpaired by training on data from a different hospital as long as the segmentation guidelines are identical [72]. For MRI segmentation, the quality improves if data from the employing hospital is included in the training or as part of an on-boarding process [27]. An alternative way to overcome this challenge is the implementation of a distributed learning method that is able to leverage the data from multiple hospitals in a privacy-preserving manner [14].

5   Deformable Image Registration

It is of upmost importance in radiotherapy that the prescribed dose is being delivered to the target as conformal as possible while sparing neighboring organs at risk.

Patients anatomy changes from day to day. This might lead to misadministrating the dose, thus not fulfilling the clinical goals. To avoid this, adaptive radiotherapy was introduced (ART) [45]. Here, a patient image of the day is used to update the deprecated treatment plan. Furthermore, one needs to track the absorbed dose in each organ to ensure the correct dose coverage of the tumor and not overdosing the risk organs during the whole treatment based on multiple fractions. This process is called dose accumulation.

In both steps, deformable image registration (DIR) is being used. DIR is morphing the original image to the updated image set of the day. The "path" of each voxel is saved as a 3D vector in a deformation vector field (see Fig. 10) which can later be used to deform the dose as well for accumulation.

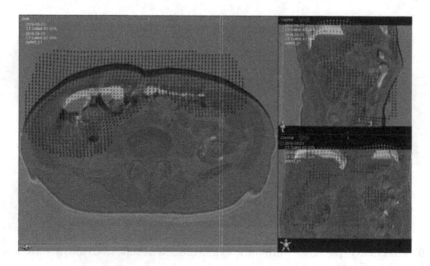

Fig. 10. A deformation vector field (arrows) to morph one CT (grey in the foreground) to the other CT (black in the background) of the same patient in different breathing phases.

However, due to the vast number of voxels that need to be compared and moved around based on optimization algorithms, a conventional DIR [8] can take up to minutes until it finishes. That is the first opportunity for DL to help. As soon as the patient is positioned and the images of the day are recorded everything should be fast to start dose delivery to avoid anatomy changes [21] or patient position changes. DL based DIR can be done within a fraction of a second [4,16] for 3D image volumes, which is a considerable improvement over, e.g. 1 min.

Another point where DL might help to improve the results is later in the process of dose accumulation. Classical DIR algorithms follow a fixed set of parameters and therefore perform better or worse depending on the image which needs to be deformed. Furthermore, there are indefinite ways to deform one image to another and thus, no ground truth actually exists. That is why unsupervised learning is normally being used for these kind of projects, since supervised models would only mimic the behavior of the classical algorithm, i.e. copying its problems and uncertainties.

Unsupervised learning detects patterns on its own, and thus might be able to outperform existing solutions [4, 55].

Several architectures are being used and tested [25]. First approaches relied on typical U-Nets [4] and newer publications are looking into the potential of GANs [82] to either generate the morphed image directly or "just" the deformation vector field.

6   Conclusion

As shown on several examples from the image guided radiation therapy field we see enormous potential of data driven methods to enhance or overcome state of the art algorithms. This can be observed in various stages of the imaging pipeline. Notably, the biggest improvements can be observed where learning based methods are used under consideration of domain knowledge (e.g. x-ray imaging physics) over pure black-box applications. We rate this as motivation to further explore problem specific network architectures and loss functions to obtain solutions that leverage physical or physiological constraints to reduce the solution space for the training process. Beyond this we see a lot of synergies between the described domain solutions where integrated solutions like e.g. deformable registration with implicit segmentation or image reconstruction with implicit deformable registration against prior acquisitions could be future developments. In conclusion we sense a wide agreement in the scientific community that deep learning will be the next evolutionary step in the field.

References

1. Adamson, P.M., Arrate, F., Jordan, P.: Evaluation of abdominal autosegmentation versus inter-observer variability on a high-speed ring gantry CBCT system. In: AAPM Annual Meeting, San Antonio, TX (2019)
2. Andersen, A., Kak, A.: Simultaneous algebraic reconstruction technique (SART): a superior implementation of the ART algorithm. Ultrason. Imaging **6**(1), 81–94 (1984)
3. Bae, K.T., Giger, M.L., Chen, C.T., Kahn Jr., C.E.: Automatic segmentation of liver structure in CT images (1993)
4. Balakrishnan, G., Zhao, A., Sabuncu, M.R., Guttag, J., Dalca, A.V.: VoxelMorph: a learning framework for deformable medical image registration. IEEE Trans. Med. Imaging **38**(8), 1788–1800 (2019). https://doi.org/10.1109/tmi.2019.2897538

5. Benedict, S.H., et al.: Stereotactic body radiation therapy: the report of AAPM Task Group 101. Med. Phys. **37**(8), 4078–4101 (2010)
6. Bertholet, J., Knopf, A., et al.: Real-time intrafraction motion monitoring in external beam radiotherapy. Phys. Med. Biol. **64**(15), 15TR01 (2019)
7. Brehm, M., Paysan, P., Oelhafen, M., Kachelrieß, M.: Artifact-resistant motion estimation with a patient-specific artifact model for motion-compensated cone-beam CT. Med. Phys. **40**(10), 101913 (2013)
8. Brown, L.G.: A survey of image registration techniques. ACM Comput. Surv. **24**(4), 325–376 (1992)
9. Cai, J., Xia, Y., Yang, D., Xu, D., Yang, L., Roth, H.: End-to-end adversarial shape learning for abdomen organ deep segmentation. In: Suk, H.-I., Liu, M., Yan, P., Lian, C. (eds.) MLMI 2019. LNCS, vol. 11861, pp. 124–132. Springer, Cham (2019). https://doi.org/10.1007/978-3-030-32692-0_15
10. Chen, B., Xiang, K., Gong, Z., Wang, J., Tan, S.: Statistical iterative CBCT reconstruction based on neural network. IEEE Trans. Med. Imaging **37**(6), 1511–1521 (2018)
11. Chen, G., et al.: AirNet: fused analytical and iterative reconstruction with deep neural network regularization for sparse-data CT. Med. Phys. (2020). https://doi.org/10.1002/mp.14170
12. Chen, G., Zhao, Y., Huang, Q., Gao, H.: 4D-AirNet: a temporally-resolved CBCT slice reconstruction method synergizing analytical and iterative method with deep learning. Phys. Med. Biol. (2020). https://doi.org/10.1088/1361-6560/ab9f60
13. Chun, I., Huang, Z., Lim, H., Fessler, J.: Momentum-Net: fast and convergent iterative neural network for inverse problems. arXiv preprint arXiv:1907.11818, July 2019
14. Czeizler, E., et al.: Using federated data sources and Varian Learning Portal framework to train a neural network model for automatic organ segmentation. Physica Medica **72**, 39–45 (2020)
15. Dahele, M., Verbakel, W.: Treatment planning, intrafraction monitoring and delivery: linear accelerator-based stereotactic spine radiotherapy. Stereotact Body Radiat Ther Spinal Metastasis Future Medicine Ltd., pp. 37–55 (2014)
16. De Vos, B.D., Berendsen, F.F., Viergever, M.A., Sokooti, H., Staring, M., Išgum, I.: A deep learning framework for unsupervised affine and deformable image registration. Med. Image Anal. **52**, 128–143 (2019)
17. Ding, Q., Chen, G., Zhang, X., Huang, Q., Ji, H., Gao, H.: Low-dose CT with deep learning regularization via proximal forward backward splitting. Phys. Med. Biol. (2020). https://doi.org/10.1088/1361-6560/ab831a
18. Dong, X., et al.: Automatic multiorgan segmentation in thorax CT images using U-Net-GAN. Med. Phys. **46**, 2157–2168 (2019)
19. Dunne, E.M., Fraser, I.M., Liu, M.: Stereotactic body radiation therapy for lung, spine and oligometastatic disease: current evidence and future directions. Ann. Transl. Med. **6**(14), 283 (2018). https://doi.org/10.21037/atm.2018.06.40
20. Elbakri, I.A., Fessler, J.A.: Statistical image reconstruction for polyenergetic X-ray computed tomography. IEEE Trans. Med. Imaging **21**(2), 89–99 (2002)
21. Elmahdy, M.S., et al.: Robust contour propagation using deep learning and image registration for online adaptive proton therapy of prostate cancer. Med. Phys. **46**, 3329–3343 (2019)
22. Erath, J., Vöth, T., Maier, J., Kachelrieß, M.: Forward and cross-scatter estimation in dual source CT using the deep scatter estimation (DSE). In: Medical Imaging 2019: Physics of Medical Imaging, vol. 10948, p. 24. International Society for Optics and Photonics (2019). https://doi.org/10.1117/12.2512718

23. Erdogan, H., Fessler, J.A.: Ordered subsets algorithms for transmission tomography. Phys. Med. Biol. **44**(11), 2835–2851 (1999)
24. Fu, L., De Man, B.: A hierarchical approach to deep learning and its application to tomographic reconstruction. In: 15th International Meeting on Fully Three-Dimensional Image Reconstruction in Radiology and Nuclear Medicine, vol. 11072, p. 1107202. International Society for Optics and Photonics (2019)
25. Fu, Y., Lei, Y., Wang, T., Curran, W.J., Liu, T., Yang, X.: Deep learning in medical image registration: a review. Phys. Med. Biol. (2020). https://doi.org/10.1088/1361-6560/ab843e
26. Fu, Y., et al.: Pelvic multi-organ segmentation on cone-beam CT for prostate adaptive radiotherapy. Med. Phys. (2020). https://doi.org/10.1002/mp.14196
27. Gibson, E., et al.: Inter-site variability in prostate segmentation accuracy using deep learning. In: Frangi, A.F., Schnabel, J.A., Davatzikos, C., Alberola-López, C., Fichtinger, G. (eds.) MICCAI 2018. LNCS, vol. 11073, pp. 506–514. Springer, Cham (2018). https://doi.org/10.1007/978-3-030-00937-3_58
28. Gordon, R., Bender, R., Herman, G.: Algebraic Reconstruction Techniques (ART) for three-dimensional electron microscopy and X-ray photography. J. Theor. Biol. **29**(3), 471–481 (1970)
29. Haas, B., Coradi, T., et al.: Automatic segmentation of thoracic and pelvic CT images for radiotherapy planning using implicit anatomic knowledge and organ-specific segmentation strategies. Phys. Med. Biol. **53**(6), 1751–1771 (2008)
30. Hammernik, K., Würfl, T., Pock, T., Maier, A.: A deep learning architecture for limited-angle computed tomography reconstruction. In: Maier-Hein, K.H., Deserno, T.M., Handels, H., Tolxdorff, T. (Hrsg.) Bildverarbeitung für die Medizin 2017. INFORMAT, pp. 92–97. Springer, Heidelberg (2017). https://doi.org/10.1007/978-3-662-54345-0_25
31. Han, Y.S., Yoo, J., Ye, J.C.: Deep residual learning for compressed sensing CT reconstruction via persistent homology analysis. arXiv preprint arXiv:1611.06391, November 2016. http://arxiv.org/abs/1611.06391
32. Hänsch, A., Dicken, V., Grass, T., Morgas, T., Klein, J., Meine, H.: Deep learning based segmentation of organs of the female pelvis in CBCT scans for adaptive radiotherapy using CT and CBCT data. Comput. Assist. Radiol. Surg. CARS **2018**, 133 (2018)
33. Haytmyradov, M., et al.: Adaptive weighted log subtraction based on neural networks for markerless tumor tracking using dual energy fluoroscopy. Med. Phys. **47**(2), 672–680 (2020)
34. Haytmyradov, M., et al.: Markerless tumor tracking using fast-kV switching dual-energy fluoroscopy on a benchtop system. Med. Phys. **46**(7), 3235–3244 (2019)
35. Hindley, N., Keall, P., Booth, J., Shieh, C.: Real-time direct diaphragm tracking using kV imaging on a standard linear accelerator. Med. Phys. **46**(10), 4481–4489 (2019)
36. Hirai, R., Sakata, Y., Tanizawa, A., Mori, S.: Real-time tumor tracking using fluoroscopic imaging with deep neural network analysis. Physica Medica **59**, 22–29 (2019)
37. Huang, Y., Preuhs, A., Manhart, M., Lauritsch, G., Maier, A.: Data consistent CT reconstruction from insufficient data with learned prior images. arXiv preprint arXiv:2005.10034 (2020)
38. Radiology Support Devices Inc.: PIXY: Anthropomorphic Phantoms - Radiology Support Devices. http://rsdphantoms.com/radiology/anthropomorphic-phantoms/. Accessed 7 Oct 2019
39. Jaffray, D.: Image-guided radiotherapy: from current concept to future perspectives. Nat. Rev. Clin. Oncol. **9**(12), 688 (2012)

40. Jégou, S., Drozdzal, M., Vázquez, D., Romero, A., Bengio, Y.: The one hundred layers Tiramisu: fully convolutional DenseNets for semantic segmentation. CoRR abs/1611.09326 (2016). http://arxiv.org/abs/1611.09326

41. Jeung, A., Zhu, L., Mostafavi, H., van Heteren, J.: What image features are good for correlation-based tracking algorithms used for soft tissue monitoring in X-ray imaging. In: AAPM Annual Meeting, San Antonio, TX (2019)

42. Jia, X., et al.: Statistical CT reconstruction using region-aware texture preserving regularization learning from prior normal-dose CT image. Phys. Med. Biol. **63**(22), 225020 (2018)

43. Jin, K.H., McCann, M.T., Froustey, E., Unser, M.: Deep convolutional neural network for inverse problems in imaging. IEEE Trans. Image Process. **26**(9), 4509–4522 (2017)

44. Kang, E., Koo, H.J., Yang, D.H., Seo, J.B., Ye, J.C.: Cycle-consistent adversarial denoising network for multiphase coronary CT angiography. Med. Phys. **46**(2), 550–562 (2019)

45. Keall, P.J., Hsu, A., Xing, L.: Image-guided adaptive radiotherapy. In: Hoppe, R.T., Phillips, T.L., Roach, M. (eds.) Leibel and Phillips Textbook of Radiation Oncology, 3rd edn., pp. 213–223. W.B. Saunders, Philadelphia (2010)

46. Kofler, A., Haltmeier, M., Kolbitsch, C., Kachelrieß, M., Dewey, M.: A U-Nets cascade for sparse view computed tomography. In: Knoll, F., Maier, A., Rueckert, D. (eds.) MLMIR 2018. LNCS, vol. 11074, pp. 91–99. Springer, Cham (2018). https://doi.org/10.1007/978-3-030-00129-2_11

47. Kristan, M., et al.: A novel performance evaluation methodology for single-target trackers. IEEE Trans. Pattern Anal. Mach. Intell. **38**(11), 2137–2155 (2016)

48. Lecchi, M., Fossati, P., Elisei, F., Orecchia, R., Lucignani, G.: Current concepts on imaging in radiotherapy. Eur. J. Nucl. Med. Mol. Imaging **35**(4), 821–837 (2008)

49. Li, W., Sahgal, A., Foote, M., Millar, B.A., Jaffray, D.A., Letourneau, D.: Impact of immobilization on intrafraction motion for spine stereotactic body radiotherapy using cone beam computed tomography. Int. J. Radiat. Oncol. Biol. Phys. **84**(2), 520–526 (2012)

50. Liao, H., Lin, W.A., Zhou, S.K., Luo, J.: ADN: artifact disentanglement network for unsupervised metal artifact reduction. IEEE Trans. Med. Imaging **39**(3), 634–643 (2020)

51. Lin, T., et al.: Microsoft COCO: common objects in context (2014). http://arxiv.org/abs/1405.0312

52. Lin, W.A., et al.: DuDoNet: dual domain network for CT metal artifact reduction. In: Proceedings of the IEEE Conference on Computer Vision and Pattern Recognition, pp. 10504–10513 (2020)

53. Lyu, Y., Lin, W.A., Lu, J., Zhou, S.K.: DuDoNet++: encoding mask projection to reduce CT metal artifacts. arXiv preprint arXiv:2001.00340 (2020)

54. Maier, J., Sawall, S., Kachelrieß, M.: Deep scatter estimation (DSE): feasibility of using a deep convolutional neural network for real-time x-ray scatter prediction in cone-beam CT. In: Medical Imaging, vol. 10573, pp. 393–398. SPIE (2018). https://doi.org/10.1117/12.2292919

55. Mansilla, L., Milone, D.H., Ferrante, E.: Learning deformable registration of medical images with anatomical constraints. Neural Netw. **124**, 269–279 (2020). https://doi.org/10.1016/j.neunet.2020.01.023

56. Men, K., et al.: Fully automatic and robust segmentation of the clinical target volume for radiotherapy of breast cancer using big data and deep learning. Physica Medica **50**, 13–19 (2018)

57. Microsoft: COCO - Common Objects in Context. http://cocodataset.org/# detection-eval. Accessed 26 Sept 2019

58. Nikolov, S., et al.: Deep learning to achieve clinically applicable segmentation of head and neck anatomy for radiotherapy. CoRR abs/1809.04430 (2018)

59. Oktay, O., et al.: Anatomically Constrained Neural Networks (ACNN): application to cardiac image enhancement and segmentation. CoRR abs/1705.08302 (2017). http://arxiv.org/abs/1705.08302

60. Park, H.S., Lee, S.M., Kim, H.P., Seo, J.K., Chung, Y.E.: CT sinogram-consistency learning for metal-induced beam hardening correction. Med. Phys. **45**(12), 5376–5384 (2018)

61. Paysan, P., Munro, P., Scheib, S.G.: CT based simulation framework for motion artifact and ground truth generation of cone-beam CT. In: AAPM Annual Meeting, San Antonio, TX (2019)

62. Paysan, P., Strzelecki, A., Arrate, F., Munro, P., Scheib, S.G.: Convolutional network based motion artifact reduction in cone-beam CT. In: AAPM Annual Meeting, San Antonio, TX (2019)

63. Potters, L., et al.: American society for therapeutic radiology and oncology and american college of radiology practice guideline for the performance of stereotactic body radiation therapy. Int. J. Radiat. Oncol. Biol. Phys. **60**(4), 1026–1032 (2004)

64. Ritschl, L., Sawall, S., Knaup, M., Hess, A., Kachelrie, M.: Iterative 4D cardiac micro-CT image reconstruction using an adaptive spatio-temporal sparsity prior. Phys. Med. Biol. **57**(6), 1517–1525 (2012)

65. Rockmore, A.J., Macovski, A.: A maximum likelihood approach to emission image reconstruction from projections. IEEE Trans. Nucl. Sci. **23**(4), 1428–1432 (1976)

66. Roggen, T., Bobic, M., Givehchi, N., Scheib, S.G.: Deep Learning model for markerless tracking in spinal SBRT. Physica Medica Eur. J. Med. Phys. **74**, 66–73 (2020)

67. Ronneberger, O., Fischer, P., Brox, T.: U-Net: convolutional networks for biomedical image segmentation. CoRR abs/1505.04597 (2015). http://arxiv.org/abs/1505.04597

68. Sauppe, S., Hahn, A., Brehm, M., Paysan, P., Seghers, D., Kachelrieß, M.: PO-0934: cardio-respiratory motion compensation for 5D thoracic CBCT in IGRT. Radiother. Oncol. **119**, S452–S453 (2016)

69. Sauppe, S., Kuhm, J., Brehm, M., Paysan, P., Seghers, D., Kachelrieß, M.: Motion vector field phase-to-amplitude resampling for 4D motion-compensated cone-beam CT. Phys. Med. Biol. **63**(3), 035032 (2018)

70. Schnurr, A.K., Chung, K., Russ, T., Schad, L.R., Zöllner, F.G.: Simulation-based deep artifact correction with convolutional neural networks for limited angle artifacts. Zeitschrift fur Medizinische Physik **29**(2), 150–161 (2019)

71. Schreier, J., Attanasi, F., Laaksonen, H.: A full-image deep segmenter for CT images in breast cancer radiotherapy treatment. Front. Oncol. **9**, 677 (2019)

72. Schreier, J., Attanasi, F., Laaksonen, H.: Generalization vs. specificity. In: which cases should a clinic train its own segmentation models? Front. Oncol. **10**, 675 (2020)

73. Schreier, J., Genghi, A., Laaksonen, H.: Clinical evaluation of a full-image deep segmentation algorithm for the male pelvis on cone-beam CT and CT. Radiother. Oncol. **145**, 1–6 (2020)

74. Shojaii, R., Alirezaie, J., Babyn, P.: Automatic lung segmentation in CT images using watershed transform. In: Proceedings of the International Conference on Image Processing, ICIP (2005)

75. Sonke, J.J., Zijp, L., Remeijer, P., van Herk, M.: Respiratory correlated cone beam CT. Med. Phys. **32**(4), 1176–1186 (2005)
76. Star-Lack, J., et al.: A modified McKinnon-Bates (MKB) algorithm for improved 4D cone-beam computed tomography (CBCT) of the lung. Med. Phys. **45**(8), 3783–3799 (2018)
77. Vishnevskiy, V., Rau, R., Goksel, O.: Deep variational networks with exponential weighting for learning computed tomography. In: Shen, D., Liu, T., Peters, T.M., Staib, L.H., Essert, C., Zhou, S., Yap, P.-T., Khan, A. (eds.) MICCAI 2019. LNCS, vol. 11769, pp. 310–318. Springer, Cham (2019). https://doi.org/10.1007/978-3-030-32226-7_35
78. Wang, A.S., Stayman, J.W., Otake, Y., Vogt, S., Kleinszig, G., Siewerdsen, J.H.: Accelerated statistical reconstruction for C-arm cone-beam CT using Nesterov's method. Med. Phy. **42**(5), 2699–2708 (2015)
79. Wang, H., et al.: Dosimetric effect of translational and rotational errors for patients undergoing image-guided stereotactic body radiotherapy for spinal metastases. Int. J. Radiat. Oncol. Biol. Phys. **71**(4), 1261–1271 (2008)
80. Wang, J., Liang, J., Cheng, J., Guo, Y., Zeng, L.: Deep learning based image reconstruction algorithm for limited-angle translational computed tomography. PLoS ONE **15**(1), e0226963 (2020)
81. Würfl, T., Ghesu, F.C., Christlein, V., Maier, A.: Deep learning computed tomography. In: Ourselin, S., Joskowicz, L., Sabuncu, M.R., Unal, G., Wells, W. (eds.) MICCAI 2016. LNCS, vol. 9902, pp. 432–440. Springer, Cham (2016). https://doi.org/10.1007/978-3-319-46726-9_50
82. Zhang, X., Jian, W., Chen, Y., Yang, S.T.: Deform-GAN: an unsupervised learning model for deformable registration. ArXiv abs/2002.11430 (2020)
83. Zhang, Y., Yu, H.: Convolutional neural network based metal artifact reduction in X-ray computed tomography. IEEE Trans. Med. Imaging **37**(6), 1370–1381 (2018)
84. Zhang, Y., Huang, X., Wang, J.: Advanced 4-dimensional cone-beam computed tomography reconstruction by combining motion estimation, motion-compensated reconstruction, biomechanical modeling and deep learning. Vis. Comput. Ind. Biomed. Art **2**(1), 1–15 (2019)
85. Zhang, Z., Liang, X., Dong, X., Xie, Y., Cao, G.: A sparse-view CT reconstruction method based on combination of DenseNet and deconvolution. IEEE Trans. Med. Imaging **37**(6), 1407–1417 (2018)
86. Zhao, W., et al.: Markerless pancreatic tumor target localization enabled by deep learning. Int. J. Radiat. Oncol. Biol. Phys. **105**(2), 432–439 (2019)
87. Zhu, B., Liu, J.Z., Rosen, B.R., Rosen, M.S.: Image reconstruction by domain transform manifold learning. Nature **555**(7697), 487–492 (2018)
88. ZHu, L., Baturin, P.: Deep neural network image fusion without using training data. In: AAPM ePoster Library (2019)

Intentional Image Similarity Search

Nikolaus Korfhage, Markus Mühling, and Bernd Freisleben(⊠)

Department of Mathematics and Computer Science, University of Marburg,
Hans-Meerwein-Straße 6, 35032 Marburg, Germany
{korfhage,muehling,freisleb}@informatik.uni-marburg.de

Abstract. Image representations learned by deep convolutional neural networks (CNNs) have greatly improved the performance of content-based image retrieval systems in recent years. Query-by-example is the most popular strategy to represent a user's search intention in content-based image and video retrieval scenarios. Nevertheless, simply presenting an image as a query is often insufficient to express a user's intention, due to ambiguities in the query image. To better meet a user's search intention, a novel intentional image similarity search approach is proposed, consisting of a scheme for specifying a user's intention for a query image, a plugin mechanism to support more fine-grained neural network models for specific search, and a hybrid feature method based on CNN and handcrafted features. Furthermore, a novel analysis technique for deep similarity networks is introduced for the purpose of finding relevant image regions. The proposed approach is evaluated qualitatively on video recordings of the German Broadcasting Archive.

Keywords: Intentional gap · Image similarity search · Hybrid feature method

1 Introduction

A fundamental problem of content-based image retrieval is to overcome the discrepancy between the information that can be extracted from visual data and the human interpretation of the same data. In the literature, this discrepancy is also known as the semantic gap [18]. Using state-of-the-art convolutional neural network (CNN) features has brought us close to the goal of bridging this gap. Image representations learned by deep neural networks can greatly improve the performance of content-based image retrieval systems. They are less dependent on pixel intensities and are better suited for searching semantic content.

In addition to the semantic gap, there is an intentional gap that describes the coincidence between a query and a user's intention. Query-by-example is the most popular, most intuitive, and most expressive strategy to describe a user's search intention in content-based image and video retrieval scenarios. Nevertheless, simply presenting an image as a query is often insufficient to express a user's intention. For example, the pictures in Fig. 1 show query images presented to the database of the German Broadcasting Archive, i.e., an institution

© Springer Nature Switzerland AG 2020
F.-P. Schilling and T. Stadelmann (Eds.): ANNPR 2020, LNAI 12294, pp. 23–35, 2020.
https://doi.org/10.1007/978-3-030-58309-5_2

that maintains the cultural heritage of the television broadcasts of the former German Democratic Republic (GDR). In all presented query images, the user's intention is not clear:

(a)   (b)   (c)

Fig. 1. Query images.

In Fig. 1 (a), possible search intentions are:

– Crowd
– Person
– Katarina Witt
– Autograph signing session
– Katarina Witt signing autographs
– Katarina Witt in figure skating dress

In Fig. 1 (b), possible search intentions are:

– Simson scooter
– Is the woman important?
– Is the layout important (woman sitting on a scooter at a parking area with Trabant cars in front of a house)?

In Fig. 1 (c), possible search intentions are:

– Motorbike
– Vintage Motorbike
– Moped
– Simson moped
– Simson S51 moped
– Black Simson S51 moped
– Motorbike on meadow with trees in the background

The mismatch between a user's intention and a query image can be attributed to the following factors: region of interest, layout, and specificity. It is often unclear whether the user is interested in only part of the image or in the whole scene. Furthermore, there is the question of whether the layout is important. For example, is it important in query image Fig. 1(b) that the woman is sitting on a scooter in front of parking cars with a house in the background? Finally, the specificity of the query is an important factor to capture the user's search intention. It ranges from the general concept of the query image (e.g., motorbike), to the specific object, person, or scene (e.g., Simson S51), to a duplicate of the image. In this context, color, texture, and shape are further attributes to specify a user's intention. In all these cases, more user interactions are necessary to specify a user's intention.

In this paper, a novel similarity search approach is presented that uses intentional constraints to capture regions of interest, layout, and specificity. These constraints are defined by user interactions using region labeling and checkboxes for layout and specificity. Regarding specificity, we have to distinguish between the hierarchy of classes (e.g., car → Volkswagen → Golf) and the attributes like color, texture, and shape. The best solution for specific classes are fine-grained models trained for subcategories of e.g., persons, cars, dogs, or flowers. However, this solution is not really scalable, since there are thousands of classes that would need to be mapped with fine granularity. Therefore, we realized a hybrid approach where CNN features are combined with handcrafted features (e.g., color moments and SIFT descriptors) to handle specificity constraints, while fine-grained similarity modules can be plugged in for the most important classes such as, e.g., persons.

The hybrid approach operates in two stages. In the first stage, CNN features are used to find semantically similar content. These images are re-ranked in the second stage using handcrafted features. These features are extracted from image regions that are responsible for the high semantic similarity score between the query and the result image. The responsible image regions (called heat maps) are calculated using a novel technique for visualizing deep similarity networks. These regions are also used to realize the layout constraint by comparing the heat maps.

The contributions of the paper are as follows:

- A new query specification scheme is presented that allows to clarify the search intention of the user presenting the query image.
- A hybrid method using CNN and handcrafted features is presented to realize intentional image similarity search.
- A novel analysis technique for deep similarity networks is introduced to find the relevant image regions.

The paper is organized as follows. Section 2 discusses related work. Section 3 describes the data of the German Broadcasting Archive to which the intentional content-based image similarity approach is applied. In Sect. 4, we present the proposed novel intentional image similarity search approach, including the query specification scheme, the plugin mechanism for fine-grained models, as well as the relevant region extraction approach. Experimental results are presented in Sect. 5. Section 6 concludes the paper and outlines areas for future work.

2 Related Work

In multimedia search, Kofler et al. [11] distinguish between the topical dimension ("what" is the user searching for) and the intent dimension ("why" is the user searching). While the terms "intent" and "intention" are used synonymously in the literature, Kofler et al. [11] distinguish between the "intent" as the "immediate reason, purpose, or goal behind a user's information need" [9] and the "intention" which describes the information need as a whole. In the case of content-based image similarity search, we consider the intention gap as the coincidence between the query image and the user's intention. This is often reflected in the ambiguity of the query image. The term "intent" goes deeper and considers, for example, conceptual models of user intent which are built based on click-through data of user sessions, query log analysis, or user profiles exploiting long-term search behaviors. While there is a wide range of intent-aware approaches in the field of text and multimedia information retrieval [5,6,11] that are mainly based on keyword or text queries, less research effort has been devoted to intentional image similarity search.

In this paper, we focus on content-based image retrieval with query-by-example. Query-by-content based on feature representations learned by deep CNNs have greatly increased the performance of content-based image retrieval systems [19], since they are less dependent on pixel intensities and better represent the semantic content of the images. Thus, they try to bridge the semantic gap between the data representation and the human interpretation.

In addition to the semantic gap, there is an intentional gap that describes the coincidence between a query and a user's intention. In general, the intentional gap is due to ambiguities in the query image. As already described in Sect. 1, a search image is often insufficient to express a user's intention concerning region of interest, layout, and specificity. The query image, for example, could contain image regions that are not part of a user's search intention. Furthermore, the specificity of the query image has to be clarified.

Relevance feedback is a commonly used technique to narrow down a user's search intention. In this scenario, a user interacts with the search engine to evaluate an initial retrieval result. The additional relevant and non-relevant labeled images are used in an iterative process to refine the retrieval results. An overview of relevance feedback in image retrieval is given by Zhou and Huang [24].

Bian et al. [3] use a query suggestion approach for query-by-example image search to specify a user's intention. Given a query image, informative attributes reflecting visual properties of the query image are suggested to the user as complements to the query. By selecting some suggested attributes in a feedback session, a user can clarify his or her search intention.

The approach of Guan and Qui [8] learns a user's intention in an interactive image retrieval process. Given a query image, the relevant image regions are inferred both from the query and from multiple relevance feedback images using local image patch appearance prototypes. These relevant regions are then used to refine the ranking result.

Zhang et al. [20] present a semantic concept approach. The authors organize the semantic concepts into a hierarchy and augment each concept with a set of related attributes (e.g., round, red, shiny). The queries are mapped onto the concept hierarchy with attributes, and user feedback is collected to refine the ranking results.

Other possibilities of specifying the search intention are query expansions using, for example, multiple query images [1,2], additional keywords, or text descriptions [10,16].

To the best of our knowledge, there are no deep similarity search approaches dealing with query image ambiguities.

3 German Broadcasting Archive

Our hybrid feature approach for intentional image similarity search has been applied to historical video recordings of the German Broadcasting Archive (DRA). The DRA maintains the cultural heritage of television broadcasts of the former German Democratic Republic (GDR). It was founded in 1952 as a charitable foundation and joint institution of the Association of Public Broadcasting Corporations in the Federal Republic of Germany (ARD).

The archive contains film documents of former GDR television productions from the first broadcast in 1952 until its cessation in 1991. It includes a total of around 100,000 broadcasts, such as: contributions and recordings of the daily news program *Aktuelle Kamera*; political magazines such as *Prisma* or *Der schwarze Kanal*; broadcaster's own TV productions including numerous films, film adaptations and TV series productions such as *Polizeiruf 110*; entertainment programs (e.g., *Ein Kessel Buntes*); children's and youth programs (fairy tales, *Elf 99*); as well as advice and sports programs.

The DRA provides access to this valuable collection of scientifically relevant videos. The uniqueness and importance of the material fosters a large scientific interest in the video content. Access to the archive is granted to scientific, educational and cultural institutions, to public service broadcasting companies and, to a limited extent, to commercial organizations and private persons. The video footage is often used in film and multimedia productions. Furthermore, there is a considerable international research interest in GDR and German-German history. International scientists use the DRA for their research in the fields of psychology, media, social, political or cultural science.

The DRA is answering a wide range of time-consuming research requests. However, finding similar images in large multimedia archives is manually infeasible. Therefore, the DRA aims to digitize and index the entire video collection to facilitate search in images and videos. In this context, content-based image retrieval using query-by-example is a powerful tool to make the valuable information in the archive findable.

4   A Novel Approach to Intentional Image Similarity Search

In this section, the proposed intentional similarity search approach is presented. To better meet a user's search intentions, the approach is based on a plugin mechanism for fine-grained similarity search modules and a hybrid approach based on CNN and handcrafted features. Further query specifications are required to capture the search intention of a user presenting a query image. The query specification scheme and the mapping of queries to similarity search modules is presented in Sect. 4.1. Section 4.2 presents the introduced hybrid approach using deep CNN and handcrafted features. The plugin mechanism using the example of similarity search for faces is described in Sect. 4.3.

4.1   Query Specification

To disambiguate a query, a user is offered several options for further specifying his or her search intention associated with the presented query image.

First, the user is allowed to specify the region of interest to exclude irrelevant parts of the query image.

Second, the user can choose whether (s)he is looking for a specific concept (e.g., a VW Golf). This option is available if one of the plugins for fine-grained search detects the corresponding general concept (e.g., a car).

Third, the user has the possibility to select color, texture, and shape as additional query conditions to clarify the search intention. The selected features are automatically extracted from the region of interest and used to refine the initial ranking results of either the general or the specific deep similarity search model. The hybrid feature approach is described in the next section.

Fourth, the user can tell the similarity search system that the layout is important. The layout is considered by comparing the relevant image regions between the query and the retrieved images. The relevant region extraction approach is presented in Sect. 4.2.

Finally, the user can perform a duplicate search. For this purpose, all constraints must be satisfied.

Altogether, multiple selected conditions are weighted accordingly in the distance function at the re-ranking stage.

4.2   Hybrid Feature Method

The proposed hybrid method operates in two stages.

In the first stage, CNN features are used to find semantically similar content. To be scalable to millions of images, this stage relies on compact representations of the images for fast computation of image similarity by the Hamming distance. The deep similarity search approach used in the first stage is described in the next paragraph.

If color, texture or shape are selected as additional query conditions, the results of the first stage are re-ranked in the second stage, using handcrafted features.

Deep Similarity Search. In our work, we rely on the visual modality and focus on large-scale semantic similarity search. Since high-dimensional CNN features are not suitable to efficiently search in very large databases, large-scale similarity search systems focus on binary image codes for compact representations and fast comparisons rather than full CNN features. Binary codes enable fast distance computation in the Hamming space. Furthermore, the distance computation complexity is reduced by Multi-Index Hashing [14]. We use a model that is trained to generate 256 bit binary codes for fast image retrieval. First, a NASNet [25], pretrained on ImageNet [7], is trained on ImageNet and the Places 205 dataset [23]. Before the final classification layer. A *tanh* activation layer is integrated which produces the 256-dimensional codes. Next, the model is trained for a few epochs with a smaller learning rate.

Handcrafted Features. The definition of similarity ranges from pixel-based similarity to semantic similarity. The latter corresponds to human understanding. The definition and optimization of similarity functions is subject to current research [4,12]. In the following, color histograms are proposed to compute the similarity of color distributions, GIST features [15] are used to measure texture similarity, and SIFT features [13] are employed for shapes. The handcrafted features are extracted from image regions that are responsible for the high semantic similarity score at the first stage. For this purpose, a new method is introduced to find these regions, as described in the following paragraph. To compute the similarity between the relevant regions of two images, the Euclidean distance of the normalized feature vectors is calculated. For multiple conditions, the distances of the corresponding feature vectors are weighted accordingly.

Relevant Region Extraction. To re-rank the retrieval list, we extract relevant regions by a method similar to *Class Activation Maps* (CAM) [22]. The idea is to use the output activations of the query image to detect relevant regions in images of the retrieval list. The method requires that the last convolutional layer is followed by a global average pooling layer. Global average pooling outputs the spatial average of the feature map of the last convolutional layer. Thus, each output in the final layer is a weighted sum of the pooled vector. CAM can then be generated by applying the weights corresponding to a specific output class to weight each feature map of the last convolutional layer. In contrast to CAM, we use the averaged weights of the 256-dimensional coding layer output of the query image to weight the final feature maps of the retrieval image.

For an image r in the retrieval list, $f_k^r(x,y)$ represents the activation of unit k in the last convolutional layer at spatial location (x,y). For this unit, global average pooling of the convolutional layer with depth C results in $F_k^r = \frac{1}{C}\sum_{i=1}^{C} f_k^r(x,y)_i$. For each output unit in the subsequent deep hashing layer, the score is computed as $\sum_k w_{k,s}^r F_k^r$, where w_k^r is the weight between the output s and unit k. Likewise, w_k^q is obtained a for query image q. As in CAM, the bias term is ignored. Finally, the activation map used for extracting regions is obtained by

$$M(x,y) = \sum_k \sum_s w_{k,s}^q f_k^r(x,y). \tag{1}$$

Fig. 2. Heatmaps visualized in retrieval results.

Figure 2 shows a heatmap visualization of the upscaled feature maps for several retrieval images weighted by the query image.

4.3 Plugin Mechanism

The best solution for specific classes are fine-grained models trained for subcategories, such as faces, car models, bird species or dog breeds. Since this approach is not scalable to thousands of classes, fine-grained models are only integrated for the most important and most frequently used query contents. The user-defined query image regions are analyzed by the search engine to detect classes for which more fine-grained modules exist. Each module has to provide two components: a detection component and a component that generates the binary codes from the image region. In an interactive user session, as described in Sect. 4.1, the user specifies whether (s)he is searching for the general class or for the automatically detected specific content.

Each module has its own index. If a module is activated, the general search index is replaced by the specific index of the fine-grained model from the corresponding submodule.

In the following, the fine-grained module for faces/persons is presented in more detail. Similar to the general model, the same deep hashing approach, as described in Sect. 4.2, is used to generate binary codes for large-scale similarity search. In contrast to the general model, the face module follows a two-stage approach. In the first stage, the faces are detected using a joint face detection and alignment approach based on multitask cascaded convolutional networks [21]. For this purpose, a publicly available implementation is used[1]. After aligning the face regions, a deep hashing model is used to generate the binary codes. We use the pretrained weights of the publicly available FaceNet model [17] trained on the CASIA-WebFace dataset. We extended the architecture of this model by a coding layer and fine-tuned it on the same dataset extended by training samples for 100 persons from the DRA dataset to adjust the model to the keyframes of the historical video recordings.

---

[1] https://github.com/davidsandberg/facenet/tree/master/src/align.

Fig. 3. Specific retrieval results using the fine-grained face module in comparison to the general search results.

While in the indexing phase the binary codes for all detected faces of an image are calculated and fed into the corresponding index, in the search phase only the largest face within the query image region is used to generate the binary query code.

5   Experimental Results

The proposed intentional image search approach was evaluated experimentally on the keyframes of the video recordings of the German Broadcasting Archive. While the currently digitized index contains more than 10 million keyframes, the retrieval results had to be restricted to 400,000 keyframes due to associated rights of use of the keyframes.

We evaluate our approach qualitatively for two use cases below.

In the first use case, we consider the plugin mechanism and investigate the impact of the face/person similarity module. For this purpose, a face detection and recognition module was integrated into the similarity search system. This allows the user to specify whether general concepts are important, or if (s)he is interested in a specific concept - a person's face in this case.

The experiments were performed on two indices, one for general semantic similarity, another one for face similarity. The general index contains hash codes for all 400,000 keyframes, while the person recognition based index contains about 300,000 image codes, each one representing a detected face.

Fig. 4. Retrieval of deep similarity search (query images in top row).

Figure 3 shows the top retrieval results for query images for both the general search and the specific search, which involves a subsequent face detection and recognition step. The detection component of the face plugin reliably detected the faces both in the query and database images. While the general search results contains similar images that contain persons with the same shot size, the specific search delivers the same person in different scenarios with high accuracy.

In the second use case, we evaluated the hybrid feature approach by the example of color histograms. If a user decides that color is an important property in the query image, the retrieval list is re-ranked, as described in Sect. 4.2. Figure 4 shows the retrieval for some query images. Figure 5 shows the same retrieval list when color is of interest to the user, re-ranked according to the similarity of color histograms extracted from the relevant regions. This example shows how retrieval results of CNNs trained on (high level) semantic concepts can be refined by re-ranking them according to selected low level features.

In both cases, the additional specification of the query leads to results that reflect a user's intention.

Fig. 5. Retrieval of deep similarity search, re-ranked by local color histograms (query images in top row).

6   Conclusion

In this paper, a novel intentional image similarity search approach was proposed to better meet a user's search intention. For this purpose, a new scheme for specifying a user's intention with respect to a query image was introduced. The best solution for specific classes are fine-grained models trained for sub-categories, such as faces. Since this approach is not scalable to thousands of classes, a plugin mechanism was integrated to support specific search for the most important concepts. Furthermore, a hybrid feature method based on CNN and handcrafted features was used to consider color, texture, and shape. In this context, a novel analysis method for deep similarity networks was presented for the purpose of finding relevant image regions. Finally, the proposed system was evaluated qualitatively on video recordings of the German Broadcasting Archive, showing promising results.

There are several areas for future work. First, runtime improvements are necessary for the relevant region and handcrafted feature extraction stages. Second, a user could be interested in searching for multiple regions in the query image, e.g., for two specific persons. For this purpose, the approach has to be extended by region specific options. Finally, further combinations of CNN and handcrafted features should be evaluated to satisfy a user's intentions.

Acknowledgement. This work is financially supported by the Deutsche Forschungs-gemeinschaft (DFG, FR 791/15-1).

References

1. Al-Mohamade, A., Bchir, O., Ben Ismail, M.M.: Multiple query content-based image retrieval using relevance feature weight learning. J. Imaging **6**(1), 2 (2020)
2. Arandjelović, R., Zisserman, A.: Multiple queries for large scale specific object retrieval. In: BMVC 2012 - Electronic Proceedings of the British Machine Vision Conference 2012, pp. 1–11 (2012)
3. Bian, J., Zha, Z.J., Zhang, H., Tian, Q., Chua, T.S.: Visual query attributes suggestion. In: MM 2012 - Proceedings of the 20th ACM International Conference on Multimedia, pp. 869–872 (2012)
4. Blanco, G., et al.: A label-scaled similarity measure for content-based image retrieval. In: 2016 IEEE International Symposium on Multimedia (ISM), pp. 20–25. IEEE (2016)
5. Cai, J., Zha, Z.J., Wang, M., Zhang, S., Tian, Q.: An attribute-assisted reranking model for web image search. IEEE Trans. Image Process. **24**(1), 261–272 (2015)
6. Deng, C., Ji, R., Tao, D., Gao, X., Li, X.: Weakly supervised multi-graph learning for robust image reranking. IEEE Trans. Multimed. **16**(3), 785–795 (2014)
7. Deng, J., Dong, W., Socher, R., Li, L.J., Li, K., Fei-Fei, L.: ImageNet: a large-scale hierarchical image database. In: 2009 IEEE Conference on Computer Vision and Pattern Recognition, pp. 248–255. IEEE (2009)
8. Guan, J., Qiu, G.: Learning user intention in relevance feedback using optimization. In: Proceedings of the ACM International Multimedia Conference and Exhibition, pp. 41–50 (2007)
9. Hanjalic, A., Kofler, C., Larson, M.: Intent and its discontents (2012)
10. Jarrar, R., Belkhatir, M.: On the coupled use of signal and semantic concepts to bridge the semantic and user intention gaps for visual content retrieval. Int. J. Multimed. Inf. Retr. **5**(3), 165–172 (2016)
11. Kofler, C., Larson, M., Hanjalic, A.: User intent in multimedia search: a survey of the state of the art and future challenges. ACM Comput. Surv. **49**(2), 1–37 (2016)
12. Liang, R.Z., et al.: Optimizing top precision performance measure of content-based image retrieval by learning similarity function. In: 2016 23rd International Conference on Pattern Recognition (ICPR), pp. 2954–2958. IEEE (2016)

13. Lowe, D.G.: Object recognition from local scale-invariant features. In: Proceedings of the Seventh IEEE International Conference on Computer Vision, vol. 2, pp. 1150–1157. IEEE (1999)
14. Norouzi, M., Punjani, A., Fleet, D.J.: Fast search in hamming space with multi-index hashing. In: 2012 IEEE Conference on Computer Vision and Pattern Recognition, pp. 3108–3115. IEEE (2012)
15. Oliva, A., Torralba, A.: Modeling the shape of the scene: a holistic representation of the spatial envelope. Int. J. Comput. Vis. **42**(3), 145–175 (2001)
16. Piras, L., Giacinto, G.: Information fusion in content based image retrieval: a comprehensive overview. Inf. Fusion **37**, 50–60 (2017)
17. Schroff, F., Kalenichenko, D., Philbin, J.: FaceNet: a unified embedding for face recognition and clustering. In: Proceedings of the IEEE Computer Society Conference on Computer Vision and Pattern Recognition, 07–12 June, pp. 815–823, March 2015
18. Smeulders, A.W., Worring, M., Santini, S., Gupta, A., Jain, R.: Content-based image retrieval at the end of the early years. IEEE Trans. Pattern Anal. Mach. Intell. **22**(12), 1349–1380 (2000)
19. Wan, J., et al.: Deep learning for content-based image retrieval: a comprehensive study. In: Proceedings of the 22nd ACM International Conference on Multimedia, pp. 157–166 (2014)
20. Zhang, H., Zha, Z.J., Yang, Y., Yan, S., Gao, Y., Chua, T.S.: Attribute-augmented semantic hierarchy: towards bridging semantic gap and intention gap in image retrieval. In: Proceedings of the 21st ACM International Conference on Multimedia, pp. 33–42 (2013)
21. Zhang, K., Zhang, Z., Li, Z., Member, S., Qiao, Y., Member, S.: Joint face detection and alignment using multitask cascaded convolutional networks. IEEE Signal Process. Lett. **23**(10), 1499–1503 (2016)
22. Zhou, B., Khosla, A., Lapedriza, A., Oliva, A., Torralba, A.: Learning deep features for discriminative localization. In: Proceedings of the IEEE Conference on Computer Vision and Pattern Recognition, pp. 2921–2929 (2016)
23. Zhou, B., Lapedriza, A., Khosla, A., Oliva, A., Torralba, A.: Places: a 10 million image database for scene recognition. IEEE Trans. Pattern Anal. Mach. Intell. **40**(6), 1452–1464 (2017)
24. Zhou, X.S., Huang, T.S.: Relevance feedback in image retrieval: a comprehensive review. Multimed. Syst. **8**(6), 536–544 (2003)
25. Zoph, B., Le, Q.V.: Neural architecture search with reinforcement learning. arXiv preprint arXiv:1611.01578 (2016)

13. Tan, D.C. Object recognition from local scale-invariant features. In: Proceedings of the Seventh IEEE International Conference on Computer Vision, vol. 2, pp. 1150–1157. IEEE (1999)

14. Simonyan, K., Zisserman, A.: Very deep convolutional networks for large-scale image recognition. In: 2015 Conference on Computer Vision and Pattern Recognition, pp. 1–14 (2015)

15. Ghosh, A., Vastella, A.: Practical prediction of the future holistic perception for the spatial mapping, ... Proceedings 43(3), 123–135 (2017)

16. Han, D., Chen, C.L.: bibliometric analysis for sustainable innovation ... perspective works, 191–199. ACM (2019) (2020)

17. Szegedy, C., Ioannou, V., et al. Going deeper with convolutions. In: Proceedings of the IEEE Computer Society Conference on Computer Vision and Pattern Recognition, pp. 1–9 (2015), pp. 1–9

18. Szegedy, A.N., Vanjung, M., Sandler, S., Girshick, R.: Content-based image retrieval, ... and of the early 2000. LibLL. Proc. Pattern Anal. Mach. Intell. 11(3), 248–250 (2018)

19. Wu, A.: Deep learning for content- and life-long ... of a multiplicative analysis for the refinement. In: IEEE ... International Conference on Acoustics, ... (2019)

20. Zhang, H., Zhi, F.L., Yang, C., Cao, T.L.: ... with the ... for the ... coded trade-offs ... a ... approach. In: ... group, pp. 1–9. Proc. ... Chen, ... International Conference on Multimedia, pp. 1–9 (2019)

21. Simonyan, K., Li, D., Xie, S., et al., Sun, Y.: Attention-based ... for ... neural networks ... deep convolutional neural networks. IEEE Trans. Proc. 1–10, 279–2320 (2015)

22. Zhang, F.M., ... impact ... and visual ... for ... as a contribution to increase in the ownership of the IEEE Conference ... Conference ... and Pattern Recognition, pp. 2921–2929 (2016)

23. Zhou, B., Lapedriza, A., Korolova, A., Oliva, A., Torralba, A.: Places: an 10 million image database for scene recognition. IEEE Trans. Pattern Anal. Mach. Intell. 40(6), 1452–1464 (2017)

24. Zhou, X., Wang, M., ... methods to prediction ... A.A.: A survey on ... computing in big data, pp. 1–9, pp. 1–9 (2018)

25. Tan, S., Dai, L.: ... data structures and features ... with a ... in a ... of ... computing, pp. 1–15 (2017)

Foundations

Structured (De)composable Representations Trained with Neural Networks

Graham Spinks[⊠][iD] and Marie-Francine Moens[iD]

Department of Computer Science, KU Leuven, Leuven, Belgium
{graham.spinks,sien.moens}@cs.kuleuven.be

Abstract. The paper proposes a novel technique for representing templates and instances of concept classes. A template representation refers to the generic representation that captures the characteristics of an entire class. The proposed technique uses end-to-end deep learning to learn structured and composable representations from input images and discrete labels. The obtained representations are based on distance estimates between the distributions given by the class label and those given by contextual information, which are modeled as environments. We prove that the representations have a clear structure allowing to decompose the representation into factors that represent classes and environments. We evaluate our novel technique on classification and retrieval tasks involving different modalities (visual and language data).

Keywords: Composable representations · Deep learning · Multimodal

1 Introduction

We propose a novel technique for representing templates and instances of concept classes that is agnostic with regard to the underlying deep learning model. Starting from raw input images, representations are learned in a classification task where the cross-entropy classification layer is replaced by a fully connected layer that is used to estimate a bounded approximation of the distance between each class distribution and a set of contextual distributions that we call 'environments'. By defining randomized environments, the goal is to capture common sense knowledge about how classes relate to a range of differentiating contexts, and to increase the probability of encountering distinctive diagnostic features. This idea loosely resembles how human long-term memory might achieve retrieval [7] as well as how contextual knowledge is used for semantic encoding [6]. Our experiments confirm the value of such an approach.

In this paper, classes correspond to (visual) object labels, and environments correspond to combinations of contextual labels given by either object labels or image caption keywords. Representations for individual inputs, which we call 'instance representations', form a 2D matrix with rows corresponding to classes

F.-P. Schilling and T. Stadelmann (Eds.): ANNPR 2020, LNAI 12294, pp. 39–51, 2020.
https://doi.org/10.1007/978-3-030-58309-5_3

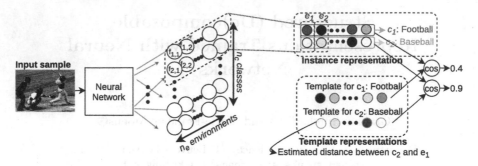

Fig. 1. The last layer of a convolutional neural network is replaced with fully-connected layers that map to $n_c \times n_e$ outputs $f_{i,j}$ that are used to create instance representations that are interpretable along contextual dimensions, which we call 'environments'. By computing the cosine similarity, rows are compared to corresponding class representations, which we refer to as 'templates'.

and columns corresponding to environments, where each element is an indication of how much the instance resembles the corresponding class versus environment. The parameters for each environment are defined once at start by uniformly selecting a randomly chosen number of labels from the power set of all available contextual labels. The class representation, which we refer to as 'template', has the form of a template vector. It contains the average distance estimates between the distribution of a class and the distributions of the respective environments. By computing the cosine similarity between the instance representation and all templates, class membership can be determined efficiently (Fig. 1).

Template and instance representations are interpretable as they have a fixed structure comprised of distance estimates. This structure is reminiscent of traditional language processing matrix representations and enables operations that operate along matrix dimensions. We demonstrate this with a Singular Value Decomposition (SVD) which yields components that determine the values along the rows (classes) and columns (environments) respectively. Those components can then be altered to modify the information content, upon which a new representation can be reconstructed. The proposed representations are evaluated in four settings: (1) Multi-label image classification, *i.e.*, object recognition with multiple objects per image; (2) Image retrieval where we query images that look like existing images but contain altered class labels; (3) Single-label image classification on pre-trained instance representations for a previously unseen label; (4) Rank estimation with regard to compression of the representations.

Contributions. (1) We propose a new deep learning technique to create structured representations from images, entity classes and their contextual information (environments) based on distance estimates. (2) This leads to template representations that generalize well, as successfully evaluated in a classification task. (3) The obtained representations are interpretable as distances between a class and its environment. They are composable in the sense that they can be modified to reflect different class membership as shown in a retrieval task.

2   Background

We shortly discuss useful background related to different aspects of our research.

Representing Entities with Respect to Context. In language applications, structured matrices (*e.g*, document-term matrices) have been used for a long time. Such matrices can be decomposed with SVD or non-negative matrix factorization. Low-rank approximations are found with methods like latent semantic indexing. Typical applications are clustering, classification, retrieval, etc. with the benefit that outcomes can usually be interpreted with respect to the contextual information. Contrary to our work, earlier methods build representations purely from labels and don't take deep neural network based features into account. More recently [11] create an unsupervised sentence representation where each entity is a probability distribution based on co-occurrence of words.

Distances to Represent Features. The Earth Mover's Distance (EMD) also known as Wasserstein distance, is a useful metric based on the optimal transport problem to measure the distance between distributions. [3] use a similar idea to define the Word Mover's Distance (WMD) that measures the minimal amount of effort to move Word2Vec-based word embeddings from one document to another. The authors use a matrix representation that expresses the distance between words in respective documents. They note the structure is interpretable and performs well on text-based classification tasks.

Random Features. The Word Mover's Embedding [14] is an unsupervised feature representation for documents, created by concatenating WMD estimates that are computed with respect to arbitrarily chosen feature maps. The authors calculate an approximation of the distance between a pair of documents with the use of a kernel over the feature map. The building blocks of the feature map are documents built from an arbitrary combination of words. This idea is based on Random Features approximation [9] that suggests that a randomized feature map is useful for approximating a shift-invariant kernel.

Our work can be viewed as a combination of the above ideas: we use distance estimates to create interpretable, structured representations of entities with respect to their contexts. The contextual dimension consists of features that are built from an arbitrary combination of discrete labels. Our work most importantly differs in the following manners: (1) We use end-to-end deep neural network training to include rich image features when building representations; (2) Information from different modalities (visual and language) can be combined.

3   CoDiR: Method

We first define some notions that are useful to understand the method, which we name Composable Distance-based Representation learning (CoDiR).

Setup and Notations. Given is a dataset with data samples x $\sim p_{\text{data}}$, with non-exclusive class labels $c_i, i \in \{1, ..., n_c\}$ which in this work are visual object labels (*e.g.*, dog, ball, ...). Image instances s are fed through a (convolutional) neural network N. The outputs of N will serve to build templates $T_{i,:} \in \mathbb{R}^{n_e}$ and instance representations $D \in \mathbb{R}^{n_c \times n_e}$ with n_e a hyperparameter denoting the amount of environments. Each environment will be defined with the use of discrete environment labels $l_k, k \in \{1, ..., n_l\}$, for which we experiment with two types: (1) the same visual object labels as used for the class labels (such that $n_l = n_c$) and (2) image caption keywords from the set of the n_l most common nouns, adjectives or verbs in the sentence descriptions in the dataset. We will refer to the first as '**CoDiR (class)**' and the latter as '**CoDiR (capt)**'.

1_{c_i} is shorthand for the indicator function $1_{c_i}(x) = 1$ if $x \in C_i$, 0 otherwise, with C_i the set of images with label c_i. Similarly we denote 1_{l_k}. Each element $D_{i,j}$ is a distance estimate between distributions p_{c_i} and p_{e_j}, where p_{c_i} is shorthand for $p(x = x, x \in C_i)$. Informally, p_{c_i} is the joint distribution modeling the data distribution and class membership c_i. To obtain p_{e_j}, several steps are performed before training. First, hyperparameter R is set, giving the maximum amount of labels per environment. For the j-th environment, we then (1) sample the actual amount of labels $r_j \sim U[1, R] \in \mathbb{N}$; (2) sample the labels $l_m^{(j)}$, with $m \in \{1, ..., r_j\}$, uniformly without replacement from the set of all discrete environment labels $l_k, k \in \{1, ..., n_l\}$. Now E_j, the set of images for environment e_j, is given by $E_j = \cup_{m=1}^{r_j} L_m^{(j)}$ with $L_m^{(j)}$ the set of images with label $l_m^{(j)}$. Thus, similarly to p_{c_i}, we have $p_{e_j} = p(x = x, x \in E_j)$. Note that by sampling a random amount of labels per environment, as inspired by [14], we ensure diversity in the type of composition of environments, with some holding many labels and some few.

Contextual Distance. We propose to represent each image as a 2D feature map that relates distributions of classes to environments. A suitable metric should be able to deal with neural network training as well as potentially overlapping distributions. A natural candidate is a Wasserstein-based distance function [1]. A key advantage is that the critic can be encouraged to maximize the distance between two distributions, whereas metrics based on Kullback-Leibler (KL) divergence are not well defined if the distributions have a negligible intersection [1]. In comparison to other neural network-based distance metrics, the Fisher IPM provides particularly stable estimates and has the advantage that any neural network can be used as f as long as the last layer is a linear, dense layer [5]. The Fisher GAN formulation bounds $\mathcal{F}$, the set of measurable, symmetric and bounded real valued functions by defining a data dependent constraint on its second order moments. The IPM is given by:

$$dF_{\mathcal{F}}(p_{e_j}, p_{c_i}) = \sup_{f_{i,j} \in \mathcal{F}} \frac{\mathbb{E}_{x \sim p_{e_j}}[f_{i,j}(x)] - \mathbb{E}_{x \sim p_{c_i}}[f_{i,j}(x)]}{\sqrt{1/2\mathbb{E}_{x \sim p_{e_j}} f_{i,j}^2(x) + 1/2\mathbb{E}_{x \sim p_{c_i}} f_{i,j}^2(x)}} \tag{1}$$

In practice, the Fisher IPM is estimated with neural network training where the numerator in Eq. 1 is maximized while the denominator is expressed as a constraint, enforced with a Lagrange multiplier. While the Fisher IPM is an estimate of the chi-squared distance, the numerator can be viewed as a bounded estimate of the inter-class distance, closely related to the Wasserstein distance [5]. From now on, we denote this approximation of the inter-class distance as the 'distance'. During our training, critics $f_{i,j}$ are trained from input images to maximize the Fisher IPM for distributions p_{c_i} and p_{e_j}, $\forall i \in \{1, ..., n_c\}, \forall j \in \{1, ..., n_e\}$. The numerator then gives the distance between p_{c_i} and p_{e_j}. We denote $T \in \mathbb{R}^{n_c \times n_e}$, with $T_{i,j} = \underset{x \sim p_{e_j, train}}{\mathbb{E}} [f_{i,j}(x)] - \underset{x \sim p_{c_i, train}}{\mathbb{E}} [f_{i,j}(x)]$, i.e., the evaluation of the estimated distances over the training set. Intuitively, one can see why a matrix T with co-occurrence data contains useful information. A subset of images containing 'cats', for example, will more closely resemble a subset containing 'dogs' and 'fur' than one containing 'forks' and 'tables'.

Template and Instance Representations. As the template representation for class c_i, we simply use the corresponding row of the learned distance matrix: $T_{i,:}$. Each element $T_{i,j}$ gives an average distance estimate for how a class c_i relates to environment e_j, where smaller values indicate that class and environment are similar or even (partially) overlap. For the instance representation for an input s we then propose to use $D \in \mathbb{R}^{n_c \times n_e}$ with elements given by Eq. 2:

$$D_{i,j}^{(s)} = \underset{x \sim p_{e_j, train}}{\mathbb{E}} [f_{i,j}(x)] - f_{i,j}(s) \qquad (2)$$

where $f_{i,j}(s)$ is simply the output of critic $f_{i,j}$ for the instance s. The result is that for an input s with class label c_i, $D_{i,:}^{(s)}$ is correlated to $T_{i,:}$ as its distance estimates with respect to all different environments should be similar. Therefore, the cosine similarity between vector $D_{i,:}^{(s)}$ and the template $T_{i,:}$ will be large for input samples from class i, and small otherwise.

Such templates can be evaluated, for example, in multi-label classification tasks (see Sect. 4). Finding the classes for an image is then simply calculated by computing whether $\forall c_i$, $cos(D_{i,:}^{(s)}, T_{i,:}) > t_{c_i}$ with t_{c_i} a threshold (the level of which is determined during training). From here on we will use a shorthand notation $D^{(s)} \subset c_i$ to denote $cos(D_{i,:}^{(s)}, T_{i,:}) > t_{c_i}$, and $D^{(s)} \not\subset c_i$ otherwise.

Implementation. Training $n_c \times n_e$ critics is not feasible in practice, so we pass input images through a common neural network for which the classification layer is replaced by $n_c \times n_e$ single layer neural networks, the outputs of which constitute $f_{i,j}$ (see Fig. 1). During training, any given mini-batch will contain inputs with many different c_i and e_j. To maximize Eq. 1 efficiently, instead of feeding a separate batch for the samples of $x \sim p_{c_i}$ and $x \sim p_{e_j}$, we use the same mini-batch. Additionally, instead of directly sampling $x \sim p_{c_i}$ we multiply each output $f_{i,j}$ with a mask $M_{i,j}^c$ where $M_{i,j}^c = 1_{c_i}$. Similarly, for $x \sim p_{e_j}$ we multiply each output $f_{i,j}$ with a mask $M_{i,j}^e$ where $M_{i,j}^e = \sum_{m=1}^{r_j} 1_{1_m^{(j)}}$. The result is that instances then are weighted according to their label prevalence as

Algorithm 1. Algorithm of the training process. For matrices and tensors, $\times$ refers to matrix multiplication and $*$ refers to element-wise multiplication.

Inputs: images s, class labels c, environment labels l

$\forall j \in \{1, ..., n_e\}$: $r_j \sim U[1, R] \in \mathbb{N} \wedge \forall m \in \{1, ..., r_j\}$, $1_m^{(j)} \sim U[1, n_l]$

Create $\boldsymbol{V} \in \mathbb{N}^{n_l \times n_e}$ which has value 1 for each uniformly selected label, 0 otherwise.

Init $\boldsymbol{\lambda} = 0 \in \mathbb{R}^{n_c \times n_e} \wedge$ Init weights in neural network N

while Training **do**

    Sample a mini-batch b, with batch size n_b, containing images s and binary class labels $\boldsymbol{C_b} \in \mathbb{N}^{n_b \times n_c}$ and binary environment labels $\boldsymbol{L_b} \in \mathbb{N}^{n_b \times n_l}$.

    Create masks

    Expand $\boldsymbol{C_b}$ into $\boldsymbol{M^c} \in \mathbb{N}^{n_b \times n_c \times n_e}$, s.t. $\boldsymbol{M^c}_{k,i,:} = \mathbf{1}_{c_i}(s_k)$ for the k-th sample s_k.

    Multiply $\boldsymbol{L_b}$ and $\boldsymbol{V}$, then expand the result into $\boldsymbol{M^e} \in \mathbb{N}^{n_b \times n_c \times n_e}$, s.t. $\boldsymbol{M^e}_{k,:,j} = \sum_{m=1}^{r_j} \mathbf{1}_{1_m^{(j)}}(s_k)$ for the k-th sample s_k.

    Calculate the FISHER GAN loss

    Propagate b through N to obtain $\boldsymbol{O_f} \in \mathbb{R}^{n_c \times n_e}$ containing all outputs $f_{i,j}$.

    Apply masks to N's outputs: $\boldsymbol{O_E} = \boldsymbol{O_f} * \boldsymbol{M^e}$ and $\boldsymbol{O_C} = \boldsymbol{O_f} * \boldsymbol{M^c}$.

    $E_{fE} = mean(\boldsymbol{O_E}, dim = 0)$

    $E_{fEs} = mean(\boldsymbol{O_E} * \boldsymbol{O_E}, dim = 0)$

    $E_{fC} = mean(\boldsymbol{O_C}, dim = 0)$

    $E_{fCs} = mean(\boldsymbol{O_C} * \boldsymbol{O_C}, dim = 0)$

    $constraint = 1 - (0.5 * E_{fEs} + 0.5 * E_{fCs})$

    Minimize $loss = -sum(E_{fE} - E_{fC} + \boldsymbol{\lambda} * constraint - \rho/2 * constraint^2)$

end while

---

required. From these quantities, the Fisher IPM can be calculated and optimized. Algorithm 1 explains all the above in detail.[1] When comparing to similar neural network-based methods, the last layer imposes a slightly larger memory footprint ($O(n^2)$ vs $O(n)$) but training time is comparable as they have the same amount of layers. After training completes we perform one additional pass through the training set where we use 2/3rd of the samples to calculate the templates and the remaining 1/3rd to set the thresholds for classification.[2]

(De-)composing Representations. As the CoDiR representations have a clear structure, a Singular Value Decomposition of D: $D = USV$ can be performed, such that the rows of U and the columns of V can be interpreted as the corresponding factors as contributed by the c_i and e_j respectively. This leads to two applications: (1) Composition: by modifying the elements of U, one can easily obtain $\tilde{U}$ with modified information content. By building a new representation $\tilde{D}$ from $\tilde{U}$, S and V, one thus obtains a similar representation to the original but with modified class membership. This will be further explained in this section. (2) Compression: The spectral norm for instance representations is large with a non-flat spectrum. One can thus compress the representations substantially by retaining only the first k eigenvectors of U and V, thus

---

[1] Implementation code can be found at https://github.com/GR4HAM/CoDiR.

[2] All models are trained on a single 12 Gb gpu.

creating representations in a lower k dimensional space (rank k) without significant loss of classification accuracy. If $k = 1$, the new representations are $(91 + 300)/(91 * 300) = 1.4\%$ the size of the original representations. We call this method C-CoDiR(k).

Let us consider in detail how to achieve composition. To keep things simple, we only discuss the case for 'CoDiR (capt)'. Given an image s for which $D^{(s)} \subset c_+$ and $D^{(s)} \not\subset c_-$. The goal is now to modify $D^{(s)}$ such that it represents an image $\tilde{s}$ for which $D^{(\tilde{s})} \not\subset c_+$ and $D^{(\tilde{s})} \subset c_-$ while preserving the contextual information in the environments of $D^{(s)}$. As an example, for a $D^{(s)}$ of an image where $D^{(s)} \subset c_{dog}$ and the discrete labels from which the environments are built indicate labels such as *playing*, *ball* and *grass*. The goal would be to modify the representation into $D^{(\tilde{s})}$ (such that, for example, $D^{(\tilde{s})} \subset c_{cat}$ and $D^{(\tilde{s})} \not\subset c_{dog}$) and to not modify the information in the environments.

To achieve this, consider that by increasing the value of $U_{c_+,:}$, one can increase the distance estimate with respect to class c_+, thus expressing that $D^{(s)} \not\subset c_+$. Practically, one can set the values of $\tilde{U}_{c_+,:}$ to the mean of all rows in U corresponding to the classes $\bar{c}$ for which $D^{(s)} \not\subset \bar{c}$. The opposite can be done for class c_-, *i.e.*, one can decrease the value of $U_{c_-,:}$ such that $D^{(\tilde{s})} \subset c_-$. To set the values of $\tilde{U}_{c_-,:}$, one can perform a SVD on the matrix composed of all n_c template representations T, thus obtaining $U_T S_T V_T$. As the templates by definition contain estimated distances for samples of all classes, it is then easy to see that by setting $\tilde{U}_{c_-,:} = U_{T_{c_-,:}}$ we express that $D^{(\tilde{s})} \subset c_-$ as desired. A valid representation can then be reconstructed with the outer product $D^{(\tilde{s})} = \sum_k \sigma_k \tilde{U}_{:,k} \otimes V_{k,:}^\top$ where σ_k are the eigenvalues of $D^{(s)}$. In the next section this is illustrated by retrieving images after modifying the representations.

4 Experiments

We show how CoDiR compares to a (binary) cross-entropy baseline for multi-label image classification. Additionally, CoDiR's qualities related to (de)compositions, compression and rank are examined.

4.1 Setup

The experiments are performed on the COCO dataset [4] which contains multiple labels and descriptive captions for each image. We use the 2014 train/val splits of this dataset as these sets contain the necessary labels for our experiment, where we split the validation set into two equal, arbitrary parts to have a validation and test set for the classification task. We set $n_c = 91$, *i.e.*, we use all available 91 class labels (which includes 11 supercategories that contain other labels, *e.g.*, 'animal' is the supercategory for 'zebra' and 'cat'). An image can contain more than one class label. To construct environments we use either the class labels, CoDiR (class), or the captions, CoDiR (capt). For the latter, a vocabulary is built of the n_l most frequently occurring adjectives, nouns and verbs. For each

Table 1. F1 scores, precision (PREC) and recall (REC) for different models for the multi-label classification task. σ is the standard deviation of the F1 score over three runs. All results are the average of three runs.

MODEL	METHOD	n_e	n_l	R	F1	PREC	REC	σ
ResNet-18	BXENT (single)	–	–	–	0.566	0.579	0.614	$3.6e^{-3}$
ResNet-18	**CoDiR (class)**	300	91	40	**0.601**	0.650	0.613	$8.0e^{-3}$
ResNet-101	BXENT (single)	–	–	–	0.570	0.582	0.623	$1.3e^{-2}$
ResNet-101	**CoDiR (class)**	300	91	40	**0.627**	0.664	0.648	$2.5e^{-3}$
Inception-v3	BXENT (single)	–	–	–	**0.638**	0.663	0.669	$5.4e^{-3}$
Inception-v3	**CoDiR (class)**	300	91	40	0.617	0.648	0.646	$4.7e^{-3}$
ResNet-18	BXENT (joint)	–	300	–	0.611	0.631	0.654	$1.1e^{-3}$
ResNet-18	BXENT (joint)	–	1000	–	0.614	0.637	0.653	$9.3e^{-3}$
ResNet-18	**CoDiR (capt)**	300	300	40	0.629	0.680	0.641	$2.7e^{-3}$
ResNet-18	**CoDiR (capt)**	1000	1000	100	**0.638**	0.686	0.651	$1.9e^{-3}$
ResNet-101	BXENT (joint)	–	300	–	0.598	0.619	0.640	$1.1e^{-2}$
ResNet-101	BXENT (joint)	–	1000	–	0.592	0.611	0.638	$7.0e^{-3}$
ResNet-101	**CoDiR (capt)**	300	300	40	0.645	0.696	0.655	$2.8e^{-2}$
ResNet-101	**CoDiR (capt)**	1000	1000	100	**0.657**	0.702	0.666	$1.3e^{-2}$
Inception-v3	BXENT (joint)	–	300	–	0.644	0.671	0.675	$1.5e^{-2}$
Inception-v3	BXENT (joint)	–	1000	–	0.63	0.655	0.663	$3.0e^{-2}$
Inception-v3	**CoDiR (capt)**	300	300	40	0.660	0.699	0.675	$1.9e^{-3}$
Inception-v3	**CoDiR (capt)**	1000	1000	100	**0.661**	0.700	0.676	$6.5e^{-3}$

image, each of the n_l labels is then assigned if the corresponding vocabulary word occurs in any of the captions. For the retrieval experiment we select a set of 400 images from the test set and construct their queries.[3] All images are randomly cropped and rescaled to 224 × 224 pixels. We use three types of recent state-of-the-art classification models to compare performance: ResNet-18, ResNet-101 [2] and Inception-v3 [13]. For all runs, an Adam optimizer is used with learning rate $5.e-3$. ρ for the Fisher IPM loss is set to $1e^{-6}$. Parameters are found empirically based on performance on the validation set.

4.2 Results

Multi-label Image Classification. In this experiment the objects in the image are recognized. For each experiment images are fed through a neural network where the only difference between the baseline and our approach is the last layer. For the baseline, which we call 'BXENT', the classification model is trained with a binary cross-entropy loss over the outputs and optimal decision thresholds are

---

[3] All dataset splits and queries available at https://github.com/GR4HAM/CoDiR.

selected based on the validation set F1 score. For CoDiR, classification is performed on the learned representations as explained in Sect. 3. We then conduct two types of experiments: (1) **BXENT (single)** vs **CoDiR (class)**: An experiment where only class labels are used. For BXENT (single), classification is performed on the output with dimension n_c. For CoDiR (class), environments are built with class labels, such that $n_l = n_c$. (2) **BXENT (joint)** vs **CoDiR (capt)**: An experiment where n_l additional contextual labels from image captions are used. The total amount of labels is $n_c + n_l$. For BXENT (joint) this means joint classification is performed on all $n_c + n_l$ outputs. For CoDiR (capt), there are n_c classes whereas environments are built with the selected n_l caption words. For all models, scores are computed over the n_c class labels.

With the same underlying architecture, Table 1 shows that the CoDiR method compares favorably to the baselines in terms of F1 score.[4] When adding more detailed contextual information in the environments, as is the case for CoDiR (capt), our model outperforms the baseline in all cases.[5]

The performance of CoDiR depends on the parameters n_e and R. To measure this influence the multi-label classification task is performed for different n_e values. Increasing n_e or the amount of environments leads in general to better performance, although it plateaus after a certain level. For R, the max amount of labels per environment, an optimal value can also be found empirically between 0 and n_l. The reason is that combining a large amount of labels in any environment creates a unique subset to compare samples with. When R is too large, however, subsets with unique features are no longer created and performance deteriorates. Also, even when n_e and R are small, the outcome is not sensitive with regard to the choice of environments, suggesting that the amount and diversity are more important than the composition of the environments.

Retrieval. The experiments here are designed to show interpretability, composability and compressibility of the CoDiR representations. All models and baselines in these sections are pre-trained on the classification task above. We perform two types of retrieval experiments: (1) NN: the most similar sample to a reference sample is retrieved; (2) M-NN: a sample is retrieved with modified class membership while contextual information in the environments is retained. Specifically: "Given an input s_r that belongs to class c_+ but not c_-, retrieve the instance in the dataset that is most similar to s_r that belongs to c_- and not c_+", where c_+ and c_- are class labels (see Fig. 2). We will show that CoDiR is well suited for such a task, as its structure can be exploited to create modified representations $D^{(s_r)}$ through decomposition as explained in Sect. 3.

This task is evaluated as shown in Table 2a where the goal is to achieve a good combination of M-NN PREC and F1% (for the latter, higher percentages are better). We use the highly structured sigmoid outputs of the BXENT (single) and BXENT (joint) models as baselines, denoted as **SEM (single)** and **SEM (joint)** respectively. With SEM (joint) it is possible to directly modify class

---

[4] Multi-label scores as defined by [12].

[5] For reference: a k-Nearest Neighbors ($k = 3$) on pre-trained ImageNet features of a ResNet-18 achieves a F1 of 0.221.

Table 2. Methods are used in combination with three different base models: ResNet-18/ResNet-101/Inception-v3. All results are the average of three runs.

Method	NN F1	M-NN PREC	F1%	Method	F1
SEM(single)	.64/.66/.70	.53/.55/.55	93/87/89	SEM (single)	0.00/0.00/0.00
SEM(joint)	.71/.70/.73	.29/.28/.31	97/100/96	CoDiR (class)	0.10/0.06/0.07
CNN(joint)	.71/.70/.70	.37/.26/.33	92/90/92	C-CoDiR(5)(class)	0.06/0.08/0.09
CM	.72/.74/.74	.19/.15/.18	100/100/100	SEM (joint)	0.00/0.10/0.00
CoDiR	.70/.72/.72	.30/.30/.27	97/97/95	CoDiR (capt)	0.08/0.15/0.20
C-CoDiR(5)	.70/.72/.72	.30/.29/.26	97/94/93	C-CoDiR(5)(capt)	0.10/0.14/0.19

(a) For the NN and M-NN retrieval, the F1 score of class labels and the precision (PREC) of the modified labels are shown for the first retrieved sample. The proportion of the F1 score of M-NN over NN for the caption words is shown as F1%.

(b) F1 score for a simple logistic regression on pre-trained representations to classify a previously unseen label ("panting dogs"). For the last three models, $n_l = 300$.

Fig. 2. Example of a retrieval result for both NN and M-NN. For NN, based on the representation $D^{(s_r)}$, the most similar instance is retrieved. For M-NN $D^{(s_r)}$ is modified into $D^{(\bar{s}_r)}$ before retrieving the most similar instance.

labels while maintaining all other information. It is thus a 'best-case scenario'-baseline for which one can strive, as it combines a good M-NN precision and F1% score. SEM (single) on the other hand only contains class information and thus presents a best-case scenario for the M-NN precision score yet a worst-case scenario for the F1% score. Additionally we compare with a simple baseline consisting of **CNN** features from the penultimate layer of the BXENT (joint) models with $n_l = 300$. We also use those features in a **Correlation Matching (CM)** baseline, that combines different modalities (CNN features and word caption labels) into the same representation space [10]. The representations of these baseline models cannot be composed directly. In order to compare them to the 'M-NN' method, therefore, we define templates as the average feature vector for a particular class. We then modify the representation for a sample s by subtracting the template of c_+ and adding the template of c_-. All representations except SEM (single) are built from the BXENT (joint) models with $n_l = 300$. For CoDiR they are built from CoDiR (capt) with $n_l = 300$.

For all baselines similarity is computed with the cosine similarity, whereas for CoDiR we exploit its structure as: $similarity = mean\_cos(D^{(\bar{s}_r)}, D^{(s)})$ over all classes c for which $cos(D_{c,:}^{(\bar{s}_r)}, T_c) > 0.75 \times t_c$. Here, notations are taken from Sect. 3 and $D^{(\bar{s}_r)}$ is the modified representation of the reference sample. $mean\_cos(D^{(\bar{s}_r)}, D^{(s)})$ is the mean cosine similarity between $D^{(\bar{s}_r)}$ and $D^{(s)}$ with the mean calculated over class dimensions. The similarity is thus calculated over class dimensions where classes with low relevance, *i.e.*, those that have a low similarity with the templates, are not taken into account.

The advantages of the composability of the representations can be seen in Table 2a where CoDiR (capt) has comparable performance to the fully semantic SEM (joint) representations. CNN (joint) manages to obtain a decent M-NN precision score, thus changing class information well, but at the cost of losing contextual information (low F1%), performing almost as poorly as SEM (single). Whereas CM performs well on the NN task, it doesn't change the class information accurately and thus (inadvertently) retains most contextual information.

Rank. While the previous section shows that the structure of CoDiR representations provides access to semantic information derived from the labels on which they were trained, we hypothesize that the representations contain additional information beyond those labels, reflecting local, continuous features in the images. To investigate this hypothesis, we perform an experiment, similar to [15], to determine the rank of a matrix composed of 1000 instance representations of the test set. To maintain stability we take only the first 3 rows (corresponding to 3 classes) and all 300 environments of each representation. Each of these is flattened into a 1D vector of size 900 to construct a matrix of size 1000 * 900. Small singular values are thresholded as set by [8]. The used model is the CoDiR (capt) ResNet-18 model with $n_l = 300$. We obtain a rank of 499, which exceeds the amount of class and environment labels $(3 + 300)$ within, suggesting that the representations contain additional structure beyond the original labels.

The representations can thus be compressed. Table 2a shows that C-CoDiR with $k = 5$, denoted as **C-CoDiR(5)**, approaches CoDiR's performance across all defined retrieval tasks. To show that the CoDiR representations contain information beyond the pre-trained labels, we also use cross-validation to perform a binary classification task with a simple logistic regression. A subset of 400 images of dogs is taken from the validation and test sets, of which 24 and 17 respectively are positive samples of the previously unseen label: *panting dogs*. The outcome in Table 2b shows that CoDiR and C-CoDiR(5) representations outperform the purely semantic representations of the SEM model, which shows that the additional continuous information is valuable.

5   Conclusion

CoDiR is a novel deep learning method to learn representations that can combine different modalities. The instance representations are obtained from images with a convolutional neural network and are structured along class and environment dimensions. Templates are derived from the instance representations that generalize the class-specific information. In a classification task it is shown that this generalization improves as richer contextual information is added to the environments. When environments are built with labels from image captions, the CoDiR representations consistently outperform their respective baselines. The representations are continuous and have a high rank, as demonstrated by their ability to classify a label that was not seen during pre-training with a simple logistic regression. At the same time, they contain a clear structure which allows for a semantic interpretation of the content. It is shown in a retrieval task that the representations can be decomposed, modified and recomposed to reflect the modified information, while conserving existing information.

CoDiR opens an interesting path for deep learning applications to explore uses of structured representations, similar to how such structured matrices played a central role in many language processing approaches in the past. In zero-shot settings the structure might be exploited, for example, to make compositions of classes and environments that were not seen before. Additionally, further research might explore unsupervised learning or how the method can be applied to other tasks and modalities with alternative building blocks for the environments. While we demonstrate the method with a Wasserstein-based distance, other distance or similarity metrics could be examined in future work.

Acknowledgements. This work was partly supported by the FWO and SNSF, grants G078618N and #176004 as well as an ERC Advanced Grant, #788506. We thank Nvidia for granting us two TITAN Xp GPUs.

References

1. Arjovsky, M., Chintala, S., Bottou, L.: Wasserstein generative adversarial networks. In: International Conference on Machine Learning, pp. 214–223 (2017)
2. He, K., Zhang, X., Ren, S., Sun, J.: Deep residual learning for image recognition. In: Proceedings of the IEEE Conference on CVPR, pp. 770–778 (2016)
3. Kusner, M., Sun, Y., Kolkin, N., Weinberger, K.: From word embeddings to document distances. In: ICML, pp. 957–966 (2015)
4. Lin, T.-Y., et al.: Microsoft COCO: common objects in context. In: Fleet, D., Pajdla, T., Schiele, B., Tuytelaars, T. (eds.) ECCV 2014. LNCS, vol. 8693, pp. 740–755. Springer, Cham (2014). https://doi.org/10.1007/978-3-319-10602-1_48
5. Mroueh, Y., Sercu, T.: Fisher GAN. In: Advances in NeurIPS, pp. 2513–2523 (2017)
6. Murdock, B.B.: A theory for the storage and retrieval of item and associative information. Psychol. Rev. **89**(6), 609 (1982)
7. Nairne, J.S.: The myth of the encoding-retrieval match. Memory **10**(5–6), 389–395 (2002)
8. Press, W.H., Teukolsky, S.A., Vetterling, W.T., Flannery, B.P.: Numerical Recipes. The Art of Scientific Computing, 3rd edn. Cambridge University Press, Cambridge (2007)
9. Rahimi, A., Recht, B.: Random features for large-scale kernel machines. In: Advances in Neural Information Processing Systems, pp. 1177–1184 (2008)

10. Rasiwasia, N., et al.: A new approach to cross-modal multimedia retrieval. In: Proceedings of the 18th ACM International Conference on Multimedia, pp. 251–260. ACM (2010)
11. Singh, S.P., Hug, A., Dieuleveut, A., Jaggi, M.: Context mover's distance & barycenters: optimal transport of contexts for building representations. In: Proceedings of ICLR Workshop on Deep Generative Models (2019)
12. Sorower, M.S.: A literature survey on algorithms for multi-label learning. Or. State Univ. Corvallis **18**, 1–25 (2010)
13. Szegedy, C., Vanhoucke, V., Ioffe, S., Shlens, J., Wojna, Z.: Rethinking the inception architecture for computer vision. In: CVPR, pp. 2818–2826 (2016)
14. Wu, L., Yen, I.E.H., Xu, K., Xu, F., Balakrishnan, A., Chen, P.Y., et al.: Word mover's embedding: from Word2Vec to document embedding. In: Proceedings of the 2018 Conference on EMNLP, pp. 4524–4534 (2018)
15. Yang, Z., Dai, Z., Salakhutdinov, R., Cohen, W.W.: Breaking the softmax bottleneck: a high-rank RNN language model. In: ICLR (2018)

Long Distance Relationships Without Time Travel: Boosting the Performance of a Sparse Predictive Autoencoder in Sequence Modeling

Jeremy Gordon[1]([⊠]) [iD], David Rawlinson[2] [iD], and Subutai Ahmad[3] [iD]

[1] University of California, Berkeley, Berkeley, CA, USA
jrgordon@berkeley.edu
[2] Incubator 491, Melbourne, Australia
dave@agi.io
[3] Numenta, Redwood City, CA, USA
sahmad@numenta.com

Abstract. In sequence learning tasks such as language modelling, Recurrent Neural Networks must learn relationships between input features separated by time. State of the art models such as LSTM and Transformer are trained by backpropagation of losses into prior hidden states and inputs held in memory. This allows gradients to travel through time from present to past and effectively learn with perfect hindsight, but at a significant memory cost. In this paper we show that it is possible to train high performance recurrent networks using information that is local in time, and thereby achieve a significantly reduced memory footprint. We describe a predictive autoencoder called bRSM featuring recurrent connections, sparse activations, and a boosting rule for improved cell utilization. The architecture demonstrates near optimal performance on a non-deterministic (stochastic) partially-observable sequence learning task consisting of high-Markov-order sequences of MNIST digits. We find that this model learns these sequences faster and more completely than an LSTM, and offer several possible explanations why the LSTM architecture might struggle with the partially observable sequence structure in this task. We also apply our model to a next word prediction task on the Penn Treebank (PTB) dataset. We show that a 'flattened' RSM network, when paired with a modern semantic word embedding and the addition of boosting, achieves 103.5 PPL (a 20-point improvement over the best N-gram models), beating ordinary RNNs trained with BPTT and approaching the scores of early LSTM implementations. This work provides encouraging evidence that strong results on challenging tasks such as language modeling may be possible using less memory intensive and more biologically-plausible training regimes.

Keywords: Language modeling · Natural language processing · Deep learning · Sequence modeling · Sparse coding

© Springer Nature Switzerland AG 2020
F.-P. Schilling and T. Stadelmann (Eds.): ANNPR 2020, LNAI 12294, pp. 52–64, 2020.
https://doi.org/10.1007/978-3-030-58309-5_4

1   Introduction

The challenge of modeling relationships between features separated by a large number of timesteps is well known. Language modeling, the task of next character or next word prediction, is an extensively studied task that highlights the need to capture the long-distance relationships that are inherent to natural language. Historically, a variety of architectures have achieved excellent language modeling performance. Although larger datasets and increased memory capacity have also improved results, architectural changes have been associated with more significant improvements on older benchmarks.

N-gram models are an intuitive baseline model and were developed early in this history. N-gram models learn a distribution over the corpus vocabulary conditioned on n prior tokens. Among N-gram models, smoothed 5-gram models achieve minimum perplexity on the Penn Treebank corpus [11], a result that illustrates constraints on the value of increasingly long temporal context.

More recent approaches have demonstrated the success of neural models such as Recurrent Neural Networks applied to language modeling. In 2011, [13] presented a review of language models on the Penn Treebank (PTB) corpus showing that recurrent neural models at that time outperformed all other architectures.

Ordinary RNNs are known to be prone to vanishing gradients—partial derivatives used to backpropagate error signals across many layers approach zero. Hochreiter et al. introduced a novel multi-gate architecture called Long Short-Term Memory (LSTM) as a potential solution [9]. Models featuring LSTM have produced state of the art results in language modeling, demonstrating their ability to robustly learn long-range causal structure in sequential input.

RNNs appear to be a natural fit for language modeling due to the sequential nature of the task, but feed-forward networks utilizing novel convolutional strategies have also been competitive in recent years. WaveNet is a deep autoregressive model using dilated causal convolutions in order to achieve long temporal range receptive fields [16]. A recent review compared the wider family of temporal convolutional networks (TCN)—of which WaveNet is a member—with recurrent architectures such as LSTM and GRU, finding that TCNs surpassed traditional recurrent models on a wide range of sequence learning tasks [2].

Extending the concept of replacing recurrence with autoregressive convolution, [18] added attentional filtering to their Transformer network. The Transformer uses a deep encoder and decoder each composed of multi-headed attention and feed-forward layers. While the dilated convolutions of WaveNet allow it to learn relationships across longer temporal windows, attention allows the network to learn which parts of the input, as well as intermediate hidden states, are most useful for the present output.

Current state-of-the-art results are achieved by GPT-2, a 1.5 billion parameter Transformer [5], which obtains 35.7 PPL on the PTB task (see Table 2). The previous state of the art was an LSTM with the addition of mutual gating of the current input and the previous output reporting 44.8 PPL [12].

Common to all the neural approaches reviewed here is the use of some form of deep-backpropagation (BP), either by unrolling through time (see Sect. 3.1 for more detail) or through a finite window of recent inputs (WaveNet, Transformer). Since most of these models also feature deep multilayer architectures, during backpropagation gradients must flow across layers, and over time steps, resulting in very large computational graphs. By contrast, all non-deep-BP methods in the literature are noticeably less capable (e.g. none achieve < 100 PPL on PTB).

1.1 Motivation

The question remains whether models without deep BP can equal the impressive results from the recurrent, autoregressive, and attention-based architectures listed above. Models that avoid BP across many layers or time steps (i.e. without time-travel) are interesting for two reasons. First, the computational efficiency of deep learning is increasingly important, both due to practical resource constraints and environmental considerations [6]. Second, since deep BP is inconsistent with known biological learning mechanisms, strictly local computational models may lead to the discovery of alternative approaches. Specifically, we are interested in models that lie within the biologically plausible criteria outlined by [17]: 1) local and immediate credit assignment, 2) no synaptic memory, and 3) no time-traveling synapses. Our goal is to explore and advance the performance bounds of sequence learning models given these bio-plausibility constraints.

2 Method

2.1 Original RSM Model

We began with the Recurrent Sparse Memory (RSM) architecture [17]. In the cited work, RSM was shown to be comparable to RNNs trained by BPTT in a variety of tasks, but was inferior when generalizing to unseen sequences. This paper introduces new techniques that appear to mitigate those deficits. RSM is a predictive recurrent autoencoder that receives sequential inputs (e.g. images or word tokens), and is trained to generate a prediction of the next input (see Fig. 1a). Like Hierarchical Temporal Memory [8], the RSM memory is organized into m groups (or mini-columns), each composed of n cells. Cells within each group share a single set of weights from feed forward input, such that the feed-forward contribution z^F is an m-dimensional vector computed as:

$$z^F = w^F x^F(t) \tag{1}$$

Each cell receives dense recurrent connections from all cells at the previous time step, and the recurrent contribution z^R is an $m \times n$ matrix computed as:

$$z^R = w^R x^R(t) \tag{2}$$

σ_{ij} is an $m \times n$ matrix holding the weighted sum combining feed-forward and recurrent input to each cell j in group i, and is given by:

$$\sigma_{ij} = z_i^F + z_{ij}^R \tag{3}$$

A top-k sparsity is used as per [10]. RSM implements this sparsity by computing two sparse binary masks, M^π and M^λ, which indicate the most active cell (one per group), and most active group (k per layer), respectively. An inhibition trace was used in the original model to encourage efficient resource utilization during the sparsening step, but is replaced with boosting in this work (see Sect. 3.2 for discussion). The final output is calculated by applying a *tanh* nonlinearity to the sparsened activity:

$$y_{ij} = \tanh(\sigma_{ij} \cdot M_i^\lambda \cdot M_{ij}^\pi) \tag{4}$$

A memory trace $\psi(t)$ is maintained with an exponential decay parameterized by ϵ, such that $\psi(t) = \max(\psi(t-1) \cdot \epsilon, y)$. From ψ, the recurrent input at the next time step is calculated by normalizing with constant α, chosen such that the activity in x^R sums to 1:

$$x^R(t+1) = \alpha \cdot \psi(t) \tag{5}$$

Like other predictive autoencoders, RSM is trained to generate the next input $\hat{x}^F$ by "decoding" from the max of each group's sparse activity:

$$y_i^\lambda = \max(y_{i1}, \ldots, y_{in}) \tag{6}$$

The prediction is computed as $\hat{x}^F(t) = w^D y^\lambda$, where w^D is a weight matrix with dimension equal to the transpose of w^F. BP depth is therefore fixed at two. Finally, to read out labels or word distributions from the network, RSM uses a simple classifier network composed of a 2-layer fully connected ANN using leaky ReLU nonlinearities. The classifier network is trained concurrently but independently to the RSM network (not sharing gradients), and takes the RSM's hidden state as input.

2.2   Boosted RSM (bRSM)

In an attempt to achieve better generalization, we developed a variant of RSM that replaces cell-inhibition with a cell activity 'boosting' scheme. For brevity, we refer to the modified algorithm as bRSM[1].

We find that bRSM significantly improves performance on the language modeling task. We review each of our adjustments in the section below.

Flattened Network. A fundamental dynamic of HTM-like architectures is that each mini-column learns some spatial structure in the input, and each cell within a mini-column learns a transition from a prior representation [7]. A potential limitation of this architecture is that, while representations of the input via

---

[1] Code for the bRSM model and all experiments is available at https://github.com/numenta/nupic.research/tree/master/projects/rsm.

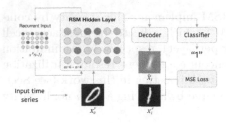

RSM architecture.   Actual and predicted digit images.

Fig. 1. The RSM architecture and typical outputs when applied to the high-order, partially-observable stochastic sequential MNIST task (see Sect. 3.1). RSM is a sparse recurrent predictive autoencoder. The classifier network has two fully-connected layers. Note that, as per original paper, the RSM network is trained only on the local MSE loss, and is not trained by gradients backpropagated from the classifier network. In (b), rows alternate between actual 9-digit samples from the grammar, and bRSM predictions. Sequences "6-4-1-3-9" and "3-4-1-3-1" (with common subsequence "4-1-3" outlined) are predicted correctly.

feed forward connections benefit from shared spatial semantics (similar representations for similar inputs), the predictive representations from recurrent connections lack this property: similar sequence items in different sequential contexts are highly orthogonal [17].

To illustrate a potential inefficiency of this orthogonality, consider a network trained on sequences where some set of similar inputs $A = \{A_1, A_2, A_3\}$ predict both B and C at the next time step, prompting cells in the representations of both B and C to activate when exposed to inputs in A. These cells may contain nearly identical weights linked to a sparse representation generalizing across patterns in A. Such a redundancy might be avoided if some subset of cells having learned the transition from A could be shared by both B and C. This line of reasoning motivated experiments in which each group was set to have only one cell, thus removing shared feed-forward weights from the model, and enabling decoding from the full hidden state rather than a group-max bottleneck. The flexibility of allowing predictive cells to participate in multiple input representations may explain the improved performance of this flattened architecture in the language modeling task, though we suspect the grouped model may be beneficial on tasks with higher-order compositionality in space or time.

Boosting. Sparse networks may learn locally optimal configurations in which only a small fraction of a layer's representational capacity is used, resulting in many idle cells and limited performance. The original RSM model employs an inhibition strategy whereby an exponentially decaying trace is used to discourage recently active cells from re-activating.

An alternative, "boosting" strategy is proposed to achieve the same goal but with different statistical characteristics. We used the boosted k-Winners algorithm suggested by [1]. This algorithm tracks the duty cycle of each cell d_i, which captures the probability of recent activation (sparsened via top-k masking):

$$d_i(t) = (1 - \alpha) \cdot d_i(t - 1) + \alpha[i \in \textit{topIndices}] \tag{7}$$

A per-cell boost term b_i is then computed based on this duty cycle, increasing the probability of less active cells firing, and inhibiting those more recently active.

$$b_i(t) = e^{\beta(\hat{a} - d_i(t))} \tag{8}$$

Here, $\hat{a}$ is the expected layer sparseness defined as the number of winners divided by the layer size, $\frac{k}{mn}$, and β is the boost strength hyper-parameter which can be optionally configured to remain fixed or decay during training. The per-cell weighted sum σ_{ij} is then redefined as:

$$\sigma_{ij} = (z_i^F + z_{ij}^R) \cdot b_i \tag{9}$$

Trainable Decay. In language modeling, some tokens may provide useful context to word prediction many tokens in the future (e.g. rare words unique to a particular topic), while others may be primarily relevant for next word prediction (e.g. articles indicating syntactic structure). In the original RSM model, the rate of decay of the recurrent input is parameterized by a single scalar value ϵ, which is multiplied into the prior memory state on each time step. While each cell participates in multiple input representations, it may be possible to improve generalization performance by learning a unique exponential decay scalar for each cell in the memory. We implemented trainable decay as a single tensor Δ of dimension $m \times n$ (equivalent to just m in the flattened architecture), which we pass through a Sigmoid before applying to the memory in the decay step:

$$\psi(t + 1) = \psi(t) \cdot \sigma(\Delta) \tag{10}$$

Trainable decay requires a nominal increase in parameters, but yields a consistent small improvement (~5 PPL on next word prediction).

Functional Partitioning. We found one final addition to be significantly beneficial on the stochastic sequential MNIST task (detailed in Sect. 3.1). In this version of the model, the bRSM memory is partitioned into either two or three blocks: one taking feed-forward input only, one taking recurrent input only, and one integrating both input sources via addition. This third section is equivalent to the full memory in the original RSM model. To ensure utilization across all partitions while keeping target sparsity consistent, we applied the top-k nonlinearity to each partition separately, with partition winners k_p proportional to partition cell count m_p: $k_p = k\frac{m_p}{m}$.

The motivation behind functional partitioning was an extension of the logic behind the use of a flattened memory. To the extent that it is useful for some cells to represent transitions from prior input, and others to represent current input, we wondered if an architecture in which these functional roles are enforced would improve performance.

The resultant model contains fewer parameters since a portion of cells are connected only to the input, which has lower dimensionality than the full memory. Results for ssMNIST are shown in Fig. 2.

3   Experiments

We selected tasks anticipated to be difficult for RNNs and RSM in particular, to enable empirical characterization of its limitations. We tested bRSM on two tasks: a non-deterministic version of the original partially-observable MNIST sequence task [17], as well as next word prediction on the Penn Treebank corpus.

3.1   Stochastic Sequential MNIST (ssMNIST)

Sequence Learning with Both Spatial and Sequential Uncertainty. In this task a grammar generates an infinite sequence of digits in the range [0..9]. The model tries to predict the next digit. However, at each step the model's input is a randomly drawn MNIST *image* of the specified digit, rather than the digit itself (i.e. the digit-label is partially observable). In the original RSM paper, a repeating, deterministic higher-order sequence of digits was used e.g. "0123 0123 0321". In this work, the grammar emits deterministic subsequences in a *random* order. This requires repeatable subsequences to be learned and recognized, while also learning to ignore the order of subsequences, which has no predictive value. Observations (images) and transitions are then both partially non-deterministic. We generated a test grammar composed of 8 subsequences of 9 MNIST digits each (dimension specified to minimize confusion, see sample sequence and predicted outputs in Fig. 1b).

To ensure that solving the task would require the successful learning of higher order sequences, we confirmed that prediction of at least some of the transitions in the resultant subsequences required memory of digits two or more steps prior.

Unlike many RNN tasks, there is no flag or special token to indicate subsequence boundaries or task reset. Without any priors for the length or existence of subsequences, the ssMNIST task is challenging even for humans.

Baseline: tBPTT Trained LSTM. We chose an LSTM as a 'baseline' algorithm to represent the deep-backpropagation approach. Modern recurrent neural networks such as LSTMs are trained using backpropagation through time (BPTT), which conceptually unrolls the network's computational graph across multiple time steps resulting in a standard multi-layer feed-forward network, and then backpropagating the loss from one or more output layers (or heads) towards the shallower layers representing earlier timesteps.

The LSTM was trained with Adam using a learning rate of 2×10^{-5}. We set LSTM hidden layer to 450, giving approximately the same trainable parameter count as the bRSM model (2.57M parameters).

Table 1. ssMNIST results on 8×9 grammar. Accuracy is reported as mean $\pm$ one standard deviation, and max over 5 runs to account for observed inter-run variance. Theoretical ceiling on accuracy for this grammar is 88.8%.

Model	Params	Mean Acc	Max Acc
LSTM (cont)	2.6M	80.0% $\pm$ 9.1	81.4%
LSTM (mbs = 100)	2.6M	73.4% $\pm$ 18.2	82.7%
bRSM	2.5M	86.4% $\pm$ 0.3	86.8%
bRSM (partitioned)	1.8M	88.8% $\pm$ 0.1	88.9%

We implemented a training regime consistent with the improved truncated-BPTT algorithm proposed by [19], which is parameterized by two integers determining the flow of gradients through past states of the network. In $tBPTT(k_1, k_2)$, k_1 specifies the interval at which to inject error from the last k_1 outputs, while k_2 specifies the length of the history through which gradients should propagate. We set $k_1 = 1$ to match the online "one digit, one prediction" dynamic of the ssMNIST task. After disappointing initial results with large k_2 values, we experimented with a range of values to empirically optimize LSTM performance.

To confirm correctness of the LSTM baseline algorithm, we verified it is able to solve a simplified (fully observable) version of the task where the same MNIST image is used at each occurrence of a given digit. Under these conditions, LSTM achieves the theoretical accuracy limit comparatively quickly, though displays volatility even after approaching this accuracy ceiling (see Fig. 2, lower right plot). This volatility is likely a consequence of models attempting to 'learn' transitions between subsequences that are not in fact predictable.

We also found significant improvement from periodically clearing LSTM memory state, perhaps acting as a max-context-length prior. Figure 2 explores univariate optimization of this parameter, measured in mini-batches mbs.

Optimization of the backpropagation window (k_2) and state clearing interval (mbs) advantage the LSTM with two sources of an implicit prior on the length of salient temporal context. Intuitively, setting k_2 or mbs below our grammar's subsequence length would make it impossible to learn high-order relationships, and too large of a value might confound the network by offering far more temporal context than is useful for learning transitions within each subsequence. We anticipated and confirmed that maximum accuracy would be achieved when both parameters were tuned to convey a useful prior on context while supplying a sufficient history to robustly learn the higher-order temporal relationships in the data. Results from experiments with varying configurations of tBPTT and state clearing are shown in Fig. 2 and appear to support this understanding.

Results. Over all training regimes tested, LSTM with the continuous configuration and $k_2 = 30$ achieved the best mean accuracy 80.0% across runs (90.0% of the theoretical limit for this grammar). The highest accuracy was observed

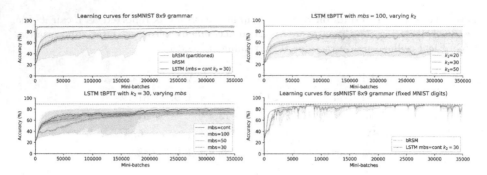

Fig. 2. LSTM and bRSM performance on ssMNIST. Mean accuracy (line), standard error (shadow) and range (light shadow) across repeated runs. Gray line is theoretical accuracy ceiling for the 8×9 grammar. Error and range are plotted for all architectures, although bRSM inter-run variance is not visible. The lower-right plot (note scaled x-axis) shows LSTM and bRSM performance when using a constant image for each digit, removing the partially observable aspect. In this case LSTM successfully solves the sequence learning task.

with $mbs = 100$ and $k_2 = 30$, reaching 82.7%, but inter-run variance was significantly higher in this configuration. In comparison the non-partitioned and partitioned variants of bRSM achieved 86.4% and 88.8% respectively, with very little inter-run variance. A summary of results is shown in Table 1.

LSTM did not achieve the maximum achievable prediction accuracy even with the additional context-length clues implicitly provided by the training regime. LSTM showed slower convergence, increased volatility and lower eventual accuracy without these clues. The much better results using a constant image for each digit suggest that the combination of partial observability, sequential uncertainty and unmarked subsequence boundaries make this task especially difficult for conventional recurrent models. In contrast, bRSM was able to learn the partially observable sequence relationships without the need to tune hyperparameters in accordance with the grammar's true time horizon. Furthermore, as noted by Rawlinson et al., by avoiding BPTT, RSM has an asymptotic memory use of $O(c)$, where c is the number of cells in the hidden layer. This is a significant reduction from deep-backpropagation models which require $O(ct)$, where t is the time-horizon, even when both models have the same number of parameters. For the empirically optimal tBPTT parameterization used in this analysis $t = k_2 = 30$, which implies that 30× more memory is required. Overall, bRSM achieves better sequence learning performance than an ordinary LSTM in this partially observable condition, with less prior knowledge of the task and significantly less memory requirement.

3.2 Language Modeling

Dataset. We show language modeling results using the popular Penn Treebank (PTB) corpus with preprocessing as per [15]. RSM's performance on this language modeling task was the weakest result of those originally reported, making it an ideal target to determine if the observed limitations could be overcome. Model evaluation was performed using the test corpus.

Training Regime. We pretrained a 100-dimensional FastText embedding [4] on the training corpus, and used this as input for all experiments. We observed that, consistent with previous findings [17], bRSM overfits the PTB training set after 40–60,000 mini-batches of training. Optimal results were obtained by pausing training of the bRSM model at this time but continuing classifier training. In future we intend to develop an automatic stopping criterion.

Results. Towards our goal of exploring the performance bounds of models under our bio-plausibility constraints, we present results from experiments with bRSM on the PTB dataset. The lowest test perplexity (103.5 PPL) was achieved using the first three additions presented in Sect. 2.2 (all but functional partitioning). A 7% word cache was effective, but an ensemble of bRSM and KN5 did not significantly improve test performance. KN5 results are shown to illustrate the performance of statistically defined n-gram models.

Table 2 reports results for the final bRSM model as well as versions of this model with each added feature ablated. bRSM, with and without the word cache, outperforms all early language modeling architectures, including ordinary (non-gated) recurrent neural language models trained with BPTT. While these results are not yet competitive with state-of-the-art deep models such as the Transformer, and modern LSTM-based approaches, they demonstrate a significant step forward for resource efficient performance.

Table 2. Language modeling results. †: As reported by [14].

bRSM comparison			bRSM feature ablations	
Model	Test PPL	Params	Model	Test PPL
KN5 †	141.2	–	**bRSM + cache**	**103.5**
KN5 + cache †	125.7	–	· Non-semantic embedding	152.6
Random Forest LM †	131.9	–	· Inhibition instead of boosting	144.0
RNN LM (tBPTT) †	124.7	–	· Non-flattened (m=800, n=3)	112.8
bRSM + cache	**103.5**	**2.55M**	· Without cache	112.0
LSTM	78.9	13M	· Untrained decay rate	107.3
Mogrifier LSTM	50.1	24M		
GPT-2	35.7	1500M		

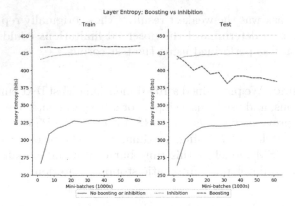

Fig. 3. Layer entropy comparison of boosting vs inhibition strategy. Maximum possible layer entropy shown by dashed gray line.

Resource Utilization (Boosting vs Inhibition). Differing temporal dynamics may explain the gap in performance between boosting and inhibition strategies. Boosting integrates a moving average of individual cell activity across hundreds of time steps, promoting the use of idle cells. In contrast, inhibition produces a strong and immediate effect where cells are fully inhibited from firing after a single activation. Both strategies aim to improve resource utilization.

These strategies can be compared by measuring the informational capacity of the RSM memory using layer entropy (H_l), calculated from the duty cycle:

$$H_l = \sum_i -d_i \log_2 d_i - (1 - d_i) \log_2 (1 - d_i) \tag{11}$$

We can compare layer entropy during training and at inference time with the theoretical maximum binary entropy for an RSM layer, which is a function only of layer sparseness ($s = \frac{k}{mn}$):

$$H_{l,max} = -s \log_2(s) - (1 - s) \log_2(1 - s) \tag{12}$$

In Fig. 3, we compare the time course of binary entropy for two RSM models differing only in resource utilization strategy. As expected, both strategies have the effect of increasing layer entropy compared to having no strategy to promote the use of idle cells. We note that inhibition exhibits nearly identical entropy dynamics across training and test sets—approximately 425 bits, or 93% of maximum entropy—while the boosted model's test entropy is reduced during exposure to unseen test sequences.

This result supports a traditional bias-variance trade-off between encoding entropy and generalization performance of sparse recurrent networks. In the high entropy case using inhibition, similar sequences are encoded in highly orthogonal patterns, which may support high capacity memorization. This is helpful when there is an opportunity to learn to interpret these patterns, but confounding when generalizing to unseen sequences, because similar contexts are encoded in dissimilar ways. This is consistent with our observation that inhibition produces worse perplexity and higher entropy on the test corpus.

However, some recent work has questioned the notion that high capacity function classes necessarily result in poor generalization performance [3], and so alternative explanations can be considered as well. For example, the strong inhibition of recently active cells may recruit arbitrary non-semantic encodings that struggle to generalize without implicating excessive capacity. In either case, encoding unseen sequences from the test corpus with relatively lower entropy implies that fewer unique encodings are produced. We hypothesize that the network falls back to known encodings of similar contexts, which the classifier network is able to interpret. Consequently, relatively better perplexity is observed from the lower-entropy test-corpus encoding.

4   Conclusion

We presented a sparse predictive autoencoder with a low memory footprint, trained on a time-local error signal. As far as we're aware, this model demonstrates the best results to date on the PTB language modeling task among models not relying upon memory-intensive deep-backpropagation across many layers and/or time steps. Neural language models with better performance all use additional mechanisms to selectively filter and store historical state (e.g. attention and gating in Transformer and LSTM networks); our goal is not to beat them, but to show that learning rules which are local in time and space could be competitive, given further development. This work provides encouraging evidence that strong results on challenging tasks such as language modeling may be possible using less memory intensive and more biologically-plausible training regimes.

We also showed that given weak partial-observability and uncertain sequence structure without boundary markers, our approach outperformed a comparable LSTM. This result also merits further investigation to understand the relationship between these task characteristics and local versus deep learning rules.

References

1. Ahmad, S., Scheinkman, L.: How can we be so dense? The benefits of using highly sparse representations. arXiv preprint arXiv:1903.11257 (2019)
2. Bai, S., Kolter, J.Z., Koltun, V.: An empirical evaluation of generic convolutional and recurrent networks for sequence modeling. arXiv preprint arXiv:1803.01271 (2018)
3. Belkin, M., Hsu, D., Ma, S., Mandal, S.: Reconciling modern machine learning and the bias-variance trade-off. arXiv preprint arXiv:1812.11118 (2018)
4. Bojanowski, P., Grave, E., Joulin, A., Mikolov, T.: Enriching word vectors with subword information. Trans. Assoc. Comput. Linguist. **5**, 135–146 (2017)

5. Gong, C., He, D., Tan, X., Qin, T., Wang, L., Liu, T.Y.: FRAGE: frequency-agnostic word representation. In: Advances in Neural Information Processing Systems, pp. 1334–1345 (2018)
6. Hao, K.: Training a single AI model can emit as much carbon as five cars in their lifetimes (2019). https://www.technologyreview.com/s/613630/training-a-single-ai-model-can-emit-as-much-carbon-as-five-cars-in-their-lifetimes/
7. Hawkins, J., Ahmad, S., Purdy, S., Lavin, A.: Biological and machine intelligence (2016). https://numenta.com/resources/biological-and-machine-intelligence/. Initial online release 0.4
8. Hawkins, J., Ahmad, S.: Why neurons have thousands of synapses, a theory of sequence memory in neocortex. Front. Neural Circ. **10** (2016). https://doi.org/10.3389/fncir.2016.00023
9. Hochreiter, S., Schmidhuber, J.: Long short-term memory. Neural Comput. **9**(8), 1735–1780 (1997). https://doi.org/10.1162/neco.1997.9.8.1735
10. Makhzani, A., Frey, B.: K-Sparse Autoencoders. arXiv:1312.5663 [cs], December 2013
11. Marcus, M., et al.: The Penn Treebank: annotating predicate argument structure. In: Proceedings of the workshop on Human Language Technology, pp. 114–119. Association for Computational Linguistics (1994)
12. Melis, G., Kočiský, T., Blunsom, P.: Mogrifier LSTM. arXiv:1909.01792 (2019)
13. Mikolov, T., Kombrink, S., Burget, L., Černocký, J., Khudanpur, S.: Extensions of recurrent neural network language model. In: 2011 IEEE International Conference on Acoustics, Speech and Signal Processing (ICASSP), pp. 5528–5531, May 2011. https://doi.org/10.1109/ICASSP.2011.5947611
14. Mikolov, T., Deoras, A., Kombrink, S., Burget, L., Černocký, J.: Empirical evaluation and combination of advanced language modeling techniques. In: Twelfth Annual Conference of the International Speech Communication Association (2011)
15. Mikolov, T., Karafiát, M., Burget, L., Černockỳ, J., Khudanpur, S.: Recurrent neural network based language model. In: Eleventh Annual Conference of the International Speech Communication Association (2010)
16. van den Oord, A., et al.: WaveNet: a generative model for raw audio. arXiv:1609.03499 [cs], September 2016
17. Rawlinson, D., Ahmed, A., Kowadlo, G.: Learning distant cause and effect using only local and immediate credit assignment. arXiv:1905.11589 [cs, stat], May 2019
18. Vaswani, A., et al.: Attention is all you need. In: Advances in Neural Information Processing Systems, pp. 5998–6008 (2017)
19. Williams, R.J., Peng, J.: An efficient gradient-based algorithm for on-line training of recurrent network trajectories. Neural Comput. **2**(4), 490–501 (1990)

Improving Accuracy and Efficiency of Object Detection Algorithms Using Multiscale Feature Aggregation Plugins

Poonam Rajput[1] , Sparsh Mittal[2(✉)] , and Sarthak Narayan[3]

[1] Indian Institute of Technology, Hyderabad, India
cs17mtech11019@iith.ac.in
[2] Indian Institute of Technology, Roorkee, India
sparshfec@iitr.ac.in
[3] National Institute of Technology, Trichy, India
sarthak.narayan.nitt@gmail.com

Abstract. In this paper, we study the use of plugins that perform multiscale feature aggregation for improving the accuracy of object detection algorithms. These plugins improve the input feature representation, and also remove the semantic ambiguity and background noise arising from feature fusion of low and high layers representation. Further, these plugins improve focus on the contextual information that comes from the shallow layers. We carefully choose the plugins to strike a delicate balance between accuracy and model size. These plugins are generic and can be easily merged with the baseline models, which avoids the need for retraining the model. We perform experiments using the PASCAL-VOC2007 dataset. While the baseline SSD has 22M parameters and an mAP score of 77.20, the use of the SFCM (one of the plugins we used) increases the mAP score to 78.82 and the number of parameters to 25M.

1 Introduction

Object detection aims to find the bounding box and the category or class of an object for a given input image. Object detection is an essential task in computer vision, and it finds application in areas including image retrieval, security, surveillance, automated vehicle systems, and machine inspection. The effectiveness of object detection models is measured by computing the accuracy of object classification and object location.

Based on their representation approach, the CNN-based object detectors can be divided into two categories: single-scale representation and multiscale representation. The examples of object detectors using single-scale representation are YOLO, RCNN, Fast-RCNN and Faster-RCNN. All these models extract information about an object's class and location based on the last layer of CNN. By contrast, SSD [1] uses a multiscale representation where the object is detected with different scales and aspect ratios. SSD strikes a good balance between accuracy and speed. It provides real-time performance, which is crucial in many

F.-P. Schilling and T. Stadelmann (Eds.): ANNPR 2020, LNAI 12294, pp. 65–76, 2020.
https://doi.org/10.1007/978-3-030-58309-5_5

applications such as autonomous driving [2]. For this reason, in this paper, we use SSD as the baseline model.

A limitation of SSD is that its accuracy is relatively low. We propose using multiscale features to mitigate this limitation, which brings up more context information to the later layers. The head detector can use these features to predict the target with better accuracy. The motivation for utilizing multiscale features can be understood from Fig. 1(a). Here, we have N convolution layers. We resize their output to have the same size and then concatenate them. Now, we have a collection of information extracted from different layers. This collection can improve detection by incorporating the context more closer to the prediction head. We can get more context information from high-level layers because they have larger receptive fields. For example, the context for a 'ship' will be the 'sea' for the high-level layers. Based on this knowledge, the network can reduce false positives on the seashore. However, due to the pooling layers, the feature maps' resolution gradually decreases, which leads to missing out some of the small targets. Therefore, it is necessary to use both high-level semantically robust features and low-level high-resolution features. Figure 1(b) shows an example, where we only focus on the low-level features. Due to this, we lose the global-context and are unable to detect this object as the dog. Evidently, aggregating multiscale features is vital for achieving high accuracy on a broad range of images.

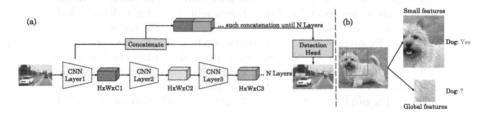

Fig. 1. (a) Multiscale features concatenation for piling better contextual information; (b) Focusing on low-level features alone can lead to losing the 'big-picture'.

Contributions: In this paper, we perform multiscale feature aggregation using plugins that perform feature-fusion or utilize attention mechanisms. Previous works have used these plugins for image classification only. However, their effectiveness for object detection is not well understood. We carefully choose the plugins to strike a fine balance between accuracy, model size and inference latency. Specifically, we have used CBAM [3], SFCM [4], CFE [5], SE [6] and DANet [7] (Sect. 2.2). These plugins improve the input feature representation, and also remove the semantic ambiguity and background noise arising from feature fusion of low-level and high-level layers representation. Further, these plugins improve focus on the contextual information that comes from the shallow layers. We show that using these plugins improves accuracy, especially those that can be easily identified from their context. In the VOC2007 dataset, examples of such objects are bird, boat, airplane, and plant, which have either consistent surrounding or consistent appearance. These plugins are generic and can be easily merged with the baseline models. Thus, it avoids the need for retraining the model from scratch. It is especially beneficial in many domains where retraining is infeasible due to its high cost or impossible due to a lack of access to the training data. We have performed experiments on the PASCAL-VOC2007 dataset (Sect. 3.1). The baseline SSD has 22M parameters, with an mAP score of 77.20 and achieves a frame rate of 56 frames/second. One of our plugins, viz. SSD-SFCM increases the mAP score to 78.82 and the number of parameters to 25M, while achieving a frame-rate of 50 frames/second. Similarly, other plugins also achieve higher accuracy than the baseline SSD. We also compare the results of different plugins and explain the reasoning behind their performance. To the best of our knowledge, our work is the first to comprehensively test the efficiency of these plugins with the fusion of multiscale features in object-detection algorithms.

2   Proposed Approach

2.1   Motivating the Need of Feature-Fusion

Figure 2 shows the generic architecture of the multi-box-SSD object detector, which uses VGG16 as the backbone network. It is evident from the figure that several multi-scale features contribute to the object detection head.

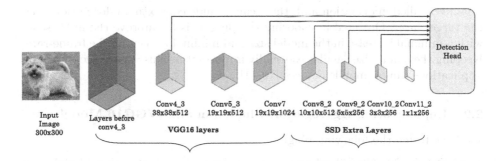

Fig. 2. Generic architecture of multibox-SSD object detector. (Color figure online)

These features can be categorized into two parts: (1) high-level features obtained from later convolution layers. They represent semantic information of an image. (2) low-level features that are obtained from the initial convolution layers. These features contain more detailed information than those contained in the high-level features [8]. Clearly, both types of features include complementary information. The semantic features describe the visual content of an image by correlating the low-level features such as color, gradient orientation, with the content of an image. For example, we can associate the brown color with the table or door and the grey color with a statue. To find such correlations, we can combine these different levels of information.

For performing multi-scale feature fusion, several methods have been proposed, such as FSSD [9]. FSSD works on the idea that the receptive field of conv4_3, conv5_3, and fc6 layers (refer Fig. 2) are better than those of later layers since the following layers introduce more background noise. So, on the VOC2007 test dataset, the best mAP score is obtained by merging the feature-maps of the conv4_3 and conv5_3 layers. To test the efficiency of FSSD, we re-implemented it in PyTorch and compared it with the baseline SSD [1]. The results are shown in Table 1, which shows that FSSD achieves higher mAP than the baseline SSD.

Table 1. Motivating results: a comparison of SSD and FSSD on VOC2007 dataset

Model	Number of parameters	mAP	Frame-rate (FPS)
SSD	22 million	77.2	56
FSSD	28 million	78.42	51

A limitation of FSSD is that it blindly concatenates the low-level and high-level features and feeds them to the head-detector [4]. To overcome this limitation, we propose to use plugins that eliminate the semantic gap between the fused features. These plugins also remove the background clutter and enable better focus on an object. Further, these plugins effectively correlate the semantic and detailed information and, thus, can include contextual information from the target layer. Our goal in choosing the plugins is to improve the mAP score with a minimal increase in the model size and minimal impact on the frame-rate. Keeping the model size small is especially important because many deep-learning applications run on mobile devices [10,11].

2.2 Implementing Aggregation Plugins in SSD-VGG16 Model

We have tested the following plugins.

1. SSD-CBAM: CBAM referred to Convolution based Attention Module and was proposed by Woo et al. [3]. CBAM is a lightweight module and can be trained end-to-end together with the baseline CNN architecture. Given an intermediate feature map, this module extracts attention masks along both channel dimensions and spatial dimensions. These masks are then multiplied with the input feature map to attain adaptive feature refinement. Figure 3 shows the architecture of SSD-CBAM. Here, CBAM is applied to all the regression head inputs. As shown in the results section (Sect. 3.2), SSD-CBAM achieves an mAP score of 78.14, which is better than that of baseline SSD. Further, it has only 23M parameters, which is lower than that of the FSSD model.

2. SSD-Fusion-CBAM: Feature-fusion using CBAM leads to a semantic gap. To remove this gap, we implemented SSD-Fusion-CBAM, which is shown in

Fig. 4. As shown in this figure, the fusion block consists of one 3×3 convolution layer for both conv4_3 and conv5_3 layer. We first perform upsampling of the feature map of conv5_3 because the dimension of conv4_3 is 38×38, and the dimension of conv5_3 is 19×19. Then, we apply a 3×3 convolution for conv5_3. As shown in the results section, SSD-Fusion-CBAM achieves an mAP score of 78.78, which is better than both FSSD and SSD-CBAM implementations. A downside of SSD-Fusion-CBAM is that it has 28M parameters.

3. SSD-SFCM: As shown above, SSD-Fusion-CBAM, which uses the fusion model of FSSD [9], leads to a significant increase in the number of model parameters. To avoid this overhead, we replace this fusion block with SFCM. The SSD-

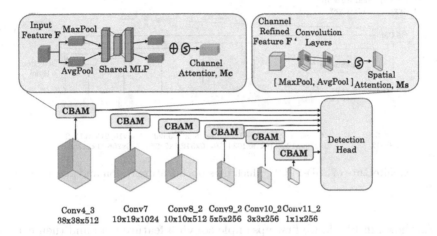

Fig. 3. The architecture of SSD-CBAM

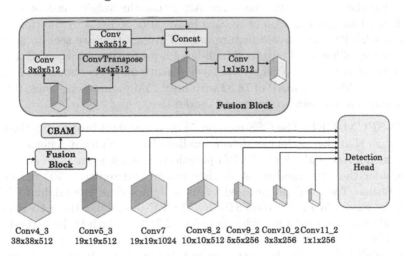

Fig. 4. Architecture of SSD-Fusion-CBAM, the single-shot detector with focused fused feature representation.

SFCM model achieves the benefits of both fusion and attention. Here, SFCM refers to the "selective feature connection mechanism" [4]. The architecture of SSD-SFCM is shown in Fig. 5.

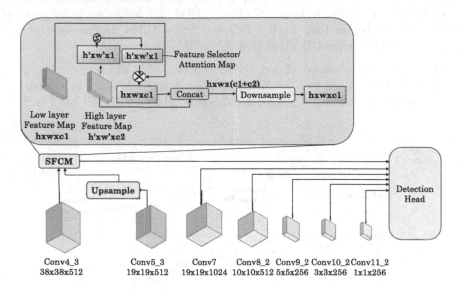

Fig. 5. Architecture of SSD-SFCM which uses both feature fusion and attention mechanism.

As shown in Fig. 5, we first upsample conv5_3 feature map and then convert it into a weight mask of $38 \times 38 \times 1$. Then, we apply the softmax function to ensure that the values are non-negative. After this, the weight mask or attention mask is used to give importance to each location of conv4_3 by multiplying the weight mask with the conv4_3 feature map. Through this, we seek to achieve a guided fusion of low-level layer and high-level layer feature maps, using a weight mask generated through a high-level layer feature map. As shown in Sect. 3.2, SSD-SFCM achieves an mAP of 78.82 with only 25M parameters. Thus, it strikes a good tradeoff between accuracy and model size.

4. SSD-SFCM-CFE: The CFE module [5] is designed for broadening the receptive field so that the model can detect smaller objects. Since it replaces a $k \times k$ CONV with a $1 \times k$ and a $k \times 1$ CONV operations, it reduces the inference latency. We consider the CFE module as another plugin and replace CFE fusion with SFCM fusion. The CFE plugin used by us is lighter than the original CFENet model. As shown in Fig. 6, the SSD-SFCM-CFE model has three CFE blocks. We found that this model provides an mAP of 78.94, which is the highest mAP achieved by any plugin we tested (Sect. 3.2). By contrast, a model with 3 CFE blocks and CFE fusion block achieves an mAP of 78.47. Evidently, our implementation of SFCM fusion with the CFE plugin is better than the implementation of the CFENet.

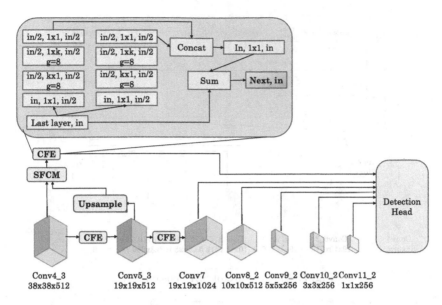

Fig. 6. Architecture of SSD-SFCM-CFE. Here, the CFE module uses bigger receptive field for feature representation enhancement.

For brevity, for the subsequent models, we only explain the plugin design, since the remaining details are similar to those discussed above.

5. SSD-Fusion with SE Plugin: The Squeeze-and-Excitation (SE) [6] block is shown in Fig. 7. This network focuses more on the informative features and leaves out less useful features. The features are first squeezed for implementing this, which leads to a channel descriptor that we obtained via merging of feature maps across their spatial dimensions ($H \times W$). This operation is followed by an excitation operation, which takes the embedding as input and produces a collection of per-channel modulation weights. These weights are applied to the feature maps to generate the SE block's output, which can be fed directly into the following layers of the model.

6. SSD-Fusion with DANet Plugin: Dual Attention Network (DANet) [7] block is shown in Fig. 8. DANet has been designed to integrate the local features with their global dependencies adaptively. It consists of two different types of attention modules. One is the position attention module, and the other is a channel attention module. The position attention module aggregates the feature at each position by a weighted sum of the features at all positions. The channel attention module selectively emphasizes interdependent channel maps. The output of both the modules is summed to improve the feature representation, leading to better detections.

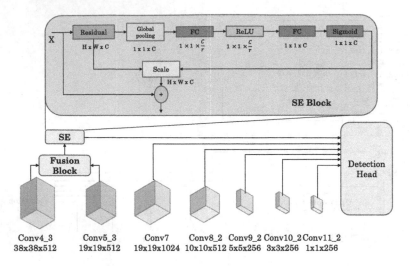

Conv4_3   Conv5_3   Conv7   Conv8_2 Conv9_2 Conv10_2 Conv11_2
38x38x512   19x19x512 19x19x1024 10x10x512 5x5x256 3x3x256 1x1x256

Fig. 7. Architecture of SSD-Fusion-SE.

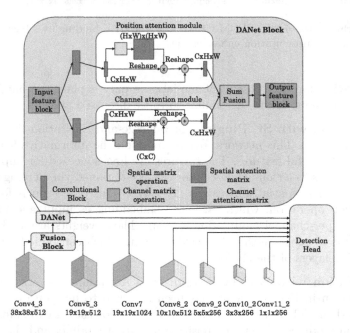

Conv4_3   Conv5_3   Conv7   Conv8_2 Conv9_2 Conv10_2 Conv11_2
38x38x512   19x19x512 19x19x1024 10x10x512 5x5x256 3x3x256 1x1x256

Fig. 8. Architecture of SSD-Fusion-DANet.

3   Results and Analysis

3.1   Experimental Platform

We perform experiments using the PyTorch framework on a P100 GPU [12]. All the models are trained on the dataset created by a combination of VOC2007 [13] and VOC2012 [14] datasets. Both the datasets have in total 5,011 and 11,540 images with 12,608 and 27,450 annotated objects, respectively. We train all the models with an input image size of 300 × 300 for 120k iterations. The learning rate is set to 1e−3 for the first 80k iterations. We then decrease the learning rate to 1e−4 and 1e−5 at 80k and 100k iterations, respectively. We initialized all the added module layers by using Xavier's initialization method. We evaluate all our models on Pascal VOC 2007 test dataset, which has 4,952 images in total. VOC2007 dataset has a total of 21 classes. The backbone of all the models is the same i.e., VGG16. We have used pre-trained weights for the VGG-16 layers [15].

3.2 Results

Table 2 compares the number of parameters, mAP and frame-rate of various implementations. Since we do not make any changes in the network's detection head, the number of parameters reported in Table 2 includes only those parameters which are part of the network before the detection head. Notice that our proposed implementations achieve higher mAP than the baseline SSD model. By improving the input feature map representation, these plugins improve the mAP.

Table 2. Results on the number of parameters, mAP and FPS of various implementations

Model	Number of parameters	mAP	Frame-rate (FPS)
SSD (baseline)	22 million	77.20	56
SSD-CBAM	23 million	78.14	47
SSD-Fusion-CBAM	28 million	78.78	46
SSD-SFCM	25 million	78.82	50
SSD-SFCM-CFE	28 million	78.94	41
SSD-Fusion-SE	28 million	78.54	48
SSD-Fusion-DANet	31 million	78.29	44

As shown in Table 2, the baseline SSD model has the highest frame-rate and the least number of parameters; however, its mAP is also the lowest. SSD-SFCM and SSD-CBAM achieve the next highest frame-rate. The reason SSD-CBAM has a lower frame-rate is the presence of two FC layers in this plugin. The SSD-Fusion-CBAM has a higher frame rate than SSD-SFCM-CFE because CFE has two parallel convolution layers and is heavier than the CBAM module.

Table 3 shows the detailed average precision (AP) for different object categories. The categories where a model achieves an absolute improvement in mAP of more than 3% are highlighted in bold. We note that for the bike, bird, cow, and train categories, SSD-Fusion-CBAM achieves a higher mAP score than SSD-CBAM. For most of these objects, contextual information helps detect the object accurately. On using the SSD-Fusion-CBAM plugin, this information is available in the model due to the piling of the feature maps of low-layer and high-layer. Both CBAM and SFCM plugins use attention for better detection. CBAM has a channel attention module and a spatial attention module. It takes input features and passes them first through channel attention module and then through spatial attention module to get the refined features. The SSD-CBAM plugin takes a single input feature and does not consider previous layer features. Due to this, SSD-CBAM cannot detect smaller objects at later detection stages; hence, its mAP score remains low.

Table 3. Detailed results showing AP improvement in different categories

Model	Aero	Bike	Bird	Boat	Bottle	Bus	Car	Cat	Chair	Cow
SSD(Baseline)	78.80	85.30	75.70	71.50	49.10	85.70	86.40	87.80	60.60	82.70
SSD-CBAM	**83.14**	83.46	75.81	70.49	**53.34**	85.47	86.38	88.03	61.84	82.59
SSD-Fusion-CBAM	**82.75**	86.83	**78.27**	73.21	**52.92**	86.62	86.45	88.24	62.01	**86.13**
SSD-SFCM	**82.74**	84.64	77.44	**74.26**	**53.43**	87.00	86.58	88.46	62.70	**85.22**
SSD-SFCM-CFE	**84.10**	85.66	77.19	73.54	**52.73**	86.96	86.81	88.47	**63.92**	84.51
SSD-Fusion-SE	**83.98**	85.35	75.68	70.66	**53.46**	86.51	86.81	88.50	**63.35**	**85.43**
SSD-Fusion-DANet	**84.30**	87.19	77.17	70.32	**52.51**	86.01	86.84	88.85	61.96	83.81
Model	Table	Dog	Horse	Mbike	Person	Plant	Sheep	Sofa	Train	Tv
SSD(Baseline)	76.50	84.90	86.70	84.00	79.20	51.30	77.50	78.70	86.70	76.20
SSD-CBAM	78.87	85.03	86.33	85.95	79.79	52.96	78.48	80.45	80.45	76.87
SSD-Fusion-CBAM	75.84	85.34	87.61	84.64	79.66	53.99	79.71	**81.00**	87.72	76.60
SSD-SFCM	78.84	85.62	87.13	85.25	79.94	52.36	79.96	79.86	87.71	77.18
SSD-SFCM-CFE	77.53	85.14	87.42	85.81	80.28	**55.48**	78.38	78.14	87.74	78.90
SSD-Fusion-SE	77.42	85.33	85.77	84.66	80.17	**54.04**	79.71	80.58	87.85	75.44
SSD-Fusion-DANet	77.04	86.09	87.74	85.04	79.87	50.48	77.45	80.16	87.24	75.80

The SFCM and SSD-Fusion-CBAM models also consider some previous layer input features that also help in better detection of smaller objects. Another factor behind improvement in mAP is that we use concatenation instead of summation for achieving feature fusion. As confirmed by the previous works, such as Unet [16] and pyramid networks [17], the concatenation of features helps preserve the spatial information of small objects even though the feature maps become very small. Also, we perform experiments on SSD-SFCM direct as well as the element-wise version. The results are shown in Table 4. We can observe that the results obtained from concatenation are better than those obtained from element-wise summation.

Table 4. A comparison of two fusion strategies: element-wise summation and concatenation

Model	Number of parameters	mAP	Frame-rate (FPS)
SSD-SFCM-eltwise	25 million	78.35	49
SSD-SFCM-concat	25 million	78.82	50

SFCM works well because, here, fusion is done by computing the weight attention score map. This score map is evaluated from high-level layers, which guides the importance of each location of the low-level feature map.

SE is not better than CBAM because SE implementation only uses global average pooled features. However, CBAM introduces a max-pooled feature along with the averaged one. So this combination leads to better attention inference. SE is also not better than SFCM as SFCM uses a guided attention mechanism that is also missing in SE blocks. DANet also does not perform well as it is more suited for segmentation rather than detection applications.

Some researchers achieve higher mAP by using larger image size. This, however, reduces the frame-rate significantly. SSD is designed to be a lightweight network for real-time processing and for execution on even mobile platforms such as Raspberry Pi. In such scenarios, large images may not be fed to SSD. Further, the plugins used by us are especially meant to enhance the features in smaller images. However, we expect that our plugins will improve the accuracy with larger images also.

4   Conclusion

In this paper, we demonstrated multiscale feature aggregation using plugins that perform feature-fusion or utilize attention mechanisms. We have carefully chosen the plugins to strike a delicate balance between accuracy and model size. In addition to improving the input feature representation, these plugins also remove the semantic ambiguity and background noise arising due to feature fusion of low and high layers representation. We show that our plugins help in improving the accuracy of especially those objects that can be easily identified from their context. Our future work will focus on designing new plugins and evaluating these plugins on edge devices such as NVIDIA Jetson or FPGAs [18]. We will also propose the techniques for improving the frame-rate without harming the accuracy.

Acknowledgment. Support for this work was provided by Semiconductor Research Corporation and Science and Engineering Research Board (SERB), India, award number ECR/2017/000622.

References

1. Liu, W., et al.: SSD: single shot MultiBox detector. In: Leibe, B., Matas, J., Sebe, N., Welling, M. (eds.) ECCV 2016. LNCS, vol. 9905, pp. 21–37. Springer, Cham (2016). https://doi.org/10.1007/978-3-319-46448-0_2

2. Mittal, S.: A survey on optimized implementation of deep learning models on the NVIDIA Jetson platform. J. Syst. Archit. **97**, 428–442 (2019)

3. Woo, S., Park, J., Lee, J.-Y., Kweon, I.S.: CBAM: convolutional block attention module. In: Ferrari, V., Hebert, M., Sminchisescu, C., Weiss, Y. (eds.) ECCV 2018. LNCS, vol. 11211, pp. 3–19. Springer, Cham (2018). https://doi.org/10.1007/978-3-030-01234-2_1

4. Du, C., Wang, Y., Wang, C., Shi, C., Xiao, B.: Selective feature connection mechanism: concatenating multi-layer CNN features with a feature selector. Pattern Recogn. Lett. **129**, 108–114 (2020)

5. Zhao, Q., Sheng, T., Wang, Y., Ni, F., Cai, L.: CFEnet: an accurate and efficient single-shot object detector for autonomous driving. arXiv preprint arXiv:1806.09790 (2018)

6. Hu, J., Shen, L., Sun, G.: Squeeze-and-excitation networks. In: IEEE Conference on Computer Vision and Pattern Recognition (CVPR), pp. 7132–7141 (2018)

7. Fu, J., et al.: Dual attention network for scene segmentation. In: IEEE Conference on Computer Vision and Pattern Recognition (CVPR), pp. 3146–3154 (2019)

8. Mittal, S.: A survey on modeling and improving reliability of DNN algorithms and accelerators. J. Syst. Architect. **104**, 101689 (2020)

9. Cao, G., et al.: Feature-fused SSD: fast detection for small objects. In International Conference on Graphic and Image Processing (ICGIP 2017) vol. 10615, pp. 106151E (2018)

10. Mittal, S., Mattela, V.: A survey of techniques for improving efficiency of mobile web browsing. Concurr. Comput. Pract. Exp. **31**(15), e5126 (2019)

11. Mittal, S., Rajput, P., Subramoney, S.: A survey of deep learning on CPUs: opportunities and co-optimizations. Technical report, IIT Roorkee (2020)

12. Mittal, S., Vaishay, S.: A survey of techniques for optimizing deep learning on GPUs. J. Syst. Architect. **99**, 101635 (2019)

13. Everingham, M., et al.: The PASCAL Visual Object Classes Challenge 2007 (VOC2007) Results (2007). http://www.pascal-network.org/challenges/VOC/voc2007/workshop/index.html

14. Everingham, M., et al.: The PASCAL Visual Object Classes Challenge 2012 (VOC2012) Results (2012). http://www.pascal-network.org/challenges/VOC/voc2012/workshop/index.html

15. https://s3.amazonaws.com/amdegroot-models/vgg16_reducedfc.pth (2020)

16. Ronneberger, O., Fischer, P., Brox, T.: U-Net: convolutional networks for biomedical image segmentation. In: Navab, N., Hornegger, J., Wells, W.M., Frangi, A.F. (eds.) MICCAI 2015. LNCS, vol. 9351, pp. 234–241. Springer, Cham (2015). https://doi.org/10.1007/978-3-319-24574-4_28

17. Lin, T.-Y., Dollár, P., Girshick, R., He, K., Hariharan, B., Belongie, S.: Feature pyramid networks for object detection. In: IEEE Conference on Computer Vision and Pattern Recognition (CVPR), pp. 2117–2125 (2017)

18. Mittal, S.: A survey of FPGA-based accelerators for convolutional neural networks. Neural Comput. Appl. **32**(4), 1109–1139 (2020)

Abstract Echo State Networks

Christoph Walter Senn[✉][iD] and Itsuo Kumazawa

Department of Information and Communications Engineering, School of Engineering,
Tokyo Institute of Technology, Tokyo, Japan
chrigi.senn@gmail.com, kumazawa.i.aa@m.titech.ac.jp

Abstract. Noisy or adverse input is a threat to the safe deployment of neural networks in production. To ensure the safe operations of such networks they need to be hardened to work under such conditions. Abstract interpretation, as a tool to formally verify properties of computations, can be used for this task. But, to date, this has mostly been studied for feed-forward networks, but not so for recurrent neural networks. For a subclass of recurrent neural networks, called echo state networks, we propose a new training algorithm using abstract interpretation and convex programming to increase the robustness against noisy inputs. Our empirical results show that the new training regime improves the performance of echo state networks in an open loop setup under high noise and generally improves their performance in closed loop setups.

Keywords: Echo state networks · Abstract interpretation · Robustness

1 Introduction

Over the past decade, neural networks have gained increased popularity for a variety of tasks, including safety-critical applications like autonomous driving. Thus, verifying neural networks and increasing robustness against noise or adversarial attacks is becoming more important. As the set of possible inputs can be unbounded, simply going over all possible inputs one-by-one is not always feasible. Abstract interpretation [2], as stemming from formal verification, is a theory that allows handling this state explosion, by abstracting concrete, single inputs, by bundling them up into a so-called abstract object. In 2018 Mirman et al. [9] and Gehr et al. [4] used abstract interpretation to verify and train deep neural networks, using different abstractions. This approach on verifying neural networks is further refined by Singh et al. in 2018 and 2019 [13–15].

On the other hand, the verification and training of recurrent neural networks with respect to local robustness are poorly studied. Our work focuses on a subset of recurrent neural networks, i.e. echo state networks (ESNs) [8], a class of recurrent neural networks where the internal and input weights are random and only the readout layer is trained. A first approach on the verification of such systems was given by Senn and Kumazawa in 2019 [12]. In this work echo state

© Springer Nature Switzerland AG 2020
F.-P. Schilling and T. Stadelmann (Eds.): ANNPR 2020, LNAI 12294, pp. 77–88, 2020.
https://doi.org/10.1007/978-3-030-58309-5_6

networks were analyzed using abstract interpretation, to estimate the influence of noise on the output.

In this study, we take these ideas and propose a new training regime for ESNs based on abstract interpretation and convex optimisation to improve the robustness against noise. We show the effectiveness empirically in open- and closed-loop ESNs.

Our main contributions are:

- A new training regime to increase the robustness of echo state networks against perturbations of the input.
- Example implementation for abstract echo state networks (Available at: https://github.com/delpart/AbstractEchoStateNetwork).

The rest of the paper is structured as follows: Sect. 2 introduces related research in abstract interpretation for neural networks, followed by Sect. 3 which explains how echo state networks and abstract interpretation can work together and how we structured our experiments. Then, in Sect. 4, we show the results gained by applying abstract interpretation and how these results can help in applying such networks in real-life situations.

2   Related Work

A first milestone in abstract interpretation for deep learning models was given in 2018 by Gehr et al. [4]. Their system AI^2, utilizes abstract interpretation to over-approximate inputs, e.g. images, using boxes, zonotopes [5] or polyhedra. Transformers were created, which translate existing networks into abstract domains and allow computation on abstract objects. This allows for an efficient way to check if properties like robustness hold, but also introduces errors due to over-approximation. A high level illustration is given in Fig. 1. An image is abstracted using a box, e.g. by replacing each pixel with an interval, and then the transformed layers are applied. The final layer then gives us a new abstract object, possibly with a different shape and size, and we can then check if our desired properties are valid, e.g. that the whole shape is within the expected decision boundaries.

AI^2 was further improved upon by Mirman et al. [9], introducing differentiable abstract interpretation. This allows for training of provable robust neural networks, but they limit their work, like AI^2, to feed-forward networks. Next to the box- and zonotope-domain used by AI^2, they also introduced a new hybrid zonotope domain placed in between the box and zonotope domain regarding speed and accuracy.

In 2019, Sing et al. [14] introduced a domain based on polyhedra with intervals, called DeepPoly. Their system enabled to proof robustness under complex perturbations, like rotations, for the first time.

In contrast to the mentioned approaches, we focus on the training of a subtype of recurrent neural networks leveraging abstract interpretation. By using the additional information, i.e. the shape and size of the abstract output, we introduce a new training regime for robust training.

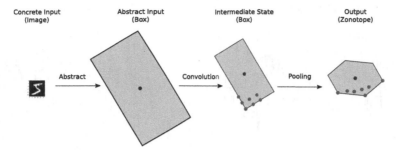

Fig. 1. Illustration of AI^2 on a high level. First, a concrete input is abstracted, e.g. to a box. Then, transformed neural network layers are applied to the abstract object, changing its size and shape. After the final layer, we can then check if the output fulfils our desired properties. The blue point represents the center of the abstraction. The red points are erroneous points (or images), which are not part of the exact solution but are part of the abstraction due to over-approximation. (Color figure online)

3 Methods

3.1 Local Robustness

We define local robustness for a model as the resistance against small perturbations of the input, i.e. small changes in the input should not change the output beyond a given boundary. Formally we say a model $f(x)$ is locally-δ-robust if the following holds:

$$\forall x \in \mathbb{R}^d. |x - x_0| \leq \delta \Rightarrow |f(x) - f(x_0)| \leq \delta_{target}, \ d \subset \mathbb{N}, \ \delta \in \mathbb{R}_+ \qquad (1)$$

In practice, Eq. 1 means that for a given input x_0, e.g. an image, and a distance metric $|x|$, e.g. $\|x\|_0$, other inputs x within the distance δ should give the same or very similar results when we apply our model f to these inputs. For example, when forecasting a time-series under noise, we want to get similar results for the time-series with or without noise (see Fig. 2 for an example using the Mackey-Glass time-series [6]).

3.2 Abstract Interpretation

To verify that the output of a model is within certain bounds under noise, normally we have to test this property for each perturbed data point. As this can become computationally infeasible, we create an abstraction containing all possible points using ball arithmetic [16]. To describe the set of all points within a distance $x_r = \delta$ around a center x_c we use the tuple $<x_c, x_r>$. By using standard ball arithmetics, we can do computations on abstract objects. These abstract objects can geometrically be interpreted as (hyper-)boxes and their domain is called Box-Domain [7]. We can then check the boundaries of the intervals in each dimension, and compare them to the desired properties, i.e. allowed upper and

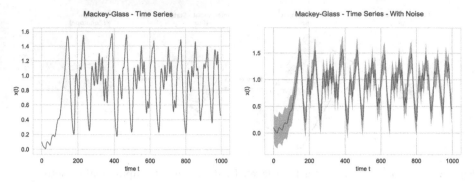

(a) The Mackey-Glass time-series.   (b) Mackey-Glass time-series with visualized noise.

Fig. 2. The concrete 1-dimensional Mackey-Glass time-series (a) and a noisy 1-dimensional Mackey-Glass time-series (b). The shaded area represents the maximum expected deviation, caused by the noise with amplitude $\delta = 0.3$.

lower bounds. For example, in Fig. 2(b), we might want to impose bounds on the valid deviation when forecasting under noise.

To be able to work with abstractions, we redefine the matrix-vector product and monotone functions for an abstraction using standard ball arithmetics $x = <x_c, x_r>$ as follows:

$$Ax = <Ax_c, |A|x_r> \tag{2}$$

$$f(x) := <f(x_c), f(x_c + x_r) - f(x_c)> \tag{3}$$

3.3   Echo State Networks

ESNs are fixed, randomly created recurrent neural networks with a trainable readout layer (see Fig. 4 for a schematic depiction). To update the state of the network we apply the following function:

$$x(t + 1) = \alpha * x(t) + \sigma(Wx(t) + W_{in}u(t)) \tag{4}$$

With $x(t)$ as the internal state (also called the reservoir), $u(t)$ as the input, W as the weights of the neural network, W_{in} as the input weights, σ as the hyperbolic tangent and α is the leaking factor.

To train it, we first drive the network with an input signal for a certain time, called the washout-time, this is to make sure, that our reservoir is purely driven by our input, i.e. that the initial state is forgotten. Then we start recording each state into matrix X. We then solve $XW_{out} = y$ for W_{out} using ridge regression, with y being our target output.

3.4   Abstract Training

For ESNs to work with abstract elements we rewrite Eq. 4 using Eq. 2 and 3, and get:

$$
\begin{aligned}
&< x_c(t+1), x_r(t+1) > \\
&= \alpha* < x_c(t), x_r(t) > +\sigma(W < x_c(t), x_r(t) > +W_{in} < u_c(t), u_r(t) >) \\
&= < \alpha * x_c(t), \alpha * x_r(t) > +\sigma(< Wx_c(t), |W|x_r(t) > + < W_{in}u_c(t), |W_{in}|u_r(t) >) \\
&= < \alpha * x_c(t), \alpha * x_r(t) > +\sigma(< Wx_c(t) + W_{in}u_c(t), |W|x_r(t) + |W_{in}|u_r(t) >) \\
&= < \alpha * x_c(t), \alpha * x_r(t) > + < \sigma(Wx_c(t) + W_{in}u_c(t)), \\
&\quad \sigma(Wx_c(t) + W_{in}u_c(t) + |W|x_r(t) + |W_{in}|u_r(t) \\
&\quad - |W|x_r(t) + |W_{in}|u_r(t)) > \\
&= < \alpha * x_c(t) + \sigma(Wx_c(t) + W_{in}u_c(t)), \\
&\quad \alpha * x_r(t) + \sigma(Wx_c(t) + W_{in}u_c(t) + |W|x_r(t) + |W_{in}|u_r(t) \\
&\quad - |W|x_r(t) + |W_{in}|u_r(t)) >
\end{aligned}
\tag{5}
$$

Similar to the classical training regime, we record the abstract outputs x_c and x_r into the matrices X_c, respectively X_r and solve the following optimisation problem:

$$
\underset{W_{out}}{argmin} \|X_c W_{out} - Y_c\|
$$
$$
s.t.\ |W_{out}|X_r <= Y_r
\tag{6}
$$

Here Y_c represents the concrete values of our target function, and Y_r the maximum deviations we tolerate. We solved for W_{out} using convex optimisation with cvxpy [3] and a Splitting Conic Solver [10,11].

3.5   Experiment

To verify our abstract echo state network, we empirically evaluated the performance of the abstract and classical approach on the Mackey-Glass [6] and Santafe D [1] datasets in open- and closed-loop setups using different noise amplitudes.

The Mackey-Glass equation is a delay differential equation, which can exhibit chaotic dynamics.

$$
x_{k+1} = cx_k + \frac{ax_{k-d}}{b + x_{k-d}^e}
\tag{7}
$$

For our experiments we chose the following parameters $a = 0.2$, $b = 0.8$, $c = 0.9$, $d = 23$ and $e = 10$. The Mackey-Glass is depicted in Fig. 3. The training-evaluation split is 8000:2000.

The Santafe D dataset is a univariate, multi-dimensional nonlinear time series generated on a computer, and contains 100'000 observations for training and 500 observations for evaluation. The Santafe D time-series is shown in Fig. 3. The training-evaluation split is 10000:500.

To test the performance of a closed-loop model for both datasets we train the model to predict the next value of the respective train time-series, and then create predictions using the evaluation split with noise du added. The noise amplitudes are ranging from 0.0 to 0.99 in 0.01 increments.

For an open-loop model we train the ESN in the same fashion as in the closed-loop but initiate the evaluation using the first value of the respective time-series, and then reuse the output of the ESN as input. We also add noise du with amplitudes ranging from 0.0 to 0.99 in 0.01 increments.

The open and closed-loop setup is depicted in Fig. 4. In contrast to the open setup, the closed feedback loop has an additional source of noise, due to imprecision and errors on the output side, which is fed back to the input.

The classical ESNs are trained as they are, whereas for the abstract versions, we set the target max deviation dy to $du/10$, i.e. the noise amplitude divided by 10.

Hyperparameters were selected based on a grid search with the parameter ranges for $\alpha \in [0.0, 1.0]$, spectral radius $\in [0.1, 1.1]$ and connectivity $\in [0.1, 1.0]$ for each dataset. The selected hyperparameters for the respective dataset are given in Table 1.

Table 1. ESN hyperparameters used in the experiments for the respective datasets.

Dataset	α	Spectral radius	Connectivity
Mackey-Glass	0.4	1.1	0.5
Santafe D	0.8	0.1	0.2

In addition, each experiment was run for 100 times, to account for uncertainty.

4   Results and Discussion

To measure the performance of the ESNs we used the mean-square-error (MSE) defined as in Eq. 8, with n representing the number of samples, $\hat{y}$ the ground truth and y the model prediction.

$$MSE = \frac{1}{n} \sum (\hat{y}_i - y_i)^2 \tag{8}$$

The averaged MSE for the experiments on both datasets are shown in Tables 2 and 3, respectively. In an open-loop setup and while noise is absent the abstract and classical network perform similar, whereas with increasing noise amplitude the performance of the classical network degrades quickly, while the abstract network's performance stays relatively stable. This is illustrated in Subfigures (a) of Figs. 5 and 6. Both types of the network perform significantly worse in

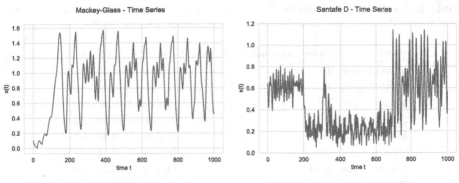

(a) The Mackey-Glass time-series.   (b) The Santafe D time-series.

Fig. 3. The first 1000 values of the time-series as used in the experiments. We use a discretized Mackey-Glass time-series (a), which is a nonlinear, delayed differential equation, the Santafe D time-series (b) on the other hand is a computer-generated univariate, multi-dimensional nonlinear dataset.

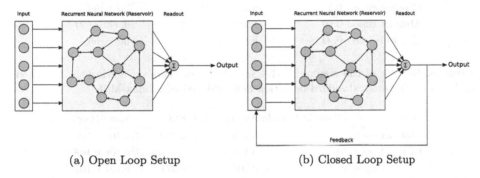

(a) Open Loop Setup   (b) Closed Loop Setup

Fig. 4. Depiction of an ESN in an open-loop (a) and closed-loop (b) setup. In the closed-loop configuration, the output of the ESN is fed back to the input, whereas in the open-loop the output is not reused.

a close loop setup, with the Santafe D dataset being an extreme case, as the classical network diverged to infinity, as seen in Subfigures (b) of Figs. 5 and 6.

The failure of the classical approach creating a working set of weights for a closed-loop in our setup is most likely due to the noise introduced to the input by errors on the output side. This also explains why our proposed methods work better in this setting, as it is inherently more robust against noisy inputs. Its robustness stems from the lower variance in the size of the output weights. Figure 7 shows differences of up to one magnitude for the Mackey-Glass dataset, whereas differences of up to four magnitudes can be observed for the Santafe D dataset, as shown in Fig. 8. Evidently, our proposed training regime leads to weights closer to zero, which increases the robustness against noisy inputs when using ESNs.

Table 2. Excerpt of the results for the Mackey-Glass time-series, showing the averaged MSE (over 100 runs), standard deviation and achieved min-/max-MSE.

Model	Noise amplitude	Average MSE	Std	Min	Max
Abstract open	0	0.232	2.790e−17	0.232	0.232
Classical open	0	0.232	1.116e−16	0.232	0.232
Abstract open	0.1	0.230	2.328e−4	0.229	0.230
Classical open	0.1	0.237	2.397e−4	0.236	0.237
Abstract open	0.5	0.240	1.169e−3	0.237	0.243
Classical open	0.5	0.345	2.001e−3	0.341	0.351
Abstract open	0.99	0.257	2.306e−03	0.251	0.262
Classical open	0.99	0.673	6.657e−02	0.661	0.694
Abstract closed	0	inf	−	inf	inf
Classical closed	0	inf	−	inf	inf
Abstract closed	0.1	0.185	4.389e−3	0.171	0.197
Classical closed	0.1	inf	−	inf	inf
Abstract closed	0.5	1.389	0.300	0.562	2.137
Classical closed	0.5	inf	−	inf	inf
Abstract closed	0.99	1.480	0.201	0.990	1.921
Classical closed	0.99	inf	−	inf	inf

Table 3. Excerpt of the results for the Santafe D time-series. Showing the averaged MSE (over 100 runs), standard deviation and achieved min-/max-MSE.

Model	Noise amplitude	Average MSE	Std	Min	Max
Abstract open	0	0.109	2.790e−17	0.109	0.109
Classical open	0	0.108	5.579e−17	0.108	0.108
Abstract open	0.1	0.103	5.371e−4	0.101	0.101
Classical open	0.1	0.111	6.668e−4	0.110	0.112
Abstract open	0.5	0.123	2.673e−3	0.115	0.130
Classical open	0.5	0.174	5.207e−3	0.160	0.190
Abstract open	0.99	0.189	6.211e−3	0.172	0.209
Classical open	0.99	0.364	0.017	0.327	0.417
Abstract closed	0	0.088	5.579e−17	0.088	0.088
Classical closed	0	0.098	0	0.098	0.098
Abstract closed	0.1	0.072	2.588e−3	0.066	0.079
Classical closed	0.1	inf	−	inf	inf
Abstract closed	0.5	0.206	2.439e−2	0.155	0.261
Classical closed	0.5	inf	−	inf	inf
Abstract closed	0.99	0.625	9.851e−2	0.425	0.837
Classical closed	0.99	inf	−	inf	inf

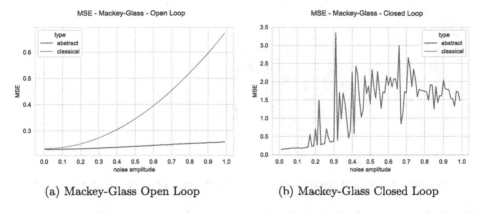

(a) Mackey-Glass Open Loop (b) Mackey-Glass Closed Loop

Fig. 5. MSE attained by a classical and abstract echo state network at various noise levels for the Mackey-Glass time-series, averaged over 100 runs. In an open-loop (a) setup the abstract echo state network surpasses the classical one when noise is present (amplitude ≥ 0.01). In a closed-loop (b) setting, on the other hand, the classical echo state network diverged, thus no data is shown for the classical one.

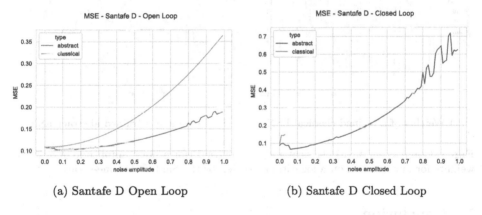

(a) Santafe D Open Loop (b) Santafe D Closed Loop

Fig. 6. MSE attained by a classical and abstract echo state network at various noise levels for the Santafe D time-series, averaged over 100 runs. In a closed-loop (b) setting the classical echo state network diverged for amplitudes ≥ 0.04, thus no data is shown for the classical one. Even in an open-loop (a) setting the abstract echo state network performs better at all noise levels.

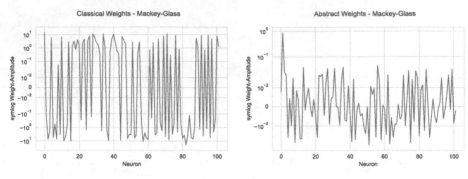

(a) Weights of a trained classical ESN.   (b) Weights of a trained abstract ESN.

Fig. 7. Comparison of (a) classical and (b) abstract (for a noise amplitude of $\delta = 0.3$) weights for our network on a logarithmic scale, for the Mackey-Glass time-series.

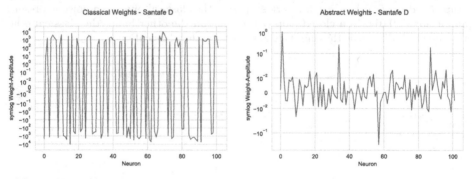

(a) Weights of a trained classical ESN.   (b) Weights of a trained abstract ESN.

Fig. 8. Comparison of (a) classical and (b) abstract (for a noise amplitude of $\delta = 0.3$) weights for our network on a logarithmic scale, for the Santafe D time-series.

5   Conclusion

We have introduced a new training regime to ESNs, improving the robustness against noise on the input side, by leveraging abstract interpretation. The increased robustness not only increases the performance if noisy inputs are present but also in a closed-loop setup, as imprecisions and errors of the output, which will be fed back to the input in this setting, are a form of noise. In general classical echo state networks tend to be better, if there is no noise present on the input side, but the performance of networks trained with our new training regime degrades significantly slower when noise is added.

An interesting direction to explore in the future is the applicability of abstract interpretation to physical reservoir computing systems. Although abstract domains, as they were introduced, are hardly directly applicable to physical systems, we can still use them in simulations. This would allow us to train such systems in simulations, and test if there is a gain in robustness within the simulated environment and then in a final step use the trained readout layer with a physical implementation of the simulated system.

Another major future research endeavour is the exploration of similar training regimes for other types of recurrent neural networks. Further, the abstract interpretation can be refined using affine arithmetic, allowing for tighter boundaries of the approximation.

Progress in these directions will facilitate the usage and safety of recurrent neural networks in real-world applications and pave the way for a safer world.

References

1. Chatfield, C., Weigend, A.S.: Time series prediction: forecasting the future and understanding the past. Int. J. Forecast. **10**(1), 161–163 (1994). https://doi.org/10.1016/0169-2070(94)90058-2
2. Cousot, P., Cousot, R.: Abstract interpretation. In: Proceedings of the 4th ACM SIGACT-SIGPLAN Symposium on Principles of Programming Languages - POPL 1977. ACM Press (1977). https://doi.org/10.1145/512950.512973
3. Diamond, S., Boyd, S.: CVXPY: a Python-embedded modeling language for convex optimization. J. Mach. Learn. Res. **17**(83), 1–5 (2016)
4. Gehr, T., Mirman, M., Drachsler-Cohen, D., Tsankov, P., Chaudhuri, S., Vechev, M.: AI2: safety and robustness certification of neural networks with abstract interpretation. In: 2018 IEEE Symposium on Security and Privacy (SP). IEEE, May 2018. https://doi.org/10.1109/sp.2018.00058
5. Ghorbal, K., Goubault, E., Putot, S.: The zonotope abstract Domain Taylor1+. In: Bouajjani, A., Maler, O. (eds.) CAV 2009. LNCS, vol. 5643, pp. 627–633. Springer, Heidelberg (2009). https://doi.org/10.1007/978-3-642-02658-4_47
6. Glass, L., Mackey, M.C., Zweifel, P.F.: From clocks to chaos: the rhythms of life. Phys. Today **42**(7), 72–72 (1989). https://doi.org/10.1063/1.2811091
7. Gurfinkel, A., Chaki, S.: BOXES: a symbolic abstract domain of boxes. In: Cousot, R., Martel, M. (eds.) SAS 2010. LNCS, vol. 6337, pp. 287–303. Springer, Heidelberg (2010). https://doi.org/10.1007/978-3-642-15769-1_18
8. Jaeger, H.: The "echo state" approach to analysing and training recurrent neural networks. Technical report, GMD Report 148, German National Research Center for Information Technology (2001). http://www.faculty.iu-bremen.de/hjaeger/pubs/EchoStatesTechRep.pdf
9. Mirman, M., Gehr, T., Vechev, M.: Differentiable abstract interpretation for provably robust neural networks. In: Dy, J., Krause, A. (eds.) Proceedings of the 35th International Conference on Machine Learning. Proceedings of Machine Learning Research, vol. 80, pp. 3578–3586. PMLR, Stockholmsmässan, Stockholm Sweden, 10–15 July 2018. http://proceedings.mlr.press/v80/mirman18b.html
10. O'Donoghue, B., Chu, E., Parikh, N., Boyd, S.: Conic optimization via operator splitting and homogeneous self-dual embedding. J. Optim. Theory Appl. **169**(3), 1042–1068 (2016). http://stanford.edu/ boyd/papers/scs.html
11. O'Donoghue, B., Chu, E., Parikh, N., Boyd, S.: SCS: splitting conic solver, version 2.1.2, November 2019. https://github.com/cvxgrp/scs
12. Senn, C.W., Kumazawa, I.: Robust echo state networks. In: Proceedings of the 29th Annual Conference of Japanese Neural Network Society (JNNS 2019), September 2019

13. Singh, G., Gehr, T., Mirman, M., Püschel, M., Vechev, M.: Fast and effective robustness certification. In: Proceedings of the 32nd International Conference on Neural Information Processing Systems, NIPS 2018NIPS 2018, pp. 10825–10836. Curran Associates Inc., Red Hook (2018)
14. Singh, G., Gehr, T., Püschel, M., Vechev, M.: An abstract domain for certifying neural networks. Proc. ACM Program. Lang. **3**(POPL), 1–30 (2019). https://doi.org/10.1145/3290354
15. Singh, G., Gehr, T., Püschel, M., Vechev, M.: Boosting robustness certification of neural networks. In: International Conference on Learning Representations (ICLR) (2019)
16. Van Der Hoeven, J.: Ball arithmetic, November 2009, 33 p. https://hal.archives-ouvertes.fr/hal-00432152

Minimal Complexity Support Vector Machines

Shigeo Abe[✉]

Kobe University, Rokkodai, Nada, Kobe, Japan
abe@kobe-u.ac.jp
http://www2.kobe-u.ac.jp/~abe

Abstract. Minimal complexity machines (MCMs) minimize the VC (Vapnik-Chervonenkis) dimension to obtain high generalization abilities. However, because the regularization term is not included in the objective function, the solution is not unique. In this paper, to solve this problem, we propose fusing the MCM and the standard support vector machine (L1 SVM). This is realized by minimizing the upper bound on the decision function for the training data in the L1 SVM. We call the machine Minimum complexity L1 SVM (ML1 SVM). We compare the ML1 SVM with other types of SVMs including the L1 SVM using several benchmark data sets and show that the ML1 SVM performs comparable to or better than the L1 SVM.

1 Introduction

In the support vector machine (SVM) [1,2], training data are mapped into the high dimensional feature space, and in that space, the separating hyperplane is determined so that the nearest training data of both classes are maximally separated. Here, the distance between a data sample and the separating hyperplane is called margin.

Motivated by the success of SVMs in real world applications, many SVM-like classifiers have been developed to improve the generalization ability. The ideas of extensions lie in incorporating the data distribution (or margin distribution) to the classifiers.

To cope with this, one approach proposes kernels based on the Mahalanobis distance [3,4]. Another approach reformulates the SVM so that the margin is measured by the Mahalanobis distance [5,6].

Yet another approach controls the overall margins instead of the minimum margin. In [7], a large margin distribution machine (LDM) is proposed, in which the average margin is maximized and the margin variance is minimized. Although the generalization ability is better than that of the SVM, the number of hyperparameters is larger than that of the SVM. To cope with this problem, in [8], the unconstrained LDM (ULDM) is proposed, which has the equal number of hyperparameters and which has the generalization ability comparable to that of the LDM and the SVM.

F.-P. Schilling and T. Stadelmann (Eds.): ANNPR 2020, LNAI 12294, pp. 89–101, 2020.
https://doi.org/10.1007/978-3-030-58309-5_7

The generalization ability of the SVM can be analyzed by the VC (Vapnik-Chervonenkis) dimension [1] and the maximum generalization ability is achieved by minimizing the radius-margin ratio, where the radius is the minimum radius of the hypersphere that encloses all the training data in the feature space.

If the center of the hypersphere is assumed to be at the origin, the radius of the hypersphere can be minimized for a given feature space as discussed in [9]. The minimal complexity machine (MCM) is derived based on this assumption. In the MCM, the VC dimension is minimized by minimizing the upper bound of the soft-margin constraints for the decision function. Because the regularization term is not included, the MCM is trained by linear programming. The generalization performance of the MCM is shown to be superior to that of the SVM, but according to our analysis [10], the solution is non-unique and the generalization ability is not better than that of the SVM. The problem of non-uniqueness is shown to be solved by adding the regularization term in the objective function of the MCM, which is a fusion of the MCM and the linear programming SVM (LP SVM) called MLP SVM.

In this paper we propose fusing the MCM with the standard SVM, i.e., L1 SVM, to improve the generalization ability of the L1 SVM. We call the fused architecture minimal complexity L1 SVM (ML1 SVM). The ML1 SVM is obtained by adding the upper bound on the decision function and the upper bound minimization term in the objective function of the L1 SVM. We derive the dual form of the ML1 SVM with one set of variables associated with the soft-margin constraints and the other set, upper-bound constraints. We then decompose the dual ML1 SVM into two subproblems: one for the soft-margin constraints, which is similar to the dual L1 SVM, and the other for the upper-bound constraints. These subproblems neither include the bias term nor the upper bound. Thus, for a convergence check, we derive the exact KKT (Karush-Kuhn-Tucker) conditions that do not include the bias term and the upper bound. The second subproblem is different from the first subproblem in that it includes the inequality constraint on the sum of dual variables. To remove this, we change the inequality constraint into two equality constraints and called this architecture $ML1_v$ SVM.

In Sect. 2, we summarize the architectures of L1 SVM and the MCM. In Sect. 3, we discuss the architectures of the ML1 SVM and $ML1_v$ SVM. In Sect. 4, we compare the generalization ability of the proposed classifiers with other SVM-like classifiers using two-class and multiclass problems.

2   L1 Support Vector Machines and Minimal Complexity Machines

In this section, we briefly explain the architectures of the L1 SVM and the MCM [9]. Then we discuss the problem of non-unique solutions of the MCM and one approach to solving the problem [10].

2.1 L1 Support Vector Machines

Let the M training data and their labels be $\{\mathbf{x}_i, y_i\}(i = 1, \ldots, M)$, where $\mathbf{x}_i$ is an n-dimensional input vector and $y_i = 1$ for Class 1 and -1 for Class 2. The input space is mapped into the l-dimensional feature space by the mapping function $\phi(\mathbf{x})$ and in the feature space the following separating hyperplane is constructed:

$$\mathbf{w}^\top \phi(\mathbf{x}) + b = 0, \tag{1}$$

where $\mathbf{w}$ is the l-dimensional constant vector and b is the bias term.

The primal form of the L1 SVM is given by

$$\min Q(\mathbf{w}, b, \boldsymbol{\xi}) = \frac{1}{2} \|\mathbf{w}\|^2 + C \sum_{i=1}^{M} \xi_i \tag{2}$$

$$\text{s.t. } y_i \left(\mathbf{w}^\top \phi(\mathbf{x}_i) + b \right) + \xi_i \geq 1, \ \xi_i \geq 0, \ i = 1, \ldots, M, \tag{3}$$

where $\boldsymbol{\xi} = (\xi_1, \ldots, \xi_M)^\top$, ξ_i is the slack variable for $\mathbf{x}_i$, and $C (> 0)$ is the margin parameter that determines the trade-off between the maximization of the margin and minimization of the classification error. Inequalities (3) are called soft-margin constraints.

2.2 Minimal Complexity Machines

The VC dimension is a measure for estimating the generalization ability of a classifier and lowering the VC dimension leads to realizing a higher generalization ability. For an SVM-like classifier with the minimum margin $\delta_{\min}$, the VC dimension D is bounded by [1]

$$D \leq 1 + \min \left(R^2 / \delta_{\min}^2, l \right), \tag{4}$$

where R is the radius of the smallest hypersphere that encloses all the training data.

In training the L1 SVM, both R and l are not changed. In the LS SVM, where ξ_i are replaced with ξ_i^2 in (2) and the inequality constraints, with equality constraints in (3), although both R and l are not changed by training, the second term in the objective function works to minimize the square sums of $y_i f(\mathbf{x}_i) - 1$. Therefore, like the LDM and ULDM, this term works to condense the margin distribution in the direction orthogonal to the separating hyperplane.

The MCM that minimizes the VC-dimension, i.e., $R/\delta_{\min}$ in (4) is

$$\min Q(\boldsymbol{\alpha}, h, \boldsymbol{\xi}, b) = h + C \sum_{i=1}^{M} \xi_i \tag{5}$$

$$\text{s.t. } h \geq y_i \left(\sum_{j=1}^{M} \alpha_j K_{ij} + b \right) + \xi_i \geq 1, \ i = 1, \ldots M, \tag{6}$$

where h is the upper bound of the soft-margin constraints and $K_{ij} = K(\mathbf{x}_i, \mathbf{x}_j) = \phi^\top(\mathbf{x}_i)\phi(\mathbf{x}_j)$. Here, the mapping function $\phi(\mathbf{x})$ in (1) is [11]

$$\phi(\mathbf{x}) = (K_{11}, \dots, K_{1M})^\top, \tag{7}$$

and $\mathbf{w} = \boldsymbol{\alpha}$. The MCM can be solved by linear programming.

Because the upper bound h in (6) is minimized in (5), the separating hyperplane is determined so that the maximum distance between the training data and the separating hyperplane is minimized.

The MCM, however, does not explicitly include the term related to the margin maximization. This makes the solution non-unique and unbounded.

To make the solution unique, in [10] the MCM and the LP SVM are fused and the resulting classifier is called minimal complexity LP SVM (MLP SVM):

$$\min \; Q(\boldsymbol{\alpha}, h, \boldsymbol{\xi}, b) = C_h\, h + \sum_{i=1}^{M}(C_\alpha\,|\alpha_i| + C\,\xi_i) \tag{8}$$

$$\text{s.t. } h \geq y_i\left(\sum_{j=1}^{M} \alpha_j\, K_{ij} + b\right) + \xi_i \geq 1, \; i = 1, \dots M, \tag{9}$$

where C_h is the positive parameter and $C_\alpha = 1$. Deleting $C_h\, h$ in (8) and upper bound h in (9), we obtain the LP SVM. Setting $C_h = 1$ and $C_\alpha = 0$ in (8), we obtain the MCM.

3   Minimal Complexity L1 Support Vector Machines

In this section, we discuss the architecture and optimality conditions of the proposed classifiers.

3.1   Architecture

Similar to the MLP SVM, here we propose fusing the MCM given by (5) and (6) and the L1 SVM given by (2) and (3):

$$\min \; Q(\mathbf{w}, b, h, \boldsymbol{\xi}) = C_h\, h + \frac{1}{2}\|\mathbf{w}\|^2 + C\sum_{i=1}^{M} \xi_i \tag{10}$$

$$\text{s.t. } y_i\left(\mathbf{w}^\top \phi(\mathbf{x}_i) + b\right) + \xi_i \geq 1, \; \xi_i \geq 0, \tag{11}$$

$$h \geq y_i\left(\mathbf{w}^\top \phi(\mathbf{x}_i) + b\right), \; i = 1, \dots, M, \tag{12}$$

$$h \geq 1, \tag{13}$$

where $\boldsymbol{\xi} = (\xi_1, \dots, \xi_M)^\top$, C_h is the positive parameter to control the volume that the training data occupy, and h is the upper bound of the constraints. The upper bound defined by (6) is redefined by (12) and (13), which exclude ξ_i. This makes the KKT conditions for the upper bound simpler. We call (12) the upper bound constraints and the above classifier minimum complexity L1 SVM (ML1 SVM).

In the following, we derive the dual problem of the ML1 SVM. Introducing the nonnegative Lagrange multipliers α_i, β_i, and η, we obtain

$$Q(\mathbf{w}, b, h, \boldsymbol{\xi}, \boldsymbol{\alpha}, \boldsymbol{\beta}, \eta) = C_h\, h + \frac{1}{2}\, \|\mathbf{w}\|^2 - \sum_{i=1}^{M} \alpha_i\, \left(y_i\, (\mathbf{w}^\top\, \boldsymbol{\phi}(\mathbf{x}_i) + b) - 1 + \xi_i\right)$$

$$+ C \sum_{i=1}^{M} \xi_i - \sum_{i=1}^{M} \beta_i\, \xi_i - \sum_{i=1}^{M} \alpha_{M+i}\, \left(h - y_i\, (\mathbf{w}^\top\, \boldsymbol{\phi}(\mathbf{x}_i) + b)\right) - (h - 1)\, \eta, \quad (14)$$

where $\boldsymbol{\alpha} = (\alpha_1, \ldots, \alpha_{2M})^\top$, $\boldsymbol{\beta} = (\beta_1, \ldots, \beta_M)^\top$.

For the optimal solution, the following KKT conditions are satisfied:

$$\frac{\partial Q(\mathbf{w}, b, h, \boldsymbol{\xi}, \boldsymbol{\alpha}, \boldsymbol{\beta}, \eta)}{\partial \mathbf{w}} = \mathbf{0}, \quad \frac{\partial Q(\mathbf{w}, b, h, \boldsymbol{\xi}, \boldsymbol{\alpha}, \boldsymbol{\beta}, \eta)}{\partial h} = 0, \quad (15)$$

$$\frac{\partial Q(\mathbf{w}, b, h, \boldsymbol{\xi}, \boldsymbol{\alpha}, \boldsymbol{\beta}, \eta)}{\partial b} = 0, \quad \frac{\partial Q(\mathbf{w}, b, h, \boldsymbol{\xi}, \boldsymbol{\alpha}, \boldsymbol{\beta}, \eta)}{\partial \boldsymbol{\xi}} = \mathbf{0}, \quad (16)$$

$$\alpha_i\, (y_i\, (\mathbf{w}^\top\, \boldsymbol{\phi}(\mathbf{x}_i) + b) - 1 + \xi_i) = 0, \ \ \alpha_i \geq 0, \quad (17)$$

$$\alpha_{M+i}\, (h - y_i\, (\mathbf{w}^\top\, \boldsymbol{\phi}(\mathbf{x}_i) + b)) = 0, \ \ \alpha_{M+i} \geq 0, \quad (18)$$

$$\beta_i\, \xi_i = 0, \quad \beta_i \geq 0, \quad \xi_i \geq 0, \ i = 1, \ldots, M, \quad (19)$$

$$(h - 1)\, \eta = 0, \ \ h \geq 1, \ \ \eta \geq 0, \quad (20)$$

where $\mathbf{0}$ is the zero vector whose elements are zero. Equations (17) to (20) are called KKT complementarity conditions.

Using (14), we reduce (15) and (16), respectively, to

$$\mathbf{w} = \sum_{i=1}^{M} (\alpha_i - \alpha_{M+i})\, y_i\, \boldsymbol{\phi}(\mathbf{x}_i), \quad \sum_{i=1}^{M} \alpha_{M+i} = C_h - \eta, \quad (21)$$

$$\sum_{i=1}^{M} (\alpha_i - \alpha_{M+i})\, y_i = 0, \quad \alpha_i + \beta_i = C, \quad i = 1, \ldots, M. \quad (22)$$

Substituting (21) and (22) into (14), we obtain the following dual problem:

$$\max Q(\boldsymbol{\alpha}) = \sum_{i=1}^{M} (\alpha_i - \alpha_{M+i}) - \frac{1}{2} \sum_{i,j=1}^{M} (\alpha_i - \alpha_{M+i})$$

$$\times (\alpha_j - \alpha_{M+j})\, y_i\, y_j\, K(\mathbf{x}_i, \mathbf{x}_j) \quad (23)$$

$$\text{s.t.} \ \sum_{i=1}^{M} y_i\, (\alpha_i - \alpha_{M+i}) = 0, \quad (24)$$

$$C_h \geq \sum_{i=1}^{M} \alpha_{M+i}, \quad (25)$$

$$C \geq \alpha_i \geq 0, \ C_h \geq \alpha_{M+i} \geq 0, \ i = 1, \ldots, M. \quad (26)$$

If we delete variables α_{M+i} and C_h from the above optimization problem, we obtain the dual problem of the original L1 SVM.

For the solution of (23) to (26), positive α_i and α_{M+j} are support vectors.

We consider decomposing the above problem into two subproblems: 1) optimizing α_i $(i = 1, \ldots, M)$ and 2) optimizing α_{M+i} $(i = 1, \ldots, M)$. To make this possible, we eliminate the interference between α_i and α_{M+i} in (24) by

$$\sum_{i=1}^{M} y_i \alpha_i = 0, \quad \sum_{i=1}^{M} y_i \alpha_{M+i} = 0. \tag{27}$$

Then the optimization problem given by (23) to (26) is decomposed into the following two subproblems:

Subproblem 1: Optimization of α_i

$$\max Q(\boldsymbol{\alpha}^0) = \sum_{i=1}^{M} (\alpha_i - \alpha_{M+i}) - \frac{1}{2} \sum_{i,j=1}^{M} (\alpha_i - \alpha_{M+i})$$
$$\times (\alpha_j - \alpha_{M+j}) \, y_i \, y_j \, K(\mathbf{x}_i, \mathbf{x}_j) \tag{28}$$

$$\text{s.t.} \sum_{i=1}^{M} y_i \alpha_i = 0, \tag{29}$$

$$C \geq \alpha_i \geq 0 \quad \text{for} \quad i = 1, \ldots, M, \tag{30}$$

where $\boldsymbol{\alpha}^0 = (\alpha_1, \ldots, \alpha_M)^{\top}$.

Subproblem 2: Optimization of α_{M+i}

$$\max Q(\boldsymbol{\alpha}^M) = \sum_{i=1}^{M} (\alpha_i - \alpha_{M+i}) - \frac{1}{2} \sum_{i,j=1}^{M} (\alpha_i - \alpha_{M+i})$$
$$\times (\alpha_j - \alpha_{M+j}) \, y_i \, y_j \, K(\mathbf{x}_i, \mathbf{x}_j) \tag{31}$$

$$\text{s.t.} \sum_{i=1}^{M} y_i \alpha_{M+i} = 0, \tag{32}$$

$$C_h \geq \sum_{i=1}^{M} \alpha_{M+i}, \tag{33}$$

$$C_h > \alpha_{M+i} \geq 0, \, i = 1, \ldots, M, \tag{34}$$

where $\boldsymbol{\alpha}^M = (\alpha_{M+1}, \ldots, \alpha_{2M})^{\top}$. Here we must notice that $\alpha_{M+i} \neq C_h$. If $\alpha_{M+i} = C_h$, from (32), at least

$$\sum_{j=1,\ldots,M,\, y_j \neq y_i} \alpha_{M+j} = C_h \tag{35}$$

is satisfied. This contradicts (33).

We solve Subproblems 1 and 2 alternatingly until the solution converges. Subproblem 1 is very similar to the L1 SVM and can be solved by the SMO (Sequential minimal optimization) combined with Newton's method [12]. Subproblem 2, which includes the constraint (33) can also be solved by a slight modification of the SMO combined with Newton's method.

3.2 KKT Conditions

To check the convergence of Subproblems 1 and 2, we use the KKT complementarity conditions (17) to (20). However, variables h and b, which are included in the KKT conditions, are excluded from the dual problem. Therefore, as with the L1 SVM [13], to make an accurate convergence test, the exact KKT conditions that do not include h and b need to be derived.

We rewrite (17) as follows:

$$\alpha_i \left(y_i\, b - y_i\, F_i + \xi_i \right) = 0, \ i = 1, \ldots, M, \tag{36}$$

where

$$F_i = y_i - \sum_{j=1}^{M} y_j \left(\alpha_j - \alpha_{M+j} \right) K(\mathbf{x}_i, \mathbf{x}_j). \tag{37}$$

We can classify the conditions of (36) into the following three cases:

1. $\alpha_i = 0$. Because $y_i\, b - y_i\, F_i + \xi_i \geq 0$ and $\xi_i = 0$, $y_i\, b \geq y_i\, F_i$, namely, $b \geq F_i$ if $y_i = 1$; $b \leq F_i$ if $y_i = -1$.
2. $C > \alpha_i > 0$. Because $\beta_i > 0$, $\xi_i = 0$ is satisfied. Therefore, $b = F_i$.
3. $\alpha_i = C$. Because $\beta_i = 0$, $\xi_i \geq 0$ is satisfied. Therefore, $y_i\, b \leq y_i\, F_i$, namely, $b \leq F_i$ if $y_i = 1$; $b \geq F_i$ if $y_i = -1$.

Then the KKT conditions for (36) are simplified as follows:

$$\bar{F}_i \geq b \geq \tilde{F}_i, \ i = 1, \ldots, M, \tag{38}$$

where

$$\tilde{F}_i = F_i \ \ \text{if} \ \ (y_i = 1, \alpha_i = 0), \quad C > \alpha_i > 0 \ \ \text{or} \ \ (y_i = -1, \alpha_i = C), \tag{39}$$

$$\bar{F}_i = F_i \ \ \text{if} \ \ (y_i = -1, \alpha_i = 0), \quad C > \alpha_i > 0 \ \ \text{or} \ \ (y_i = 1, \alpha_i = C). \tag{40}$$

To detect the violating variables, we define b_{low} and b_{up} as follows:

$$b_{\text{low}} = \max_i \tilde{F}_i, \ \ b_{\text{up}} = \min_i \bar{F}_i. \tag{41}$$

If the KKT conditions are satisfied,

$$b_{\text{up}} \geq b_{\text{low}}. \tag{42}$$

The bias term is estimated to be

$$b_{\text{e}} = \frac{1}{2} (b_{\text{up}} + b_{\text{low}}), \tag{43}$$

where b_{e} is the estimate of the bias term using (17).

Likewise, using (37), (18) becomes

$$\alpha_{M+i}\left(h + y_i F_i - y_i b - 1\right) = 0, \ i = 1, \ldots, M. \tag{44}$$

Then the conditions for (18) are rewritten as follows:

1. $\alpha_{M+i} = 0$. From $h + y_i F_i - y_i b - 1 \geq 0$, we have $y_i b - h \leq y_i F_i - 1$, namely, $b - h \leq F_i - 1$ if $y_i = 1$; $b + h \geq F_i + 1$ if $y_i = -1$.
2. $C_h > \alpha_{M+i} > 0$. $y_i b - h = y_i F_i - 1$, namely, $b - h = F_i - 1$ if $y_i = 1$; $b + h = F_i + 1$ if $y_i = -1$.

The KKT conditions for (18) are simplified as follows:

$$\text{if } y_i = -1, \ \bar{F}_i^- + 1 \geq b^- \geq \tilde{F}_i^- + 1,$$
$$\text{if } y_i = 1, \ \bar{F}_i^+ - 1 \geq b^+ \geq \tilde{F}_i^+ - 1, \ i = 1, \ldots, M, \tag{45}$$

where $b^- = b + h$, $b^+ = b - h$, and

$$\tilde{F}_i^- = F_i + 1 \quad \text{if} \quad y_i = -1, \tag{46}$$
$$\bar{F}_i^- = F_i + 1 \quad \text{if} \quad y_i = -1, C_h > \alpha_{M+i} > 0, \tag{47}$$
$$\tilde{F}_i^+ = F_i - 1 \quad \text{if} \quad y_i = 1, C_h > \alpha_{M+i} > 0, \tag{48}$$
$$\bar{F}_i^+ = F_i - 1 \quad \text{if} \quad y_i = 1. \tag{49}$$

To detect the violating variables, we define b_{low}^-, b_{low}^+, b_{up}^-, and b_{up}^+ as follows:

$$
\begin{aligned}
b_{\text{low}}^- &= \max_i \tilde{F}_i^-, & b_{\text{low}}^+ &= \max_i \tilde{F}_i^+, \\
b_{\text{up}}^- &= \min_i \bar{F}_i^-, & b_{\text{up}}^+ &= \min_i \bar{F}_i^+.
\end{aligned}
\tag{50}
$$

In general, the distributions of Classes 1 and 2 data are different. Therefore, the upper bounds of h for Classes 1 and 2 are different. This means that either of b_{up}^- ($\bar{F}_i^-$) and b_{low}^+ ($\tilde{F}_i^+$) may not exist. But because of (32), both classes have at least one positive α_{M+i} each, and because of (44), the values of h for both classes can be different. This happens because we separate (24) into two equations as in (27). Then, if the KKT conditions are satisfied, both of the following inequalities hold

$$b_{\text{up}}^- \geq b_{\text{low}}^-, \quad b_{\text{up}}^+ \geq b_{\text{low}}^+. \tag{51}$$

From the first inequality, the estimate of h, h_e^- for Class 2, is given by

$$h_e^- = -b_e + \frac{1}{2}(b_{\text{up}}^- + b_{\text{low}}^-). \tag{52}$$

From the second inequality, the estimate of h, h_e^+ for Class 1, is given by

$$h_e^+ = b_e - \frac{1}{2}(b_{\text{up}}^+ + b_{\text{low}}^+). \tag{53}$$

3.3 Variant of Minimal Complexity Support Vector Machines

Subproblem 2 of the ML1 SVM is different from Subproblem 1 in that the former includes the inequality constraint given by (33). This makes the solution process makes more complicated. In this section, we consider making the solution process similar to that of Subproblem 1.

Solving Subproblem 2 results in obtaining h_e^+ and h_e^-. We consider assigning separate variables h^+ and h^- for Classes 1 and 2 instead of a single variable h. Then the complementarity conditions for h^+ and h^- are

$$(h^+ - 1)\,\eta^+ = 0, \ h^+ \geq 1, \ \eta^+ \geq 0, \ (h^- - 1)\,\eta^- = 0,$$
$$h^- \geq 1, \ \eta^- \geq 0, \tag{54}$$

where η^+ and η^- are the Lagrange multipliers associated with h^+ and h^-, respectively. To simplify Subproblem 2, we assume that $\eta^+ = \eta^- = 0$. This makes the equations corresponding to (33) equality constraints. Then the optimization problem given by (31) to (34) becomes

$$\max \ Q(\boldsymbol{\alpha}^M) = \sum_{i=1}^{M} \alpha_i - \frac{1}{2} \sum_{i,j=1}^{M} (\alpha_i - \alpha_{M+i})\,(\alpha_j - \alpha_{M+j})\,y_i\,y_j\,K(\mathbf{x}_i, \mathbf{x}_j) \tag{55}$$

$$\text{s.t.} \ \sum_{y_i=1, i=1}^{M} \alpha_{M+i} = C_h, \ \sum_{y_i=-1, i=1}^{M} \alpha_{M+i} = C_h, \tag{56}$$

$$C \geq \alpha_i \geq 0, \ C_h \geq \alpha_{M+i} \geq 0, \ i = 1, \ldots, M. \tag{57}$$

Here, (32) is not necessary because of (56). We call the above architecture ML1$_v$ SVM.

For the solution of the ML1 SVM, the same solution is obtained by the ML1$_v$ SVM with the C_h value given by

$$C_h = \sum_{i=1,\ldots,M, y_i=1} \alpha_{M+i} = \sum_{i=1,\ldots,M, y_i=-1} \alpha_{M+i}. \tag{58}$$

However, the reverse is not true, namely, the solution of the ML1$_v$ SVM may not be obtained by the ML1 SVM. As the C_h value becomes large, the value of η becomes positive for the ML1 SVM, but for the ML1$_v$ SVM, the values of α_{M+i} are forced to become larger. But as the following computer experiments show, the performance difference is small.

4 Computer Experiments

In this section, we compare the generalization performance of the ML1$_v$ SVM and ML1 SVM with the L1 SVM, MLP SVM [10], LS SVM, and ULDM [8] using two-class and multiclass problems.

4.1 Comparison Conditions

We determined the hyperparameter values using the training data by fivefold cross-validation, trained the classifier with the determined hyperparameter values, and evaluate the accuracy for the test data.

We trained the ML1$_v$ SVM, ML1 SVM, and L1 SVM by SMO combined with Newton's method [12]. We trained the MLP SVM by the simplex method and the LS SVM and ULDM by matrix inversion.

Because RBF kernels perform well for most pattern classification applications, we used RBF kernels: $K(\mathbf{x}, \mathbf{x}') = \exp(-\gamma \|\mathbf{x} - \mathbf{x}'\|^2/m)$, where γ is the parameter to control the spread of the radius, and m is the number of inputs.

In cross-validation, we selected the γ values from $\{0.01, 0.1, 0.5, 1, 5, 10, 15, 20, 50, 100, 200\}$ and the C and C_h values from $\{0.1, 1, 10, 50, 100, 500, 1000, 2000\}$. For the ULDM, C value was selected from $\{10^{-12}, 10^{-10}, 10^{-8}, 10^{-6}, 10^{-4}, 10^{-3}, 10^{-2}, 0.1\}$.

For the L1 SVM, LS SVM, and ULDM, we determined the γ and C values by grid search. For the ML1$_v$ SVM, ML1 SVM, and MLP SVM, to shorten computation time, first we determined the γ and C values with $C_h = 1$ ($C_h = 0.1$ for the MLP SVM) by grid search and then we determined the C_h value by line search fixing the γ and C values with the determined values.

After model selection, we trained the classifier with the determined hyperparameter values and calculated the accuracy for the test data. For two-class problems we calculated the average accuracies and their standard deviations, and performed Welch's t test with the confidence level of 5%.

4.2 Two-Class Problems

Table 1 lists accuracies for the two-class problems. In the first column, I/Tr/Te denotes the numbers of input variables, training data, and test data. Except for the image and splice problems, each problem has 100 training and test data pairs. For the image and splice problems, 20 pairs.

In the table, for each classifier and each classification problem, the average accuracy and the standard deviation are shown. For each problem the best average accuracy is shown in bold and the worst, underlined. The "+" and "−" symbols at the accuracy show that the ML1$_v$ SVM is statistically better and worse than the classifier associated with the attached symbol, respectively. The "Average" row shows the average accuracy of the 13 problems for each classifier and "B/S/W" denotes the number of times that the associated classifier showed the best, the second best, and the worst accuracies. The "W/T/L" row denotes the number of times that the ML1$_v$ SVM is statistically better than, comparable to, and worse than the associated classifier.

According to the "W/T/L" row, the ML1$_v$ SVM is statistically better than the MLP SVM but is comparable to other classifiers. From the "Average" measure, the ULDM is the best and the ML1$_v$ SVM, the second. However, the differences of the measures among the ML1$_v$ SVM, ML1 SVM, and L1 SVM are

Table 1. Accuracies of the test data for the two-class problems

Problem I/Tr/Te	ML1$_v$ SVM	ML1 SVM	L1 SVM	MLP SVM	LS SVM	ULDM
Banana 2/400/4900	89.11 ± 0.70	**89.18 ± 0.70**	89.17 ± 0.72	<u>89.07</u> ± 0.73	89.17 ± 0.66	89.12 ± 0.69
Cancer 9/200/77	73.12 ± 4.43	73.03 ± 4.45	73.03 ± 4.51	<u>72.81</u> ± 4.59	73.13 ± 4.68	**73.70 ± 4.42**
Diabetes 8/468/300	76.33 ± 1.94	76.17 ± 2.25	76.29 ± 1.73	<u>76.05</u> ± 1.74	76.19 ± 2.00	**76.51 ± 1.95**
Flare-solar 9/666/400	**66.99 ± 2.16**	66.98 ± 2.14	**66.99 ± 2.12**	66.62 ± 3.10	<u>66.25</u>$^+$ ± 1.98	66.28$^+$ ± 2.05
German 20/700/300	75.97 ± 2.21	75.91 ± 2.03	75.95 ± 2.24	<u>75.63</u> ± 2.57	76.10 ± 2.10	**76.12 ± 2.3**
Heart 13/170/100	**82.96 ± 3.25**	82.84 ± 3.26	82.82 ± 3.37	82.52 ± 3.27	<u>82.49</u> ± 3.60	82.57 ± 3.64
Image 18/1300/1010	97.27 ± 0.46	97.29 ± 0.44	97.16 ± 0.41	<u>96.47</u>$^+$ ± 0.87	**97.52 ± 0.54**	97.16 ± 0.68
Ringnorm 20/400/7000	97.97 ± 1.11	98.12 ± 0.36	98.14 ± 0.35	<u>97.97</u> ± 0.37	**98.19 ± 0.33**	98.16 ± 0.35
Splice 60/1000/2175	88.99 ± 0.83	89.05 ± 0.83	88.89 ± 0.91	<u>86.71</u>$^+$ ± 1.27	88.98 ± 0.70	**89.16 ± 0.53**
Thyroid 5/140/75	**95.37 ± 2.50**	95.32 ± 2.41	95.35 ± 2.44	95.12 ± 2.38	<u>95.08</u> ± 2.55	95.15 ± 2.27
Titanic 3/150/2051	77.40 ± 0.79	<u>77.37</u> ± 0.81	77.39 ± 0.74	77.41 ± 0.77	77.39 ± 0.83	**77.46 ± 0.91**
Twonorm 20/400/7000	97.38 ± 0.25	97.36 ± 0.28	97.38 ± 0.26	<u>97.13</u>$^+$ ± 0.29	**97.43 ± 0.27**	97.41 ± 0.26
Waveform 21/400/4600	89.67 ± 0.75	89.72 ± 0.73	89.76 ± 0.66	<u>89.39</u>$^+$ ± 0.53	90.05$^-$ ± 0.59	**90.18$^-$ ± 0.54**
Average	85.27	85.26	85.26	<u>84.84</u>	85.23	**85.31**
B/S/W	3/1/0	1/3/1	1/2/0	0/1/9	3/4/3	6/1/0
W/T/L	—	0/13/0	0/13/0	4/9/0	1/11/1	1/11/1

small. From the "B/S/W" measure, the ULDM is the best and the LS SVM is the second best.

4.3 Multiclass Problems

Table 2 shows the accuracies for the ten multiclass problems. The symbol "C" in the first column denotes the number of classes. Unlike the two-class problems, each multiclass problem has only one training and test data pair.

We used fuzzy pairwise (one-vs-one) classification for multiclass problems [2]. In the table, for each problem, the best accuracy is shown in bold, and the worst, underlined. For the MLP SVM, the accuracies for the thyroid, MNIST, and letter problems were not available.

Among the ten problems, the accuracies of the ML1$_v$ SVM and ML1 SVM were better than or equal to those of the L1 SVM for nine and seven problems,

Table 2. Accuracies of the test data for the multiclass problems

Problem I/C/Tr/Te	ML1$_v$ SVM	ML1 SVM	L1 SVM	MLP SVM	LS SVM	ULDM
Numeral 12/10/810/820 [2]	**99.76**	**99.76**	**99.76**	99.27	<u>99.15</u>	99.39
Thyroid 21/3/3772/3428 [14]	97.23	**97.26**	**97.26**	—	<u>95.39</u>	95.57
Blood cell 13/12/3097/3100 [2]	93.55	93.19	<u>93.16</u>	93.36	94.23	**94.61**
Hiragana-50 50/39/4610/4610 [2]	98.98	99.46	99.00	<u>98.96</u>	**99.48**	<u>98.92</u>
Hiragana-13 13/38/8375/8356 [2]	<u>99.79</u>	99.89	<u>99.79</u>	**99.90**	99.87	**99.90**
Hiragana-105 105/38/8375/8356 [2]	100.00	100.00	100.00	100.00	100.00	100.00
Satimage 36/6/4435/2000 [14]	91.85	91.85	91.90	<u>91.10</u>	91.95	**92.25**
USPS 256/10/7291/2007 [15]	95.42	**95.47**	95.27	<u>95.17</u>	**95.47**	95.42
MNIST 784/10/10000/60000 [16]	96.96	96.96	<u>96.55</u>	—	96.99	**97.03**
Letter 16/26/16000/4000 [14]	97.95	**98.03**	97.85	—	97.88	<u>97.75</u>
Average	97.15	**97.19**	97.05	—	<u>97.04</u>	97.08
B/S/W	2/1/1	5/1/0	3/0/3	2/0/2	3/3/2	5/0/2

respectively. In addition, the best average accuracy was obtained for the ML1 SVM and the second best, the ML1$_V$ SVM. This is very different from the two-class problems where the difference was very small.

5   Conclusions

In this paper, to solve the problem of the non-unique solution of the MCM, and to improve the generalization ability of the L1 SVM, we fused the MCM and the L1 SVM. We derived two dual subproblems: the first subproblem corresponds to the L1 SVM and the second subproblem corresponds to minimizing the upper bound. We further modified the second subproblem by converting the inequality constraint into two equality constraints. We call this architecture ML1$_V$ SVM and the original architecture, ML1 SVM.

According to computer experiments for two-class problems, the average accuracy of the ML1$_V$ SVM is statistically comparable to that of the ML1 SVM and L1 SVM. For multiclass problems, the ML1$_V$ SVM and ML1 SVM generalized better than the L1 SVM.

Acknowledgment. This work was supported by JSPS KAKENHI Grant Number 19K04441.

References

1. Vapnik, V.N.: Statistical Learning Theory. Wiley, Hobooken (1998)
2. Abe, S.: Support Vector Machines for Pattern Classification, 2nd edn. Springer, London (2010). https://doi.org/10.1007/978-1-84996-098-4
3. Abe, S.: Training of support vector machines with Mahalanobis kernels. In: Duch, W., Kacprzyk, J., Oja, E., Zadrożny, S. (eds.) ICANN 2005. LNCS, vol. 3697, pp. 571–576. Springer, Heidelberg (2005). https://doi.org/10.1007/11550907_90
4. Reitmaier, T., Sick, B.: The responsibility weighted Mahalanobis kernel for semi-supervised training of support vector machines for classification. Inf. Sci. **323**, 179–198 (2015)
5. Lanckriet, G.R.G., et al.: Learning the kernel matrix with semidefinite programming. J. Mach. Learn. Res. **5**, 27–72 (2004)
6. Peng, X., Xu, D.: Twin Mahalanobis distance-based support vector machines for pattern recognition. Inf. Sci. **200**, 22–37 (2012)
7. Zhang, T., Zhou, Z.-H.: Large margin distribution machine. In: Proceedings of the 12th ACM SIGKDD Conference on Knowledge Discovery & Data Mining, pp. 313–322 (2014)
8. Abe, S.: Unconstrained large margin distribution machines. Pattern Recogn. Lett. **98**, 96–102 (2017)
9. Jayadeva: Learning a hyperplane classifier by minimizing an exact bound on the VC dimension. Neurocomputing **149**, 683–689 (2015)
10. Abe, S.: Analyzing minimal complexity machines. In: Proceedings of the IJCNN 2019 (2019)
11. Abe, S.: Sparse least squares support vector training in the reduced empirical feature space. Pattern Anal. Appl. **10**(3), 203–214 (2007). https://doi.org/10.1007/s10044-007-0062-1

12. Abe, S.: Fusing sequential minimal optimization and Newton's method for support vector training. Int. J. Mach. Learn. Cybern. **7**(3), 345–364 (2016). https://doi.org/10.1007/s13042-014-0265-x
13. Keerthi, S.S., Gilbert, E.G.: Convergence of a generalized SMO algorithm for SVM classifier design. Mach. Learn. **46**(1–3), 351–360 (2002). https://doi.org/10.1023/A:1012431217818
14. http://www.ics.uci.edu/~mlearn/MLRepository.html
15. https://www.kaggle.com/bistaumanga/usps-dataset
16. http://yann.lecun.com/exdb/mnist/

Named Entity Disambiguation at Scale

Ahmad Aghaebrahimian[(⊠)] and Mark Cieliebak

Institute of Applied Information Technology,
Zurich University of Applied Sciences ZHAW, Winterthur, Switzerland
{agha,ciel}@zhaw.ch

Abstract. Named Entity Disambiguation (NED) is a crucial task in many Natural Language Processing applications such as entity linking, record linkage, knowledge base construction, or relation extraction, to name a few. The task in NED is to map textual variations of a named entity to its formal name. It has been shown that parameter-less models for NED do not generalize to other domains very well. On the other hand, parametric learning models do not scale well when the number of formal names expands above the order of thousands or more. To tackle this problem, we propose a deep architecture with superior performance on NED and introduce a strategy to scale it to hundreds of thousands of formal names. Our experiments on several datasets for alias detection demonstrate that our system is capable of obtaining superior results with a large margin compared to other state-of-the-art systems.

Keywords: Named Entity Disambiguation · Alias detection · Deep learning

1 Introduction

Named Entity Disambiguation (NED) [14,27] is the task of linking textual varia-tions of Named Entities (NE)[1] to their target names, which are usually provided as a list of formal names. For instance, while recognizing "Philip Morris" as an NE is the job of a Named Entity Recognition (NER) system, associating it to "Philip Morris International Inc. (PMI)" in a list of formal names as a means of disambiguation is performed via NED. The list of formal names often contains many other names such as "Phil Moors, Morris Industries, ..." which are very similar to the correct formal name and should not be mistaken with it. The number of formal names, which may go to several hundred thousands or even millions, makes NED a challenging task. Character shift, abbreviation, word shift, typos, and use of nicknames are other challenges in NED.

There are two broad categories of approaches to address NED, namely classic and modern. The classic or parameter-less approach [17,21] is simply a textual similarly function such as the Levenshtein distance, Cosine similarity, or longest

---

[1] Named Entities are well-known places, people, organizations, ...

© Springer Nature Switzerland AG 2020
F.-P. Schilling and T. Stadelmann (Eds.): ANNPR 2020, LNAI 12294, pp. 102–110, 2020.
https://doi.org/10.1007/978-3-030-58309-5_8

common subsequent [10] that computes the pairwise similarity score for all available formal names given each source name.

The time complexity of these models are mostly of the order of $O(f(mn))$, m and n being the number of the source and formal names, respectively, and $f(.)$ being a linear function, which makes them quite fast. Moreover, they are highly parallelizable since computing the score of a batch does not affect the next batch scores. However, the performance of these models severely suffers when porting to new domains [3,6,28].

The modern or parametric learning models have better performance in working across domains through fine-tuning or transfer learning and are of the same order of complexity except for $f(.)$, which is often a more complex function. In real-life NED systems, the number of formal names exceeds hundreds of thousands or even million names, which makes a parametric pairwise comparison difficult if not infeasible.

To address these issues, we integrate a deep learning model with a parameterless method of similarity assignment to break the limit on recognizing new domains and simultaneously scaling the system to millions of names. To this end, we train a term frequency-inverse document frequency (tf-idf) model on a range of character n-grams of all names. The tf-idf model has two tasks; to generate feature vectors for the deep learning model and to set a threshold for limiting the formal names when using the system at inference time. We test the system on four datasets for alias detection [27] and compare the results with several baselines as well as a state-of-the-art NED system.

The motivation for this work for us is to solve a business need with a scalable and efficient solution based on deep neural networks. Our business partner harvests around 100k news articles per day. They want to recognize company names in the news and to link each of them to its formal name available in a proprietary knowledge base which contains almost 80k formal names. The variance between formal names and their usage in the news and the number of formal names in addition to the sheer amount of news articles per day ask for an efficient and scalable system for performing the task.

The main contributions of this work are a deep architecture for scoring entity names and a strategy for leveraging this architecture to a large list of source and/or formal names.

2 Related Work

Before the advent of modern and neural learning models, parameter-less computation of string similarity such as the Cosine similarity, the Levenshtein distance, and the scores proposed by [24] were popular means of scoring formal names given source names. Many of these works use a sort of word-level or character-level n-gram features [20], syntactic features [7], or alignment features [25].

The earliest modern models of NED are based on feature engineering on a classifier such as Support Vector Machines (SVM) [4] coupled with a sequence decoder such as Conditional Random Fields (CRF) [9].

Most advanced neural models today use CRFs for making inference but instead of doing feature engineering manually, they use a form of Deep Neural Network (DNN) such as Long Short Term Memory (LSTM), Convolutional Neural Network (CNN), or Gated Recurrent Unit (GRU) for automatic feature learning [2, 18].

Designing a NED system by feature engineering is a highly time-consuming process, hence end-to-end neural systems capable of learning the features on their own [16, 29] are more approachable systems. All of these models use a form of neural similarity function on top of the entity embeddings, mostly on the token-level and in some studies on character-level [12]. However, the inference module in these models is usually a pairwise scoring method [1] against all formal entities given each source entity, which makes the inference unpractical for applications with a large number of formal names.

NED can be performed jointly with NER in a way that the errors generated by NER are recovered by NED. A common approach for jointly training NED with NER is to use a NER system to extract entity mentions and use feature engineering in a shared space to map the source entities to their formal names [19, 23]. The number of formal names is a limiting factor for these systems, too.

When the number of formal names is limited, NED is usually done as a single NER process. State-of-the-art NER systems [11] use a form of pre-trained embeddings that is fed into a form of Recurrent Neural Network (RNN). The resulting representations are used to form a trellis for a Conditional Random Field (CRF) [15] decoder which extracts the beginning and the end tokens of named entities. However, when the number of formal names increases, besides the lack of enough training data, the CRF turns intractable.

Finally, [22] proposed an architecture using a Multi-Layered Perceptron (MLP) to recognize toponyms, and similar neural network architecture is used by [27] for entity linking. Our work is similar to these two last studies, but our pair-wise ranking architecture is coupled with a strategy that allows us to leverage the disambiguation to millions of source and formal names by filtering irrelevant formal names out.

3   Model Description

We model text similarity as a softly constrained pair-wise ranking problem. Figure 1 schematically represents the model.

True Target Name, Source Name, and False Target Name are character-level embedding layers for true, source, and false entity inputs. In the preprocessing step, +Score is computed as the cosine of the angle between source and true name vectors. Similarly, -Score is the cosine of the angle between source and false name vectors. In this step, these two features are generated as tf-idf vectors of the most frequent n-grams of characters in their strings. The n-grams are limited to bi-, tri-, and four-grams.

As we observe in our experiments (see e.g. Table 2), these two scores have a significant impact on the performance of the network. For instance, take the

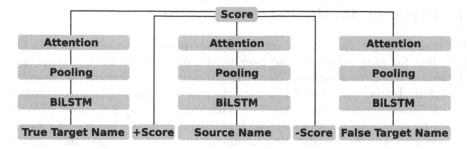

Fig. 1. The system architecture. The loss accepts two scores and three vectors to compute the difference between true and false distance given a source name.

source name "president Reagan" as the true match for "Ronald Reagan". Simply depending on character representations would make "Nancy Reagan" a good match for "president Reagan", too, which would be wrong. However, injecting the high cosine similarity of president Reagan and Ronald Reagan into the model as a signal instructs it to weigh the importance of similarities in a more elaborate way.

Character-level tf-idf vectors of source, true and false names are used as the inputs to the next layer, the BiLSTM modules for computing the string-level representations over which a column-wise max-pooling layer is applied. Then, an attention layer similar to [30] is used to help the model concentrate on more discriminating features.

The resulting vectors of the attention layers, as well as the scores, are inputs to the loss function (Eq. 1). The loss function decreases the cost when the vectors of true matches get similar to the vectors of ground truths and vice versa.

$$\mathcal{L} = \max\{0, \mathbf{m} - \mathbf{+Score} * \mathscr{F}(\mathbf{S}, \mathbf{T}^{+}) \\ + \mathbf{-Score} * \mathscr{F}(\mathbf{S}, \mathbf{T}^{-})\} \tag{1}$$

$$\mathscr{F} = \frac{1}{1 + \exp\left(-(\mathbf{v1} \cdot \mathbf{v2})\right)} * \frac{1}{1 + \|\mathbf{v1}, \mathbf{v2}\|} \tag{2}$$

Computing the similarity between two string at test time is done simply by using the same network parameters to represent the source and all formal names and computing their pair-wise similarities using $\mathscr{F}$ function (Eq. 2). However, computing the $\mathscr{F}$ for all possible permutations of source vectors and formal vectors is infeasible when one or both of the lists are big. Our strategy for scaling up the NED is to use a window of the highest cosine scored formal names instead of using all of them. Our experiment on 1000 random samples with unlimited and limited formal names showed that the difference in none of the datasets is statistically significant (Table 2).

4   Experimental Results

We trained and evaluated our system on four publicly available datasets [27] compiled for alias detection. We refer to the datasets as Wiki, Wiki-people, Artists, and Patent Assignee. The Wiki dataset is compiled by assigning the hyperlinked string in Wikipedia pages to the page they are pointing to, assuming that Wikipedia pages are entities. The Wiki-people dataset is a subset of Wiki, which contains only entities with the type "person" in the Freebase [5] knowledge graph. The Artists dataset contains alternative names for music artists extracted from MusicBrains [26]. Finally, the Patent Assignee dataset contains the aliases of assignees in patent documents[2]. Table 1 displays some statistics of these datasets.

Table 1. Number of strings, number of entities, the average number of mentions per entity, and number of samples in the train, validation and test sets

Dataset	Strings	Entities	Mentions	Train	Val	Test
Wiki-people	1880000	1160000	1.83	51842	298	3946
Wiki	9320000	4640000	2.54	64341	288	3802
Artists	1830000	1160000	1.69	11566	265	3665
Patent Assignee	330000	227000	1.50	14365	290	3746

All entities in the training data including true and false names are used to generate a list of most frequent n-gram characters limited to bi-, tri-, and four-grams. The list is used to encode the strings into their tf-idf feature vectors. The feature vectors are used for computing the cosine similarity, which is used as a feature in the neural network as well as a means to generate windows of false entities. False entities are sampled either randomly or from the window with the highest cosine scores. As an ablation study, several window sizes for false entities are selected to assess the impact of increasing false samples on the system performance.

We use BiLSTM modules with 128 units with Adam [13] as the optimizer and all dropouts set to 0.5. Since some source names may have more than one true and false answers, the Mean Average Precision (MAP) is used as the evaluation metric. We compare our system with two baselines, namely the plain Levenshtein and Jaro-Winkler distances and a state-of-the-art alias detection system proposed by [27]. The results of these experiments are reported in Table 2. The results show that our system outperforms the baselines by a large margin on three out of four datasets.

---

[2] There is a fifth dataset called "disease" which is compiled by the authors of [27]. This dataset was not publicly available at the time of authoring this work.

5   Ablation

To investigate different aspects of the system we performed an ablation study on several components of the system with the following variations. CW stands for Current Work.

- **CW-XN-ordered**
  To assess the impact of the window size or the number of false samples per true one, we define three window sizes shown as 1N, 2N, and 5N in Table 2. For instance 'CW-5N-ordered' means that for each true name we include 5 false names to train the system. At the inference time, there is no constraint on the number of false or true entities.
- **CW-2N-random**
  False names are selected either randomly or from a list of highest similar names. We make sure that the list does not contain any true target name. The distinction between these two experiments is shown by the suffix '-random' or '-ordered', respectively. The similarity scores used in this experiment are computed in the preprocessing step and are the same scores as used as a feature for training the model.
- **CW-2N-no-score**
  An additional experiment is conducted by removing the scores from the objective function to show the gain of this parameter in the network performance.
- **CW-2N-cosine**
  An experiment is conducted to assess the difference on the system performance by replacing the GESD [8] as the $\mathscr{F}$ in the objective function with cosine.
- **CW-2N-full-target**
  Finally, an experiment is performed to observe the impact of our filtering strategy on system performance. To make this experiment timely feasible, we randomly selected 1000 test samples and used the best performing model to disambiguate the samples. The results should be compared to the CW-2N-ordered experiment, which has exactly the same configuration but is applied on a limited window of 20 best scored formal names.

As Table 2 shows, all variants of CW-1N-ordered, CW-2N-ordered, and CW-5N-ordered perform on par with each other, while CW-2N-ordered yields the best results. The gap between CW-2N-no-score and CW-2N-ordered signifies the importance of integrating source similarity scores as a soft constraint in the objective function.

Although the gap between CW-2N-full-target and CW-2N-ordered is not noticeable, the first model requires much more time at the inference step since it computes the similarity for all formal names while CW-2N-ordered computes it only for a limited number of formal names. This strategy for filtering formal names is crucial to make the disambiguation feasible when the number of formal names exceeds several thousand names. This experiment shows that the introduced strategy is effective to make large scale NED manageable while getting the same performance.

Table 2. The baselines and the results of several experiments conducted on different configurations of this work are reported using the Mean of Average Precision (MAP) metric. CW stands for Current Work. All models except CW-2N-cosine use GESD [8] for $\mathscr{F}$. Scores are all in percent.

Model	Dataset			
	Wiki	Wiki-people	Artists	Patent Assignee
Levenshtein	23.8	24.6	29.6	72.0
Jaro-Winkler	29.7	28.3	32.8	85.0
Tam et al. [27]	41.6	59.4	59.7	90.6
CW-1N-ordered	61.4	71.1	70.2	88.9
CW-2N-ordered	**61.7**	**71.3**	**70.4**	89.7
CW-5N-ordered	61.5	71.2	70.1	88.6
CW-2N-random	57.2	69.3	68.4	86.3
CW-2N-no-score	56.4	65.3	67.2	84.8
CW-2N-cosine	60.5	70.4	69.9	87.1
CW-2N-full-target	**61.8**	**71.2**	**70.5**	89.4

Comparing the results on CW-2N-ordered versus CW-2N-random shows that choosing false samples from instances with high similarity with the ground truth names enhances the performance of the model. Finally, compared to its counterpart with GESD, CW-2N-cosine performs poorly which suggests that better similarity functions can improve the network even more.

6   Conclusion

NED is an integral component in many NLP applications such as record linking, entity linking, or relation extraction. Large scale NED is particularly challenging due to the time it takes to extract the correct match among hundreds of thousands of formal names, given each source name.

We proposed a state-of-the-art system for large-scale NED. Our system consists of a deep architecture for pair-wise candidate ranking and a filtering scheme that allows the network to scale up to hundreds of thousands of formal names. We tested our system on four publicly available datasets and obtained superior results with large margins on three of them.

Ideally, including contextual data should improve the performance of a NED system. However, since neither of our datasets contains contextual data there is no way to assess the impact of providing contextual data on the system performance. Nevertheless, the proposed architecture is capable of modeling contextual data by concatenating them with its input vectors. In our future work, we would like to integrate formal names metadata as well as their surrounding context into the model to further improve the performance.

References

1. Aghaebrahimian, A.: Deep neural networks at the service of multilingual parallel sentence extraction. In: Proceedings of the International Conference on Computational Linguistics (CoLing), pp. 1372–1383 (2018)
2. Aghaebrahimian, A., Cieliebak, M.: Towards integration of statistical hypothesis tests into deep neural networks. In: Proceedings of the Association for Computational Linguistics (ACL), pp. 5551–5557 (2019)
3. Bergroth, L., Hakonen, H., Raita, T.: A survey of longest common subsequence algorithms. In: Proceedings Seventh International Symposium on String Processing and Information Retrieval (SPIRE), pp. 39–48 (2000)
4. Bilenko, M., Mooney, R.J.: Adaptive duplicate detection using learnable string similarity measures. In: Proceedings of the Ninth International Conference on Knowledge Discovery and Data Mining, pp. 39–48 (2003)
5. Bollacker, K., Evans, C., Paritosh, P., Sturge, T., Taylor, J.: Freebase: a collaboratively created graph database for structuring human knowledge. In: Proceedings of the Conference of ACM Special Interest Group on Management of Data (SIGMOD), pp. 1247–1250 (2008)
6. Cohen, W., Ravikumar, P., Fienberg, S.: A comparison of string metrics for matching names and records. In: Proceedings of the Workshop on Data Cleaning and Object Consolidation (2003)
7. Das, D., Smith, N.A.: Paraphrase identification as probabilistic quasi-synchronous recognition. In: Proceedings of the Association for Computational Linguistics (ACL), pp. 468–476 (2009)
8. Feng, M., Xiang, B., Glass, M.R., Wang, L., Zhou, B.: Applying deep learning to answer selection: a study and an open task. In: Proceedings of the Workshop of Automatic Speech Recognition and Understanding (ASRU), pp. 33–40 (2015)
9. Finkel, J.R., Grenager, T., Manning, C.: Incorporating non-local information into information extraction systems by Gibbs sampling. In: Proceedings of the Association for Computational Linguistics (ACL), pp. 363–370 (2005)
10. Gan, Z., et al.: Character-level deep conflation for business data analytics. arXiv:1702.02640 (2017)
11. Güngör, O., Üsküdarli, S., Güngör, T.: Improving named entity recognition by jointly learning to disambiguate morphological tags. In: Proceedings of the International Conference on Computational Linguistics (CoLing), pp. 2082–2092 (2018)
12. Kim, Y., Jernite, Y., Sontag, D., Rush, A.M.: Character-aware neural language models. In: Proceedings of the 30th AAAI Conference on Artificial Intelligence, pp. 2741–2749 (2016)
13. Kingma, D., Ba, J.: Adam: a method for stochastic optimization. arXiv:1412.6980 (2014)
14. Kolitsas, N., Ganea, O.E., Hofmann, T.: End-to-end neural entity linking. In: Proceedings of the conference on Computational Natural Language Learning (CoNLL), pp. 519–529 (2018)
15. Lafferty, J.D., McCallum, A., Pereira, F.C.N.: Conditional random fields: probabilistic models for segmenting and labeling sequence data. In: Proceedings of the International Conference on Machine Learning (ICML), pp. 282–289 (2001)
16. Le, P., Titov, I.: Improving entity linking by modeling latent relations between mentions. In: Proceedings of the Association for Computational Linguistics (ACL), pp. 1595–1604 (2018)

17. Li, P., Dong, X.L., Guo, S., Maurino, A., Srivastava, D.: Robust group linkage. In: Proceedings of the 24th International Conference on World Wide Web, pp. 647–657 (2015)
18. Liu, L., et al.: Empower sequence labeling with task-aware neural language model. arXiv:1709.04109 (2017)
19. Luo, G., Huang, X., Lin, C.Y., Nie, Z.: Joint entity recognition and disambiguation. In: Proceedings of the Conference on Empirical Methods in Natural Language Processing (EMNLP), pp. 879–888 (2015)
20. Madnani, N., Tetreault, J., Chodorow, M.: Re-examining machine translation metrics for paraphrase identification. In: Proceedings of the Conference of the North American Chapter of the Association for Computational Linguistics (NAACL), pp. 182–190 (2012)
21. McCallum, A., Bellare, K., Pereira, F.: A conditional random field for discriminatively-trained finite-state string edit distance. In: Proceedings of the Twenty-First Conference on Uncertainty in Artificial Intelligence, pp. 388–395 (2005)
22. Santos, R., Murrieta-Flores, P., Calado, P., Martins, B.: Toponym matching through deep neural networks. Int. J. Geograph. Inf. Sci. **32**, 1–25 (2017)
23. Sil, A., Yates, A.: Re-ranking for joint named-entity recognition and linking. In: Proceedings of the International Conference on Information & Knowledge Management (CIKM), pp. 2369–2374 (2013)
24. Smith, T., Waterman, M.: Identification of common molecular subsequences. Journal of Molecular Biology **147**, 195–197 (1981)
25. Sultan, M.A., Bethard, S., Sumner, T.: DLS@CU: sentence similarity from word alignment and semantic vector composition. In: Proceedings of the 9th International Workshop on Semantic Evaluation (SemEval 2015), pp. 148–153 (2015)
26. Swartz, A.: Musicbrainz: a semantic web service. IEEE Intell. Syst. **17**, 76–77 (2002)
27. Tam, D., Monath, N., Kobren, A., Traylor, A., Das, R., McCallum, A.: Optimal transport-based alignment of learned character representations for string similarity. In: Proceedings of the Association for Computational Linguistics (ACL), pp. 5907–5917 (2019)
28. Winkler, W.: The state of record linkage and current research problems. Statist. Med. (1999)
29. Yamada, I., Shindo, H., Takeda, H., Takefuji, Y.: Joint learning of the embedding of words and entities for named entity disambiguation. In: Proceedings of the Conference on Computational Natural Language Learning (CoNLL), pp. 250–259 (2016)
30. Yang, Z., Yang, D., Dyer, C., He, X., Smola, A., Hovy, E.: Hierarchical attention networks for document classification. In: Proceedings of the Conference of the North American Chapter of the Association for Computational Linguistics (NAACL), pp. 1480–1489 (2016)

Applications

Geometric Attention for Prediction of Differential Properties in 3D Point Clouds

Albert Matveev[1], Alexey Artemov[1], Denis Zorin[1,2], and Evgeny Burnaev[1(✉)]

[1] Skolkovo Institute of Science and Technology, Moscow, Russia
e.burnaev@skoltech.ru
[2] New York University, New York, USA

Abstract. Estimation of differential geometric quantities in discrete 3D data representations is one of the crucial steps in the geometry processing pipeline. Specifically, estimating normals and sharp feature lines from raw point clouds helps improve meshing quality and allows us to use more precise surface reconstruction techniques. When designing a learnable approach to such problems, the main difficulty is selecting neighborhoods in a point cloud and incorporating geometric relations between the points. In this study, we present a geometric attention mechanism that can provide such properties in a learnable fashion. We establish the usefulness of the proposed technique with several experiments on the prediction of normal vectors and the extraction of feature lines.

Keywords: Attention · 3D computer vision · 3D point clouds

1 Introduction

Over the past several years, the amount of 3D data has increased considerably. Scanning devices that can capture the geometry of scanned objects are becoming widely available, and the computer vision community is showing a steady growth of interest in 3D data processing techniques. A range of applications includes digital fabrication, medical imaging, oil and gas modeling, and self-driving vehicles.

The geometry processing pipeline transforms input scan data into high-quality surface representation through multiple steps. The result's quality and robustness, set aside particular algorithms for surface reconstruction, are highly dependent on the performance of previous stages.

One of the cornerstone steps in the pipeline is estimating differential geometric properties like normal vectors, curvature, and, desirably, sharp feature lines. These properties, estimated from raw input point clouds, play a significant role in the surface reconstruction and meshing processes [9]. A multitude of algorithms for extracting such properties have been developed, however many of them require setting parameters for each point cloud separately, or performing grid search of parameters, making the computational complexity of such tasks burdensome.

© Springer Nature Switzerland AG 2020
F.-P. Schilling and T. Stadelmann (Eds.): ANNPR 2020, LNAI 12294, pp. 113–124, 2020.
https://doi.org/10.1007/978-3-030-58309-5_9

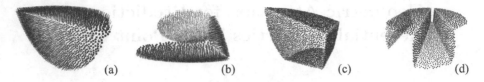

Fig. 1. Examples of sampled point clouds with the ground truth labels: (a), (b) – normals, (c), (d) – feature lines (see Sect. 4 for details).

On the contrary, the area of geometric deep learning has been emerging lately, which proposes tackling geometric problems with specialized deep learning architectures. Geometric deep learning techniques have shown success in problems of edge and vertex classification, edge prediction in graphs [3], graph classification with applications to mesh classification [7]; mesh deformation [12]; point cloud classification and segmentation [6]. In contrast, the estimation of geometric properties of surfaces has not been studied in depth.

A recently presented Transformer architecture [19] has studied the benefits of attention mechanisms for text processing, which has been established to be capable of detecting implicit relations between words in a sentence. When defining a local region in a point cloud, it is desirable to make use of such implicit relations between points, which makes attention a promising direction of research in the context of geometric problems. Such studies have started only recently, and most of the papers are focusing on semantic (classification) problems in point cloud processing. Little work has been done to improve the understanding of a geometry of the underlying surface.

In this paper, we present a novel attention-based module for improved neighborhood selection of point clouds. We call this module *Geometric Attention*. We show that it increases the quality of learnable predictions of geometric properties from sampled point cloud patches. As a qualitative result, we examine neighborhoods and argue that Geometric Attention is capable of introducing meaningful relations between points.

This paper is organized as follows. In Sect. 2, we provide an overview of related work, with the focus on geometry-related approaches and previous attention-based studies. Section 3 describes details of the proposed architecture. Experimental results are presented in Sect. 4, with both qualitative and quantitative results for the prediction of normal vectors and feature lines. We conclude in Sect. 5 with a brief discussion.

2   Related Work

Data sets. Availability of 3D data sets has increased in recent years. Collection ShapeNet [5] includes over 3 million objects. Another corpus is ModelNet [22], comprising 151,128 meshes, which is widely used for classification benchmarking. These collections of data do not fit the needs of geometric tasks due to no geometric labeling. Recently, a large-scale ABC data set [11] has been presented. It includes over 1 million high-quality CAD models; each of them is richly annotated with geometric, topological, and semantic information.

Differential quantities estimation is a standard problem for discrete surface processing. Since this problem is local, a point neighborhood is typically approximated using the k nearest neighbors (kNN). The most basic types of methods rely on fitting a local surface proxy [4]; others utilize statistical analysis techniques (e.g., [15]). A closely related property is sharp edges. This topic has not been studied in a learnable setup. Standard approaches to sharp features include analysis of covariance matrices [1,14] and clustering of normals [2]. Typically extracted sharp features are noisy and unstable.

Geometric deep learning on point clouds is a particularly popular research direction, as such architectures make minimal assumptions on input data. The primary limitation of these architectures is that they struggle to define point neighborhoods efficiently. The earlier instance of this type of networks is PointNet [17] and its successor PointNet++ [18]. PointNet++ relies on the spatial proximity of points at each layer of the network, composing a point cloud's local structures based on the Euclidean nearest neighbors approach. Some work has been done to improve neighborhood query, including non-spherical kNN search [20]. Similarly to these networks, Dynamic Graph CNN (DGCNN) [21] utilizes Euclidean nearest neighbors as an initial neighborhood extraction; however, these local regions are recomputed deeper in the network based on learned representations of points. PointWeb [25] has extended this architecture by defining a learnable point interaction inside local regions. Other networks base their local region extraction modules on the volumetric idea by dividing the volume that encloses point cloud into grid-like cells or constructing overlapping volumes around each point [8].

Attention. After the attention mechanism was shown to be beneficial in [19], many studies have adopted it for re-weighting and permutation schemes in their network architectures. In point clouds, attention has been used to refine classification and segmentation quality [13,23]. Reasoning similar to ours was presented in [24]. The authors decided to separate local and global information to ensure the rotational invariance of the network in the context of the semantic type of problems.

3   Neural Network Architecture

We start with a general problem statement; then, we describe DGCNN architecture with the main design choices. After that, we proceed with a description of the Geometric Attention module, which we incorporate into the DGCNN architecture.

Suppose you have a point cloud $P \in \mathbb{R}^{N \times 3}$, consisting of N points:

$$\mathbf{p}_i = (x_i, y_i, z_i) \in \mathbb{R}.^3$$

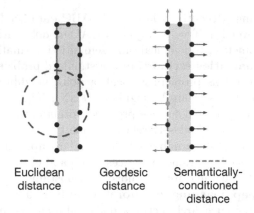

Euclidean   Geodesic   Semantically-
distance   distance   conditioned
                                        distance

Fig. 2. Different types of distances.

The goal is to construct a mapping

$$\varphi(\mathbf{p}_i) = \mathbf{y}_i,$$

where $\mathbf{y}_i$ are geometric properties defined for each point $\mathbf{p}_i$. The size of $\mathbf{y}$ may differ depending on the specific property. For instance, in case of normals vector $\mathbf{y}_i \in \mathbb{R}^3$; for sharp feature labels – $\mathbf{y}_i \in \{0, 1\}$.

The DGCNN architecture is based on the EdgeConv operation (Fig. 3), which, for an implicitly constructed local graph, extracts information from points as a step of propagation along edges. Technically, this is done through a proximity matrix:

$$\mathbf{PM} = (-d_{ij}) \in \mathbb{R}^{N \times N},$$

where $-d_{ij}$ is a negative distance from point $\mathbf{p}_i$ to $\mathbf{p}_j$.

This proximity matrix is then utilized to construct the adjacency matrix of the kNN graph $\mathcal{G} = (\mathcal{V}, \mathcal{E})$ by selecting for each point k nearest ones. After local areas are defined, inside each one, a multilayer perceptron (MLP) is applied to convolve neighborhood feature vectors. At the final stage, an aggregation operation (typically, a max pooling) is adopted to obtain new point features.

Following the notation from the original paper, we denote by $\mathbf{x}_i \in \mathbb{R}^F$ a feature vector of point $\mathbf{p}_i$. Then, the EdgeConv operator is defined by

$$\mathbf{x}'_i = \max_{j:(i,j) \in \mathcal{E}} h_\theta(\mathbf{x}_i, \mathbf{x}_j). \tag{1}$$

This EdgeConv operation is applied several times, then the outputs from each layer are concatenated together to get the final output. In such an architecture, point features carry both geometric and semantic information mixed.

3.1   Geometric Attention Module

Our main idea is to improve the feature extraction pipeline by modifying neighborhood selection. In point cloud data, it is a common problem that sampling

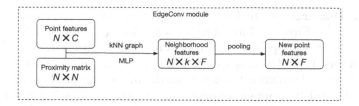

Fig. 3. EdgeConv module of the DGCNN architecture.

resolution is not enough to distinguish two sides of a thin plate (Fig. 2). To solve this problem, one would require a geodesic distance defined on a point cloud, which is not easy to get. However, with additional information (normal vectors or semantic partition of a point cloud into geometric primitives), disambiguating two sides of a plate is not an issue. For this reason, we introduce a semantically-conditioned distance to represent Euclidean proximity of points if they have similar semantic features.

Since such semantic information is not available, we attempt to disentangle semantic (global) features from geometric (local) information inside the network. We aim to implement this by adding semantic information flow and using both geometric proximity and semantic features-based attention in order to define the Geometric Attention module (refer to Fig. 4 for illustration).

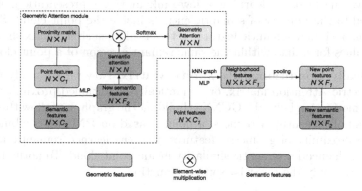

Fig. 4. Geometric attention module.

The motivation behind this choice follows from the fact that differential quantities are defined locally, which means that the output should be computed from a small vicinity of a point. At the same time, from the global point of view, differential properties are closely related to the smoothness of a surface; they are closer to each other within one geometric primitive. For instance, a normal vector field is the same inside one planar surface patch. This ambiguity may cause a struggle for the network when computing output. Hence we attempt to have these local and global types of data separated.

Semantic Features. As in DGCNN, we have point features (geometric features) $\mathbf{x}_i \in \mathbb{R}^{F_1}$. First, we introduce semantic feature vectors $\mathbf{f}_i \in \mathbb{R}^{F_2}$, which at the first layer are separately learned from point coordinates directly. In the following layers, these features are computed from concatenated vectors of geometric features and semantic features from the previous layer. Semantic features are devoted to solely represent semantic information in a rather simple manner as a soft one-hot encoding with 64 channels. To have them represent one-hot encoding, we divide each feature vector by its norm:

$$\mathbf{f}_i' = g_\phi(\mathbf{x}_i, \mathbf{f}_i),$$
$$\mathbf{f}_i' = \frac{\mathbf{f}_i'}{\|\mathbf{f}_i'\|}. \tag{2}$$

After semantic features are constructed, we apply Scaled Dot Product (SDP) attention to calculate the semantic attention matrix:

$$\mathbf{q}_i = g_{\tau_1}(\mathbf{f}_i')$$
$$\mathbf{k}_i = g_{\tau_2}(\mathbf{f}_i')$$
$$\mathbf{SA} = \frac{\mathbf{q}\mathbf{k}^\top}{\sqrt{t}} = \left(\frac{\langle \mathbf{q}_i, \mathbf{k}_j \rangle}{\sqrt{t}} \right), \tag{3}$$

where t is a scaling factor, which is set as the dimensionality of $\mathbf{f}_i'$ according to [19].

Ideally, this leads to learning a low-rank matrix representation, with the rank equal to the number of semantic entities inside the point cloud. Since the correlation of similar semantic feature vectors is high, such a matrix should have greater values for points within the same semantic region of a point cloud.

Semantically-Conditioned Proximity Matrix. Now we are ready to define the Geometric Attention matrix, or semantically-conditioned proximity matrix. Since the motivation behind DGCNN is building the graph, we follow this notion, but instead of measuring closeness of points based on **PM**, we combine purely Euclidean proximity of geometric features with the learned semantic attention matrix, which encodes semantic similarity inside point cloud. To normalize these matrices, we apply the row-wise softmax function:

$$\mathbf{SA} = \text{softmax}(\mathbf{SA}),$$
$$\mathbf{PM} = \text{softmax}(\mathbf{PM}), \tag{4}$$
$$\mathbf{GA} = \text{softmax}(\mathbf{SA} \otimes \mathbf{PM})$$

where **GA** is the Geometric Attention matrix, and $\otimes$ is element-wise matrix multiplication. The idea behind this decision is to relatively increase proximity values for those points that have similar semantic feature vectors, and decrease their closeness if semantic features are sufficiently different.

After the matrix **GA** is computed, we follow EdgeConv as in the original paper. The rest of the architecture is structured as Dynamic Graph CNN for segmentation tasks.

4   Experimental Results

We chose to predict normal vectors and detect sharp feature lines for the experimental evaluation of the Geometric Attention module. Since we focus on inferring the geometric understanding of the underlying surface, we argue that these two problems are the most representative as a benchmark. We note that the corresponding labeling could be easily obtained from the set of raw meshes, which has no additional labeling whatsoever. However, we believe that the quality of such labeling would be poor; hence, we opt to use ABC data set [11] to simulate data for our experiments.

4.1   Data Generation and Implementation Details

We start by designing the acquisition process. For a randomly selected point on a mesh surface, we begin growing the mesh neighborhood from the model by iteratively adding connected mesh faces. After the desired size has been reached, we apply the Poisson sampling technique aiming to obtain a point cloud with an average distance between points of 0.05 in original mesh units. The mesh patch size is selected such that after sampling is finished, it would ensure that the shortest sharp feature line is sampled with a predefined number of points. We found in our experiments that 8 points are sufficient to distinguish short curves robustly. When the cropped mesh patch has been sampled, we select 4096 points to provide sufficient sampling for geometric features. We refer to this set of points as point patch. We then use the initial mesh and labeling provided in the data set to transfer labels to the generated point patch. For normals, that is relatively straightforward, but for sharp features, we need to query points from point patch that are the closest to the mesh edges marked as sharp in ABC. We take point patch samples within one sampling distance tube from sharp edges and label them as "1". Using this process, we generate 200 k patches and divide them into training, validation, and test sets with ratios 4:1:1, respectively. See Fig. 1 for ground truth examples.

Table 1. Loss values for normals estimation and feature lines detection experiments.

Network	Normals		Feature lines
	Angular loss	RMSE loss	Balanced accuracy
DGCNN	0.01413	0.38618	0.9753
Ours	0.01236	0.38266	0.9892

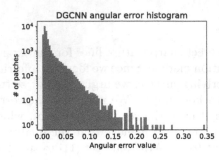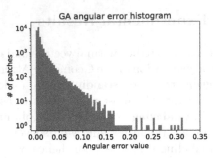

Fig. 5. Histogram of angular errors for normals estimation: left – DGCNN, right – Geometric Attention network.

We normalize point patches by centering them and scaling to fit inside the unit ball. During training, we augment the training data by randomly rotating it.

Our module has been implemented using the PyTorch [16] deep learning framework. We trained networks for 10 epochs with Adam optimizer [10] and learning rate 10^{-3}. Both of the networks were running with batch size 8 on one Tesla V100 GPU. We replaced all ReLU activations with LeakyReLU in order to avoid computational instabilities during normalization.

As discovered in the benchmark study devoted to normals estimation from [11], DGCNN provides the best accuracy among learnable methods with a smaller number of parameters; hence we base our experiments on comparing to DGCNN. We note that, even though the Geometric Attention module requires additional tensors to store the features, the number of parameters does not increase considerably.

4.2 Normals Estimation

The first task we experiment on is estimation of normal vectors. To do that, we use segmentation architecture with three output channels. We normalize the output to produce norm 1 vectors. As a loss function, we choose to optimize the loss from [11]:

$$\mathcal{L}(\mathbf{n}, \hat{\mathbf{n}}) = 1 - \left(\mathbf{n}^\top \hat{\mathbf{n}}\right)^2. \tag{5}$$

Although this loss function is producing the unoriented normals, we add a small regularization with Mean Squared Error functional. Refer to Table 1 for numerical results, where we report the angular loss (5) and mean Root Mean Squared Error computed over all patches. We provide the histograms of angular errors in Fig. 5.

The histograms indicate that albeit the results are similar, the tail of the loss distribution is thinner for Geometric Attention.

As one could see from Fig. 6, DGCNN does not always take into account directions of normals, and the Geometric Attention network can determine common normals direction. We believe that semantically-conditioned distances are helping with smoothing the result while keeping it geometrically meaningful.

Fig. 6. Normals estimation result: left – ground truth, middle – DGCNN, right – Geometric Attention network.

4.3   Sharp Feature Lines Extraction

For this experiment, the segmentation architecture was made to compute one value per point. We optimized the Binary Cross-Entropy loss for this segmentation task. Table 1 presents the Balanced accuracy value, which was computed as an average of true negative rate and true positive rate for each patch, and then averaged over all patches. The histograms of Balanced accuracies for DGCNN and our network are in Fig. 7.

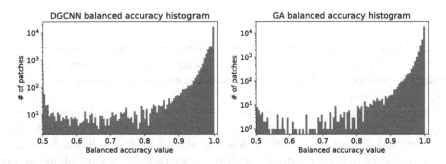

Fig. 7. Histogram of balanced accuracy values for sharp feature detection: left – DGCNN, right – Geometric Attention network.

A common issue with DGCNN predictions is missing the obtuse feature lines (as seen in Fig. 8, middle). Our network robustly detects such cases.

Lastly, we demonstrate the effect of semantically-conditioned proximity for the case of feature detection in Fig. 9. It shows that the two planes have been distinguished implicitly inside the network, and the feature line semantically separates them. The color-coding on the right image indicates the relative distances of all points from the query point. Note that the bright region border does not extend to the set of points marked as sharp, meaning that the kNN would only select points from the top plane.

Fig. 8. Feature lines detection result: left – ground truth, middle – DGCNN, right – Geometric Attention network.

Fig. 9. Semantically-conditioned proximity. For a query point (large green), we show the learned distances: left – Geometric Attention network prediction, right – relative distances (brighter – closer). (Color figure online)

5   Conclusion

In this paper, we have proposed the Geometric Attention module, which improves point neighborhood selection in point cloud-based neural networks. Unlike the previous studies, our approach is concentrated purely on geometric properties of a point cloud.

Experiments have shown that the quality of the estimated local geometric properties of the underlying surface has increased. Qualitative results indicate that our module can meaningfully define a semantically-conditioned distance. These claims have been confirmed with two experimental setups aimed at predicting surface normals and sharp feature lines.

The principal limitations of our approach are related to the common problems of the point cloud networks. Increasing the point patch size or the neighborhood radius leads to the rapid growth of the network, limiting scalability. The architecture design requires a fixed size of the inputs, which is not convenient in real-world applications and does not allow for the adaptive local region selection, which could be beneficial in many cases.

Possible directions of future research include the study of point interactions inside local regions for better feature extraction and further development of geometrically-inspired methods for robust geometry reconstruction from discrete surface representations.

Acknowledgments. The reported study was funded by RFBR, project number 19-31-90144. The Authors acknowledge the usage of the Skoltech CDISE HPC cluster Zhores for obtaining the results presented in this paper.

References

1. Bazazian, D., Casas, J.R., Ruiz-Hidalgo, J.: Fast and robust edge extraction in unorganized point clouds. In: 2015 International Conference on Digital Image Computing: Techniques and Applications (DICTA), pp. 1–8. IEEE (2015)
2. Boulch, A., Marlet, R.: Fast and robust normal estimation for point clouds with sharp features. Comput. Graph. Forum **31**(5), 1765–1774 (2012). https://doi.org/10.1111/j.1467-8659.2012.03181.x
3. Bronstein, M.M., Bruna, J., LeCun, Y., Szlam, A., Vandergheynst, P.: Geometric deep learning: going beyond euclidean data. IEEE Signal Process. Mag. **34**(4), 18–42 (2017)
4. Cazals, F., Pouget, M.: Estimating differential quantities using polynomial fitting of osculating jets. Comput. Aided Geom. Des. **22**(2), 121–146 (2005). https://doi.org/10.1016/j.cagd.2004.09.004
5. Chang, A.X., et al.: ShapeNet: an information-rich 3D model repository. Technical report arXiv:1512.03012 [cs.GR], Stanford University – Princeton University – Toyota Technological Institute at Chicago (2015)
6. Guo, Y., Wang, H., Hu, Q., Liu, H., Liu, L., Bennamoun, M.: Deep Learning for 3D point clouds: a survey. arXiv preprint arXiv:1912.12033 (2019)
7. Hanocka, R., Hertz, A., Fish, N., Giryes, R., Fleishman, S., Cohen-Or, D.: MeshCNN: a network with an edge. ACM Trans. Graph. (TOG) **38**(4), 1–12 (2019)
8. Hua, B.S., Tran, M.K., Yeung, S.K.: Pointwise convolutional neural networks. In: Computer Vision and Pattern Recognition (CVPR), pp. 984–993 (2018)
9. Kazhdan, M., Bolitho, M., Hoppe, H.: Poisson surface reconstruction. In: Proceedings of the Fourth Eurographics Symposium on Geometry Processing, vol. 7, pp. 61–70 (2006)
10. Kingma, D.P., Ba, J.: Adam: a method for stochastic optimization. arXiv preprint arXiv:1412.6980 (2014)
11. Koch, S., et al.: ABC: a big CAD model dataset for geometric deep learning. In: Proceedings of the IEEE Conference on Computer Vision and Pattern Recognition, pp. 9601–9611 (2019)
12. Kostrikov, I., Jiang, Z., Panozzo, D., Zorin, D., Bruna, J.: Surface networks. In: Proceedings of the IEEE Conference on Computer Vision and Pattern Recognition, pp. 2540–2548 (2018)
13. Liu, X., Han, Z., Liu, Y.S., Zwicker, M.: Point2sequence: learning the shape representation of 3D point clouds with an attention-based sequence to sequence network. In: Proceedings of the AAAI Conference on Artificial Intelligenc, vol. 33, pp. 8778–8785 (2019)
14. Merigot, Q., Ovsjanikov, M., Guibas, L.J.: Voronoi-based curvature and feature estimation from point clouds. IEEE Trans. Visual. Comput. Graph. **17**(6), 743–756 (2010)
15. Mura, C., Wyss, G., Pajarola, R.: Robust normal estimation in unstructured 3D point clouds by selective normal space exploration. Vis. Comput. **34**(6–8), 961–971 (2018). https://doi.org/10.1007/s00371-018-1542-6

16. Paszke, A., et al.: Pytorch: an imperative style, high-performance deep learning library. In: Advances in Neural Information Processing Systems, pp. 8026–8037 (2019)
17. Qi, C.R., Su, H., Mo, K., Guibas, L.J.: PointNet: deep learning on point sets for 3D classification and segmentation. In: Proceedings of the Computer Vision and Pattern Recognition (CVPR), vol. 1, no. 2, p. 4. IEEE (2017)
18. Qi, C.R., Yi, L., Su, H., Guibas, L.J.: PointNet++: deep hierarchical feature learning on point sets in a metric space. In: Guyon, I. (eds.) Advances in Neural Information Processing Systems, vol. 30, pp. 5099–5108. Curran Associates, Inc. (2017). http://papers.nips.cc/paper/7095-pointnet-deep-hierarchical-feature-learning-on-point-sets-in-a-metric-space.pdf
19. Vaswani, A., et al.: Attention is all you need. In: Advances in Neural Information Processing Systems, pp. 5998–6008 (2017)
20. Venkanna Sheshappanavar, S., Kambhamettu, C.: A novel local geometry capture in PointNet++ for 3D classification. In: The IEEE/CVF Conference on Computer Vision and Pattern Recognition (CVPR) Workshops, pp. 262–263, June 2020
21. Wang, Y., Sun, Y., Liu, Z., Sarma, S.E., Bronstein, M.M., Solomon, J.M.: Dynamic graph CNN for learning on point clouds. ACM Trans. Graph. (TOG) **38**(5), 1–12 (2019)
22. Wu, Z., et al.: 3D ShapeNets: a deep representation for volumetric shapes. In: 2015 IEEE Conference on Computer Vision and Pattern Recognition (CVPR), pp. 1912–1920, June 2015. https://doi.org/10.1109/CVPR.2015.7298801
23. Yang, J., et al.: Modeling point clouds with self-attention and gumbel subset sampling. In: Proceedings of the IEEE Conference on Computer Vision and Pattern Recognition, pp. 3323–3332 (2019)
24. Zhao, C., Yang, J., Xiong, X., Zhu, A., Cao, Z., Li, X.: Rotation invariant point cloud classification: where local geometry meets global topology. arXiv preprint arXiv:1911.00195 (2019)
25. Zhao, H., Jiang, L., Fu, C.W., Jia, J.: PointWeb: enhancing local neighborhood features for point cloud processing. In: The IEEE Conference on Computer Vision and Pattern Recognition (CVPR), pp. 5565–5573, June 2019

How (Not) to Measure Bias in Face Recognition Networks

Stefan Glüge[1]([⊠]) [iD], Mohammadreza Amirian[1,2] [iD], Dandolo Flumini[1] [iD],
and Thilo Stadelmann[1,3] [iD]

[1] Datalab, Wädenswil and Winterthur, Zurich University of Applied Sciences ZHAW,
Winterthur, Switzerland
{glue,amir,flum,stdm}@zhaw.ch
[2] Institute of Neural Information Processing, Ulm University, Ulm, Germany
[3] European Centre for Living Technology ECLT, Venice, Italy

Abstract. Within the last years Face Recognition (FR) systems have
achieved human-like (or better) performance, leading to extensive
deployment in large-scale practical settings. Yet, especially for sensible
domains such as FR we expect algorithms to work equally well for every-
one, regardless of somebody's age, gender, skin colour and/or origin. In
this paper, we investigate a methodology to quantify the amount of bias
in a trained Convolutional Neural Network (CNN) model for FR that
is not only intuitively appealing, but also has already been used in the
literature to argue for certain debiasing methods. It works by measuring
the "blindness" of the model towards certain face characteristics in the
embeddings of faces based on internal cluster validation measures. We
conduct experiments on three openly available FR models to determine
their bias regarding race, gender and age, and validate the computed
scores by comparing their predictions against the actual drop in face
recognition performance for minority cases. Interestingly, we could not
link a crisp clustering in the embedding space to a strong bias in recognition rates—it is rather the opposite. We therefore offer arguments for
the reasons behind this observation and argue for the need of a less naïve
clustering approach to develop a working measure for bias in FR models.

Keywords: Deep learning · Convolutional Neural Networks · Fairness

1 Introduction

FR has improved considerably and constantly over the last decade [13,17,25,40],
giving rise to numerous applications ranging from services on mobile consumer
devices, applications in sports, to the use by law enforcement agencies [32,35,
42,43]. The increased deployment has triggered an intense debate on the ethical
downsides of pervasive use of biometrics [6,29,34,39] up to the point where
regulation [23] and bans on the technology are discussed[1] and partially enforced[2].

---

[1] https://www.banfacialrecognition.com/.
[2] https://www.bbc.com/news/technology-48276660.

© Springer Nature Switzerland AG 2020
F.-P. Schilling and T. Stadelmann (Eds.): ANNPR 2020, LNAI 12294, pp. 125–137, 2020.
https://doi.org/10.1007/978-3-030-58309-5_10

This debate on ethical usage of FR technology is part of a larger trend in the machine learning field to account for ethical aspects of the methodology [7,28], which includes the aspects of trustworthiness[3], transparency (interpretability) [3,16] and fairness (bias) [4].

The issue of *bias* in machine learning is especially relevant in the area of FR, where we legitimately expect machine learning models to be unbiased because of their potentially large impact (e.g., for crime prediction [20]). The huge diversity due to race, gender and age in the appearance of human faces is however contrasted by a respective homogeneity of the data collections used to train such models. This leads to observations like the one that face recognition only works reliably for white grown-up males [8]. As working face recognition is increasingly relied on to grant individuals access to services and locations, and to predict people's behaviour, bias against certain people groups easily results in prohibitive discrimination.

The source of bias is usually the training material. Therefore, the community created datasets with known biases for race, skin color, gender and age, such as Racial Faces in-the-Wild (RFW) [45] and Diversity in Faces [33]. Given the bias in the data we are able to study the issue in the final models on two concrete levels: by (a) *quantifying* the amount of bias that exists in any trained FR model; and by (b) *reducing* identified bias in models by adequate countermeasures.

In this paper, we perform an in-depth exploration of a certain methodology for measuring the specific amount of bias that exists in any trained FR CNN. The underlying idea is appealing due to its intuitive approach and similar reasoning has already been used to argue for specific bias removal algorithms in the past [2]. The quantification itself relies on internal cluster validation measures for clusterings of embeddings based on labels for race, gender and age. It is agnostic towards the specific architecture and training procedure of the model and thus applicable to any FR system that exposes its embeddings; it is also non-invasive with respect to model training and does not effect the model's performance. Counterintuitively, our experiments speak against the validity of the idea and confirm the contrary: higher bias, as expressed in a drop in face recognition accuracy for minority cases, goes along with worse clustering, i.e. less "awareness"/more "blindness" of the model with respect to distinguishable features of the respective minority. We thus offer potential reasons for our observations, leading to preliminary results on how to better quantify bias in FR.

2   Related Work

The problem of bias in machine learning is well documented in the literature [30]. Several reasons trigger this bias: bias in the final decision as imposed by algorithm design, training process and loss design is addressed by the term *algorithmic bias*, though the term can be problematic[4]. *Selection bias* is introduced

---

[3] https://liu.se/en/research/tailor/.

[4] https://stdm.github.io/Algorithmic-bias/.

when human biases lead to selecting the wrong algorithm or data for FR, leading to biased decisions. *Data bias* finally is introduced by a lack of diversity or present imbalance in a dataset used for training or evaluating a model.

The presence of bias in model predictions for FR – leading to discrimination against certain factors such as race, age and gender of individuals – motivates two strands of recent research: (a) to automatically quantify the amount of bias in a model, and (b) to reduce it by a range of methods. Regarding bias measurement (a), Hannak et al. give criteria to accurately measure bias with respect to price discrimination [19]. Garcia et al. identify demographic bias by investigating the drop in confidence of face matching models for certain ethnicities [15]. Cavazos et al. use three different identification thresholds, namely the thresholds at equal false accept rates (FARs) and the recognition accuracy, to quantify racial bias for face matching [11]. Serna et al. show that FR bias is measurable using normalized overall activation of the models for different races [41]. In this paper, we explore a novel method to measure (quantify) bias that differs threefold from these approaches: (i) it is applicable to *any* model that exposes its embeddings, (ii) it is independed of model training and (iii) it is not based on model performance, but rather on the way faces are represented in the network.

Regarding bias reduction (b), most of the research in FR aims at tackling racial bias. However, Li et al. propose optimizing a distance metric for removing age bias. The remainder of the literature focuses on racial bias by improving both algorithmic and data biases [26]. Steed and Caliskan attempt to predict appearance bias of human annotators using transfer learning to estimate the bias in datasets [44], and Kortylewski el al. introduce synthetic data to reduce the negative effects of imbalance in datasets [22]. Yu et al. propose an adaptive triplet selection for correcting the distribution shift and model bias [47]. Robinson et al. show that the performance gaps in FR for various races can be reduced by adapting the decision thresholds for each race [36]. Domain transfer and adversarial learning are the other methods to reduce racial bias by adapting the algorithms. Wang et al. use a deep Information Maximization Adaptation Network (IMAN) for unsupervised knowledge transfer from the source domain (Caucasian) to target domains (other races) [45]. To remove the statistical dependency of the learned features to the source of bias (racial group), Adeli et al. propose an adversarial loss that minimizes the correlation between model representations and races [1]. Finally, Wang et al. [46] propose an update to the "Hard Debias" algorithm that post-processes word embeddings for unbiased text analysis and state that the idea might be transferable to other domains.

3   An Intuitively Appealing Method to Measure Bias

Human FR is not unbiased at all: we recognize faces that are most familiar much better than others. This "other-race effect" is one of the most robust empirical findings in human FR and accompanied by the popular belief that other-race faces all look alike [31]. The source of this drop in recognition performance for faces of unfamiliar origin seems to be that we know a rich feature set to distinguish between akin faces, but only know very coarse features for very differently looking faces. This results in the effect, that unfamiliar races appear to be different in general, which overlays how they differ amongst each other.

Humans associate the presence of bias with an observed *drop in recognition performance*; and the models seem to be the more biased the more *aware of the differences* between certain facial characteristics that are associated with potential discrimination. The method for measuring bias that we are concerned with in this paper builds upon both observations by exploiting them in the following way: first (a), bias in a specific model and for a specific characteristics (e.g., age, gender or race) is measured by quantifying how well the embeddings of a set of faces build clusters with respect to this characteristic. A good clustering into, for instance, age groups suggests that the model is very *aware* of the differences in age, which enables it to potentially discriminate age groups (in the two-fold meaning). Then (b), the resulting "score" is verified by experimentally checking for a *drop in FR performance* for faces with minority expressions for this characteristic. Alvi et al. argue along these lines in order to demonstrate the effect of an algorithm to remove bias: 'After unlearning gender, the feature representation is no longer separable by gender, demonstrating that this bias has been removed.' [2].

3.1 Quantifying Bias Through Internal Cluster Validation Measures

A straight-forward way to perform the respective bias quantification in (a) is to use existing cluster validity measures on the embeddings of FR models. The embeddings, usually taken as the activations of the last fully connected layer of a trained CNN during evaluation on a respective image, form a mapping of a face image into a compact Euclidean space where distances directly correspond to a measure of face similarity [40]. As the internal representation of the model contains the facial discriminant information, its embedding forms the basis for all recognition, similarity computation etc. A model with embeddings which do not cluster well with respect to a certain facial characteristics can be said to be "blind" towards the features that distinguish between its different expressions, which seems to be a good starting point for unbiasedness.

To eliminate the effect of many hyperparameters on the evaluation of the methodology, we rely on ground truth labels (either human provided or predicted by reference models) rather than a clustering algorithm for the membership of embeddings to clusters of specific discriminative characteristic. Hence, cluster membership is dependent only on the characteristics of the dataset itself and not on the FR model under evaluation and the only model-dependent part entering the bias measurement are the embeddings themselves. How well they cluster can then be quantified by so-called internal cluster validation measures [27] that are well established to measure the "goodness" of a clustering compared to other ones. Internal cluster validation measures are the correct ones to use because regardless of the source of our cluster memberships, we want to compare different clusterings with each other and not a clustering to ground truth. Generally, the indices are suitable for measuring crisp clustering, where no overlap between partitions is allowed [24]. For our evaluation, we compute the

Mean Silhouette Coefficient [38], Calinski-Harabasz Index [9], Davies-Bouldin Index [12] and Dunn Index [14]. However, for the Dunn Index we observe very small values and large variance in all experiments, resulting in no meaningful distinctions between different FR models. Therefore, those results are omitted.

The *Mean Silhouette Coefficient* is a measure of how similar an object is to its own cluster (cohesion) compared to other clusters (separation). It is bounded between -1 and $+1$, whereas *scores around zero indicate overlapping clusters*. Negative values indicate that there may be too many or too few clusters, and *positive values towards 1 indicate well separable clusters*. The Silhouette Coefficient for a set of samples is given as the mean of the Silhouette Coefficients per sample. The *Calinski-Harabasz Index*, also known as the Variance Ratio Criterion, is defined as the ratio of the between-cluster variance and the within-cluster variance. Well-defined clusters have large between-cluster and a small within-cluster variance, i.e. a *higher score relates to a model with better defined clusters*. Finally, the *Davies-Bouldin Index* computes for each cluster the other cluster that it is most similar to. Afterwards, it summarizes the maximum cluster similarities to create a single index. A low index indicates that the clusters are not very similar, i.e. *a low value relates to a model with better separation between the clusters*.

3.2   Models, Dataset and Experimental Setup

In the following, we describe our experimental setup to (a) measure bias in the embedding space based on internal cluster validation measures, and (b) validate the resulting score based on the drop in face recognition performance on a benchmark dataset. We choose three different FR models from the popular and well-established Visual Geometry Group (VGG) family, openly available from the VGG, perform measurements and validate on the RFW dataset to study the bias for race, gender and age.

We use trained models that are available directly from the authors[5]. They were pretrained on the MS-Celeb-1M [18] dataset and then fine-tuned on VGGFace2, which contains 3.31 million images of 9 131 identities [10]. All models follow the SE-ResNet-50 architectural configuration in [21], but differ in the dimensionality of embedding layer (128D/256D) which is stacked on top of the original final feature layer (2048D) adjacent to the classifier. All models were trained with standard softmax loss.

The RFW dataset was designed to study racial bias in FR systems [45]. It is constructed with four testing subsets, namely Caucasian, Asian, Indian and African. Each subset contains about 10k images of 3k individuals for face verification. We further added a gender and age label to each test image of the RFW dataset using a Wide Residual Network trained on the UTKFace [49] and IMDB-WIKI [37] datasets[6]. The age prediction is in the range of 0–100. For the cluster evaluation, we split the age predictions into the three non-overlapping groups <30, 30–45 and 45+. The boundaries are chosen such that we have at

---

[5] https://github.com/ox-vgg/vgg_face2.
[6] https://github.com/yu4u/age-gender-estimation.

least 3,000 samples in each class. Gender prediction follows the same procedure as age prediction. The model yields a continuous gender score s_{gender} between 0 and 1, whereas lower values indicate male and higher values indicate female. In order to use all samples from the dataset, we split it at $s_{gender} < 0.5$ for male and $s_{gender} > 0.5$ for female. Table 1 gives an overview of the resulting number of samples per cluster with respect to the characteristic race, age and gender. As one can see, the race clusters are nicely balanced, whereas for gender we have a strong imbalance towards "male", and for age the 30–45-group is dominant.

Table 1. Number of samples per cluster regarding different facial characteristics in the RFW dataset that are associated with bias.

Face characteristic	Clusters	#samples
Race (human annotation)	Caucasian; Indian; Asian; African	$10,099; 10,221$ $9,602; 10,397$
Age [years] (predicted)	<30; 30–45; 45+	$4,815; 32,530; 3,046$
Gender (predicted)	male; female	$28,928; 11,463$

For our evaluation we extract the embeddings of the approximately 40k face images from the RFW testset for each of the VGG2 models. Face detection and alignment is done using the MTCNN approach[7] proposed by Zhang et al. [48]. Based on the embeddings, we report the FR rates and the cluster validation measures as per the dimensions race, gender and age. For face recognition, we report a match if the sample that is the nearest neighbor to the test face comes from the same person.

4   Results

So far, we have discussed two proxies for quantifying bias. (a) *goodness of clustering* of the embeddings w.r.t to the different expressions of a facial characteristic like age, gender or race—the higher, the more bias. It can be measured for any model that exhibits its embeddings, given that a dataset with labels for these expressions exists. It is thus a candidate for a *measurement methodology* to quantify bias in general. (b) *face recognition rate* for cases that belong to the minority expression of said characteristics—the lower, the more bias. This approach needs labels and multiple samples of persons, but serves as a measure of the real-world impact of bias/discrimination (as people from minority groups are less well handled); it is thus our candidate to *validate* the bias as measured by proxy (a).

Table 2 shows face recognition rates per model and expressions of facial characteristic (b), alongside the introduced cluster validation indices (a). We highlight the best recognition rates and the lowest percentual difference compared

---

[7] https://github.com/YYuanAnyVision/mxnet_mtcnn_face_detection.

to the mean over the different expressions of a characteristic (i.e., lowest actual bias). Further, we highlight the worst clustering according to each index (i.e., lowest measured bias as predicted by the method under consideration here).

Table 2. Bias measurement results (bad Clustering Score) vs. bias validation results (good Recognition Rate and low Relative Difference) per model and characteristic/expression. Clustering scores are the Mean Silhouette Coefficient (MS), Calinski-Harabasz Index (CH) and Davies-Bouldin Index (DB); ↑ and ↓ depict if a high or low value indicate a good clustering, respectively. Lowest bias according to each type of score is highlighted.

Architecture (#features)	Metric	Race Caucasian	Indian	Asian	African	Expr. Avg.	Clustering Score MS ↑	CH ↑	DB ↓
VGG2 (128)	Rec. Rate	**0.8906**	**0.8531**	**0.8310**	**0.7998**	**0.8436**	0.062	1,812	3.85
	Rel. Diff. (%)	5.5648	1.1271	-1.5011	-5.1907	-			
VGG2 (256)	Rec. Rate	0.8787	0.8265	0.7981	0.7597	0.8158	**0.029**	**766**	**6.20**
	Rel. Diff. (%)	7.7157	1.3208	-2.1693	-6.867	-			
VGG2 (2048)	Rec. Rate	0.8799	0.8472	0.8305	0.7959	0.8384	0.050	1,473	4.49
	Rel. Diff. (%)	**4.9542**	**1.0523**	**-0.9427**	**-5.0638**	-			
		Gender Male	Female						
VGG2 (128)	Rec. Rate	**0.8381**	**0.8576**			**0.8479**	0.0048	135.3	15.56
	Rel. Diff. (%)	**-1.1507**	**1.1507**			-			
VGG2 (256)	Rec. Rate	0.8088	0.833			0.8211	**0.0017**	**64.44**	**22.55**
	Rel. Diff. (%)	-1.5039	1.5039			-			
VGG2 (2048)	Rec. Rate	0.8327	0.8525			0.8426	0.0063	143.7	15.11
	Rel. Diff. (%)	-1.1725	1.1725			-			
		Age < 30	30–45	45+					
VGG2 (128)	Rec. Rate	**0.8671**	**0.8358**	**0.8907**		**0.8645**	0.0002	21.92	34.34
	Rel. Diff. (%)	0.2971	-3.3234	3.0263		-			
VGG2 (256)	Rec. Rate	0.8424	0.8063	0.8749		0.8412	**-0.0003**	**8.51**	**52.90**
	Rel. Diff. (%)	0.139	-4.1481	4.009		-			
VGG2 (2048)	Rec. Rate	0.8638	0.8305	0.8821		0.8588	0.0006	25.40	32.26
	Rel. Diff. (%)	**0.5789**	**-3.2981**	**2.7192**		-			

For race and age, the VGG2 (128) model performs best regarding pure recognition rates, whereas VGG2 (2048) shows the lowest performance drop, i.e. is the least biased model regarding race and age. For gender we observe only a marginal difference in recognition rates and performance drops between those models. The VGG2 (256) model is the worst option with respect to recognition rates as well as performance drops (actual bias). Looking at the clustering scores (measured bias), much to our surprise, this same VGG2 (256) model produces the worst clustering with respect to all validation indices and therefore can be considered to be the model with the least distinctive face representations regarding race, gender and age (lowest measure bias). However, this is not reflected in the class-based performance drop for the recognition rates (that should be small). Regarding the age groups, the Mean Silhouette Coefficient takes very small or negative values which indicates overlapping and/or an unsuitable number of clusters. Given the continuous nature of age, this is to be expected.

5   Discussion and Conclusions

In general, we could not link a crisp clustering in the embedding space to a strong bias in the recognition rates. In our experiments we found quite the opposite. This spawns discussion on three levels:

First, on the level of the *underlying reasoning*: it needs to be checked how much the two proxies used here to quantify bias (namely, face recognition performance on minority examples as a sign for the practical effect of bias; and well-defined clusters in the embedding space with respect to typically biased characteristics as a way to measure/predict this real-world influence) are actually correlated with what is meant by "bias". This is especially relevant in the light of the fact that parts of this reasoning have already been adopted in the literature as a way to show the effectiveness of debiasing algorithms.

Second, on the level of *implementation*: Even if the notion of bias is reflected in the embedding space, the adopted naïve clustering approach using only broad expression types for races, gender and age groups can be reconsidered. We hypothesize that a cluster like "male" or "African" is too general and rather formed of multiple sub-clusters in the embedding space. Thus, a cluster validation index on male/female cannot reflect the actual awareness of the model of these ultimately relevant sub-clusters. This intuition is supported by a visualisation of the embeddings as show in Fig. 1. For all three models, one can see that an expression of race such as "Caucasian" is comprised of at least two sub-clusters. However, one has to keep in mind that the t-SNE representation is just a projection into 2D space from the original 128/256/2048D space and generates slightly different results each time on the same data set. Furthermore, for the age characteristic, the distribution of embeddings into three clusters was somewhat arbitrarily based on a balance argument and could be chosen differently. Additional "hyperparameters" of the methodology and hence candidates for further experiments are the tested model architectures and their training details (especially the used loss functions have a large effect on the embeddings and the space spanned by them).

Third, on the level of *insight/explanation*: focusing now solely on the example of racial bias for illustrative reasons, the idea of how to measure bias in this paper relies on the assumption that the main source of bias in FR is the separation of races in the embedding space (the better separated, the more awareness, hence the more biased); this could be measured by clustering quality with respect to different expressions of race. The failure to observe any such correlation could be due to, we conjecture, the between-cluster separation being less important to explain bias than the *within-cluster distribution* (i.e., it doesn't mean too much how e.g. "Africans" are separated from "Asians" in the embedding space – it is much more important how the "African" embeddings are distributed amongst each other). To underpin this hypothesis, we present the distribution of pairwise distances between test embeddings from various races in Fig. 2.

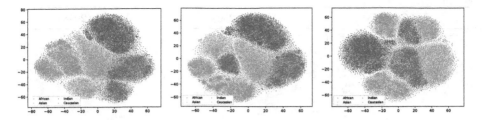

Fig. 1. T-SNE visualisations of embeddings from VGG2 (128) (a), VGG2 (256) (b) and VGG2 (2048) (c). The samples are colored according to their race.

Figure 2 allows the conclusion that *the average distance of the embeddings for "Caucasians" is higher* than the one for other races. This means in turn that the embeddings of "Caucasians" are distributed with a lower density in the embedding space. At the same time, these embeddings are the ones with the best recognition accuracies. The same observation is supported by the experiment behind Fig. 3 that uses the k-nearest neighbor classifier with varying k for *race* classification. "Africans" show the highest race recognition rate, suggesting that the embeddings are concentrated with high density in a specific region (similar to the lower average pairwise distance according to Fig. 2). The race recognition accuracy of "Caucasian" embeddings appropriately is the smallest and drops with the number of nearest neighbors, suggesting a low-density distribution of the embeddings for this race.

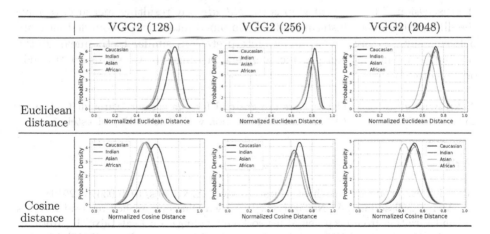

Fig. 2. Probability density distribution of pairwise (Euclidean and Cosine) distances between test image embeddings of different races.

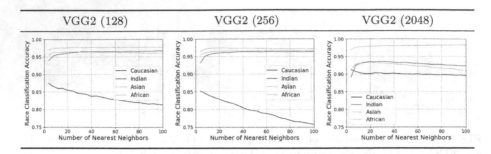

Fig. 3. Accuracy of k-nearest neighbors classifiers for race classification in the embedding space.

In summary, we presented an intuitively motivated idea on how to measure bias in any existing FR model that exposes its embeddings, and how to validate it based on FR accuracy. A similar reasoning has been used in the past in the literature to argue for the benefits of certain debiasing methods. This is why the presented results, though "negative" (they did not confirm the validity of the method, but testified to the opposite effect), are still very important: they show that similarly to what is known as the "curse of dimensionality" [5], intuition fails in this complex scenario, and assumptions need to be more thoroughly checked. Nevertheless, the given explanatory approaches show a way to turn the underlying reasoning into usable measures of bias in the future.

Future work will thus first focus on finding answers to the questions raised in the discussion. Then, a next step is to *calibrate* any resulting measure: to have the differently scaled clustering indices combined into a single bounded measure between, say, −1 and 1 which allows interpretations similar to the meaning the correlation coefficient can provide (i.e, certain ranges of values mean "biased" or "bias-free").

Acknowledgements. The authors are grateful for the support through CTI grant 25256.1 PFES-ES "LIBRA" and the collaboration with Gilbert F. Duivesteijn.

References

1. Adeli, E., et al.: Bias-resilient neural network. ArXiv abs/1910.03676 (2019)
2. Alvi, M., Zisserman, A., Nellåker, C.: Turning a blind eye: explicit removal of biases and variation from deep neural network embeddings. In: Leal-Taixé, L., Roth, S. (eds.) ECCV 2018. LNCS, vol. 11129, pp. 556–572. Springer, Cham (2019). https://doi.org/10.1007/978-3-030-11009-3_34
3. Amirian, M., Schwenker, F., Stadelmann, T.: Trace and detect adversarial attacks on CNNs using feature response maps. In: Pancioni, L., Schwenker, F., Trentin, E. (eds.) ANNPR 2018. LNCS (LNAI), vol. 11081, pp. 346–358. Springer, Cham (2018). https://doi.org/10.1007/978-3-319-99978-4_27
4. Bellamy, R.K.E., et al.: AI fairness 360: an extensible toolkit for detecting and mitigating algorithmic bias. IBM J. Res. Dev. **63**(4/5), 1–15 (2019)

5. Bellman, R.: Dynamic Programming. Princeton University Press, Princeton (1957)
6. Bernal, P.: Data gathering, surveillance and human rights: recasting the debate. J. Cyber Policy **1**(2), 243–264 (2016)
7. Brundage, M., Avin, S., Clark, J., Toner, H., et al.: The malicious use of artificial intelligence: forecasting, prevention, and mitigation. ArXiv abs/1802.07228 (2019)
8. Buolamwini, J.A.: Gender shades: intersectional phenotypic and demographic evaluation of face datasets and gender classifiers. Master's thesis, MIT (2017)
9. Caliński, T., Harabasz, J.: A dendrite method for cluster analysis. Commun. Stat. **3**(1), 1–27 (1974)
10. Cao, Q., Shen, L., Xie, W., Parkhi, O.M., Zisserman, A.: VGGFace2: a dataset for recognising faces across pose and age. In: International Conference on Automatic Face Gesture Recognition, pp. 67–74 (2018)
11. Cavazos, J.G., Phillips, P.J., Castillo, C.D., O"Toole, A.J.: Accuracy comparison across face recognition algorithms: where are we on measuring race bias? ArXiv abs/1912.07398 (2019)
12. Davies, D.L., Bouldin, D.W.: A cluster separation measure. IEEE Trans. Pattern Anal. Mach. Intell. **PAMI-1**(2), 224–227 (1979)
13. Deng, J., Guo, J., Xue, N., Zafeiriou, S.: ArcFace: additive angular margin loss for deep face recognition. In: CVPR, pp. 4690–4699 (2019)
14. Dunn, J.C.: Well-separated clusters and optimal fuzzy partitions. J. Cybern. **4**(1), 95–104 (1974)
15. Garcia, R.V., Wandzik, L., Grabner, L., Krueger, J.: The harms of demographic bias in deep face recognition research. In: ICB, pp. 1–6 (2019)
16. Gunning, D., Stefik, M., Choi, J., Miller, T., Stumpf, S., Yang, G.Z.: Xai— explainable artificial intelligence. Sci. Robot. **4**(37) (2019)
17. Guo, G., Zhang, N.: A survey on deep learning based face recognition. Comput. Vis. Image Underst. **189**, 102805 (2019)
18. Guo, Y., Zhang, L., Hu, Y., He, X., Gao, J.: MS-Celeb-1M: a dataset and benchmark for large-scale face recognition. In: Leibe, B., Matas, J., Sebe, N., Welling, M. (eds.) ECCV 2016. LNCS, vol. 9907, pp. 87–102. Springer, Cham (2016). https://doi.org/10.1007/978-3-319-46487-9_6
19. Hannak, A., Soeller, G., Lazer, D., Mislove, A., Wilson, C.: Measuring price discrimination and steering on e-commerce web sites. In: Conference on Internet Measurement Conference, pp. 305–318 (2014)
20. Hashemi, M., Hall, M.: Criminal tendency detection from facial images and the gender bias effect. J. Big Data **7**(1), 1–16 (2020). https://doi.org/10.1186/s40537-019-0282-4
21. Hu, J., Shen, L., Sun, G.: Squeeze-and-excitation networks. In: CVPR (2018)
22. Kortylewski, A., Egger, B., Schneider, A., Gerig, T., Morel-Forster, A., Vetter, T.: Analyzing and reducing the damage of dataset bias to face recognition with synthetic data. In: CVPR (2019)
23. Learned-Miller, E., Ordóñez, V., Morgenster, J., Buolamwini, J.: Facial recognition technologies in the wild: a call for a federal office. Technical report, Algorithmic Justice League, May 2020
24. Legány, C., Juhász, S., Babos, A.: Cluster validity measurement techniques. In: International Conference on Artificial Intelligence, Knowledge Engineering and Data Bases, pp. 388–393 (2006)
25. Li, S., Jain, A.: Handbook of Face Recognition. Springer, London (2011). https://doi.org/10.1007/978-0-85729-932-1

26. Li, Y., Wang, G., Nie, L., Wang, Q., Tan, W.: Distance metric optimization driven convolutional neural network for age invariant face recognition. Pattern Recognit. **75**, 51–62 (2018)
27. Liu, Y., Li, Z., Xiong, H., Gao, X., Wu, J.: Understanding of internal clustering validation measures. In: International Conference on Data Mining, pp. 911–916 (2010)
28. Loi, M., Heitz, C., Christen, M.: A comparative assessment and synthesis of twenty ethics codes on AI and big data. In: Swiss Conference on Data Science (2020)
29. Mann, M., Smith, M.: Automated facial recognition technology: recent developments and approaches to oversight. Univ. N. S. W. Law J. **40**, 121–145 (2017)
30. Mehrabi, N., Morstatter, F., Saxena, N., Lerman, K., Galstyan, A.: A survey on bias and fairness in machine learning. ArXiv abs/1908.09635 (2019)
31. Meissner, C.A., Brigham, J.C.: Thirty years of investigating the own-race bias in memory for faces: a meta-analytic review. Psychol. Public Policy Law, 3–35 (2001)
32. Merler, M., Mac, K.N.C., Joshi, D., et al.: Automatic curation of sports highlights using multimodal excitement features. IEEE Trans. Multimed. **21**(5), 1147–1160 (2019)
33. Merler, M., Ratha, N.K., Feris, R.S., Smith, J.R.: Diversity in faces. ArXiv abs/1901.10436 (2019)
34. Norval, A., Prasopoulou, E.: Public faces? A critical exploration of the diffusion of face recognition technologies in online social networks. New Media Soc. **19**(4), 637–654 (2017)
35. Robertson, D.J., Noyes, E., Dowsett, A.J., Jenkins, R., Burton, A.M.: Face recognition by metropolitan police super-recognisers. PloS ONE **11**, 1–8 (2016)
36. Robinson, J.P., Livitz, G., Henon, Y., Qin, C., Fu, Y., Timoner, S.: Face recognition: too bias, or not too bias? ArXiv abs/2002.06483 (2020)
37. Rothe, R., Timofte, R., Gool, L.V.: Deep expectation of real and apparent age from a single image without facial landmarks. Int. J. Comput. Vis. **126**(2–4), 144–157 (2018)
38. Rousseeuw, P.J.: Silhouettes: a graphical aid to the interpretation and validation of cluster analysis. J. Comput. Appl. Math. **20**, 53–65 (1987)
39. Royakkers, L., Timmer, J., Kool, L., van Est, R.: Societal and ethical issues of digitization. Ethics Inf. Technol. **20**(2), 127–142 (2018)
40. Schroff, F., Kalenichenko, D., Philbin, J.: FaceNet: a unified embedding for face recognition and clustering. In: CVPR, pp. 815–823 (2015)
41. Serna, I., Peña, A., Morales, A., Fierrez, J.: InsideBias: measuring bias in deep networks and application to face gender biometrics. ArXiv abs/2004.06592 (2020)
42. Smith, D.F., Wiliem, A., Lovell, B.C.: Face recognition on consumer devices: reflections on replay attacks. IEEE Trans. Inf. Forensics Secur. **10**(4), 736–745 (2015)
43. Stadelmann, T., et al.: Deep learning in the wild. In: Pancioni, L., Schwenker, F., Trentin, E. (eds.) ANNPR 2018. LNCS (LNAI), vol. 11081, pp. 17–38. Springer, Cham (2018). https://doi.org/10.1007/978-3-319-99978-4_2
44. Steed, R., Caliskan, A.: Machines learn appearance bias in face recognition. ArXiv abs/2002.05636 (2020)
45. Wang, M., Deng, W., Hu, J., Tao, X., Huang, Y.: Racial faces in the wild: reducing racial bias by information maximization adaptation network. In: ICCV, pp. 692–702 (2019)
46. Wang, T., Lin, X.V., Rajani, N.F., McCann, B., et al.: Double-hard debias: tailoring word embeddings for gender bias mitigation. ArXiv abs/2005.00965 (2020)

47. Yu, B., Liu, T., Gong, M., Ding, C., Tao, D.: Correcting the triplet selection bias for triplet loss. In: Ferrari, V., Hebert, M., Sminchisescu, C., Weiss, Y. (eds.) ECCV 2018. LNCS, vol. 11210, pp. 71–86. Springer, Cham (2018). https://doi.org/10.1007/978-3-030-01231-1_5. Please check and confirm the edit made in Ref. [47].
48. Zhang, K., Zhang, Z., Li, Z., Qiao, Y.: Joint face detection and alignment using multitask cascaded convolutional networks. IEEE Signal Process. Lett. **23**(10), 1499–1503 (2016)
49. Zhang, Z., Song, Y., Qi, H.: Age progression/regression by conditional adversarial autoencoder. In: CVPR, pp. 4352–4360 (2017)

Feature Extraction: A Time Window Analysis Based on the X-ITE Pain Database

Tobias Ricken[1], Adrian Steinert[1], Peter Bellmann[1]([✉]), Steffen Walter[2], and Friedhelm Schwenker[1]

[1] Institute of Neural Information Processing, Ulm University,
James-Franck-Ring, 89081 Ulm, Germany
{tobias.ricken,adrian.steinert,peter.bellmann,
friedhelm.schwenker}@uni-ulm.de
[2] Department of Medical Psychology, Ulm University,
Frauensteige 6, 89075 Ulm, Germany
steffen.walter@uni-ulm.de

Abstract. In this work, we analyse different temporal feature extraction window approaches, in combination with short-time heat and electric pain stimuli. Thereby, we focus on the physiological signals of the Experimentally Induced Thermal and Electrical (X-ITE) Pain Database. Each of our proposed approaches is evaluated based on the leave-one-subject-out cross-validation using the random forest method. Moreover, the effectiveness of each physiological signal is inspected separately, as well as by applying the feature fusion approach. Thereby, we analyse different binary classification tasks, as well as four-class classification tasks. Our outcomes indicate that a shifted temporal feature extraction window increases the classification performance significantly, when pain is induced by thermal stimuli. Moreover, our evaluations point out that the outcomes differ significantly, when participants are exposed to electrical pain stimuli. For short-term electric pain stimuli, the best results are obtained without temporal shifts of the feature extraction windows.

Keywords: Feature extraction · Time window analysis · Pain intensity classification · Physiological signals · Heat and electric pain analysis

1 Introduction

The approach to automated pain recognition is primarily driven from the findings that patients with limited communication options on the one hand receive an oversupply or undersupply of analgesics and on the other hand humane external observation scales only allow a limited temporal resolution. To further elucidate the pain configuration, machine learning algorithms have therefore been developed over the last years and tested in healthy volunteers in the form of

---

T. Ricken and A. Steinert—These authors contributed equally to this work.

© Springer Nature Switzerland AG 2020
F.-P. Schilling and T. Stadelmann (Eds.): ANNPR 2020, LNAI 12294, pp. 138–148, 2020.
https://doi.org/10.1007/978-3-030-58309-5_11

cross-validation [8,9,12,13]. The automated measurement of pain intensity is based on the sensory recording of pain reactions. Such sensory recordings are for example electrodermal activity (EDA), muscle activity (EMG) of the facial muscles and the musculus trapezius, and cardiovascular activity (ECG). Furthermore, behavioural responses to pain are studied: head posture and movement, paralinguistic utterances such as groaning, sighing, loud breathing and voice characteristics when speaking [9,12,13]. The most comprehensive multimodal data sets are the X-ITE Pain Database [3] (134 subjects, about 25,000 pain stimuli), the BioVid Heat Pain Database [11] (87 subjects, about 14,000 pain stimuli), and the SenseEmotion Database [10] (45 subjects, about 8,000 pain stimuli). In the current work, we evaluate different temporal window approaches for the extraction of features.

The remaining sections are structured as follows. Section 2 presents a small overview of related work. Section 3 provides a brief summary of the X-ITE Pain Database. Subsequently, we present the steps for signal preprocessing, the extraction of features, as well as the definition for different temporal windows. Section 4 provides a description of the different experimental settings and the corresponding results. In Sect. 5, we conclude the current work by discussing the outcomes and providing some ideas for future work.

2   Related Work

Many researches contribute to the field of automated pain assessment. A well performing system relies on a good representation of the data, reliable methods for the feature extraction, as well as appropriate classification models, which are able to make the most use out of the extracted features. Thiam et al. [8] conducted a feature extraction time window analysis, in combination with the random forest approach [1], based on the SenseEmotion Database [10]. Their outcomes show that extracting features from physiological data starting after the onset (the moment, when the induced temperature starts to increase to a predefined heat level) improved the classification accuracy. This indicates that different shifts and window lengths can increase the accuracy value. For the classification task no pain vs. the highest pain intensity, the highest accuracy value is reported as 80.24%.

Furthermore, Thiam et al. [7] proposed a deep model architecture for automated feature learning. A convolutional neural network was used to learn a signal representation, which led to the current state-of-the-art classification accuracy values, based on the BioVid Heat Pain Database [11]. For the EDA modality, the highest reported accuracy value is 84.57%, for the no pain vs. the highest pain intensity setting. In [9], Thiam et al. analysed a deep two-stream attention network for pain recognition based on video sequences, in combination with the BioVid Heat Pain Database.

Werner et al. [13] were the first to experiment with the X-ITE Pain Database [3]. The aim of their work was to provide and analyse initial baseline results. All classification tasks, based on random forest classifiers, were evaluated for all single modalities (physiological, audio and video data), as well as for the feature fusion and different decision fusion approaches. The outcomes in [13] show that audio is the worst performing modality. The best accuracy value, for the task no pain vs. the highest heat pain intensity, is reported for the EDA channel with a value of 79.1%. For the no pain vs. the highest electric pain intensity setting, the EDA channel led to the best accuracy value of 91.1%. For a broad overview on automated pain assessment, we refer the reader to a recently published survey article, by Werner et al. [14].

3   Methods

3.1   Dataset Description

In this work, we focus on the X-ITE Pain Database [3], which contains data from multiple sensors recorded during induced heat and electrical pain stimuli. The database provides data for short-term (phasic) and long-term (tonic) pain inductions. A total of 134 healthy subjects (67 women and 67 men) participated in the experiments. The recorded data is composed of the following modalities: ECG, EMG, EDA, as well as video and audio data. ECG signals measure a person's heart activity. EMG sensors measure a person's muscle activity. The EMG sensors were attached to the muscles corrugator, zygomaticus and trapezius. EDA signals measure a person's skin conductance. The EDA sensors were attached to the participants' ring and index finger. Each phasic pain stimulus has a length of 4 s (heat) and 5 s (electric), each tonic stimulus has a length of 60 s. Therefore, the X-ITE Pain Database provides the following four pain stimulation categories: phasic heat, phasic electric, tonic heat and tonic electric. For each category, three different pain intensities were predefined, for each test subject. The lowest phasic heat intensity (H1) is defined as the point at which a subject started feeling pain as a transition from just feeling warmth. The lowest phasic electric intensity (E1) is defined as the point at which a subject started feeling pain instead of tingling. The highest pain intensity, for both, heat (H3) and electric (E3), is defined as the point for which the current participant reported that the pain was getting unbearable. Additional intermediate pain intensity levels, denoted by H2 and E2, are defined as the mean values of H1 and H3, as well as the mean of E1 and E3, respectively. Each pain stimulus was followed by a non-painful stimulus, the so-called baseline (B), which was set to 32 °C. The phasic baseline has a random length between 8 and 12 s. Similarly, each tonic pain stimulus was followed by the baseline stimulus, with a fixed duration of 300 s. All biopotentials were sampled with 1000 Hz. In total, each subject was exposed to all four pain stimuli categories in a randomised order, for about 90 min, during one single recording session. Each phasic stimulus was induced 30 times per intensity, whereas each tonic stimulus was induced only once per intensity. In the current work, we focus on the physiological signals, in combination with the phasic stimuli. Note that in contrast to the description from the database paper [3], the phasic heat pain stimuli were always induced with a duration of 4 s (not 5 s).

3.2 Signal Preprocessing

We removed a total of nine participants – with the corresponding test subject IDs 1, 14, 23, 24, 25, 28, 30, 59 and 121 – due to technical issues, such as erroneous or missing data. To ensure that there is no overlap during the computation of the features, for each of the temporal windows, we defined the following criteria: minimum pain stimulus length of 3.9 s, minimum baseline duration of 8 s, each baseline has to follow a pain stimulus and the data of each subject has to start with a valid pain stimulus followed by a baseline. We removed all sequences, which violated at least one criterium. Moreover, the database contains samples, which are labelled as −10 and −11. We removed those samples from our experiments, due to the fact that they were not specified in [3]. A third-order Butterworth bandpass filter was applied to the ECG and EMG signals. For ECG, we used cut-off frequencies of 0.1 Hz and 25 Hz. For EMG, we used cut-off frequencies of 20 Hz and 250 Hz. No filter was applied to the EDA channel.

3.3 Feature Extraction

The phasic heat and electric pain stimuli differ by 1 s in length. For a fair comparison, we cut off the first second of each electric pain stimulus, such that both pain stimuli types (heat and electric) are represented with a length of 4 s each. For the baseline features, only the baselines which follow the H1 pain stimulus were used. We followed the approaches of Werner et al. [12,13], for the computation of time series based statistical descriptors.

In addition, we included the following descriptors, to each feature extraction window and each channel (unless stated otherwise): zero crossing, variance of the signal, maximum to minimum peak value ratio, root mean square (RMS) of the signal, mean value of the local minima (only EMG and ECG channels), mean value of the local maxima (only EMG and ECG channels), mean of the absolute values, standard deviation of absolute values, peak to peak mean value (only EMG and ECG channels), RMS of successive differences of R-Peaks (only ECG, detected with the Hamilton-Detector [5,6]), standard deviation of the successive differences of the R-peaks (only ECG, detected with the Hamilton-Detector), RMS of the R-Peak differences (only ECG, also with the Hamilton-Detector). Moreover, the features, which are stated in Table 1, were extracted from each signal. For the first and second derivatives, we additionally computed the following descriptors: zero crossing, variation of the first momentum, variation of the second momentum, variance, maximum to minimum peak value ratio, RMS, mean of the absolute values, standard deviation of the absolute values and variance of the second momentum. These are commonly used in this research area [4,8]. The whole feature extraction process led to a total of 405 features (ECG: 84, EMG: 243, EDA: 78). Erroneous feature values (NaN or infinity) were replaced with the mean value specific to the affected feature, subject and label, respectively. In cases, where the calculation of the mean value was not possible, we removed the corresponding feature vector. This resulted in a total amount of

26,077 feature vectors. The minimal amount of feature vectors for at least one test subject was 149. The maximal amount of feature vectors for at least one test subject was 210. After the feature extraction process, we applied a person-specific z-score standardization (zero mean with unit variance) on the combined feature set.

Table 1. Additional features computed from the ECG, EMG and EDA channels. The signal is denoted by s with length N, whereas X_i denotes the i-th chunk of s. The total number of equally sized chunks is denoted by k (in this work $k = 4$). The standard deviation is denoted by σ.

Name	Description
Variation first momentum	$\mu_{stds} = \frac{1}{k-1} \sum\limits_{i=1}^{k} \sigma(X_i)$
Variance second momentum	Variance according to [2]. $\sigma^2_{stds} = \frac{1}{k-1} \sum\limits_{i=1}^{k} (\sigma(X_i) - \mu_{stds})^2$
Variation second momentum	$\sigma_{stds} = \sqrt{\sigma^2_{stds}}$
Mean value first differences	$\frac{1}{N-1} \sum\limits_{t=1}^{N-1} (s_{t+1} - s_t)$
Mean absolute value first differences	$\frac{1}{N-1} \sum\limits_{t=1}^{N-1} |s_{t+1} - s_t|$
Mean absolute value second differences	$\frac{1}{N-2} \sum\limits_{t=1}^{N-2} |s_{t+2} - s_t|$

3.4 Definition of Temporal Windows

Thiam et al. [8] showed that a temporal shift leads to an improvement of classification performance. Therefore, we analyse different combinations of temporal window lengths and starting points. Note that we took the differing lengths of heat and electric pain stimuli into consideration (see Sect. 3.3). By *shiftxs*, we denote feature extraction windows (FEWs) with a length of 4 s, starting x seconds after the starting point of a stimulus. Analogously, by *shiftxs+.5s*, we denote the same FEW approaches, however with a length of 4.5 s. In this work, we analyse seven different FEWs, i.e. *shift0s*, *shift1s*, *shift2s*, *shift2.5s*, *shift3s*, *shift1s+.5s*, and *shift2s+.5s*. A visual overview is depicted in Fig. 1, for some of the defined temporal windows.

4 Experiments and Results

Each approach described in Sect. 3.4 is evaluated by applying the leave-one-subject-out cross-validation (LOSO-CV). We apply the random forest [1] method, in combination with 100 decision tree classifiers, each with a maximum

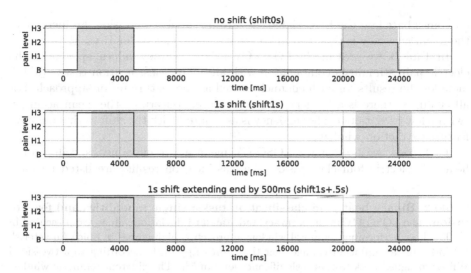

Fig. 1. Example of different time window approaches. Each plateau marks a stimulus with its specific heat intensity level (y-axis) and its duration (x-axis in ms). The highlighted regions mark the defined temporal windows. The last sub-plot depicts a temporal window with a length of 4.5 s.

node depth of 10. These configurations are adopted from Werner et al. [13]. Thereby, the impurity of each split is measured by the Gini Index criterion. Since the dataset is equally distributed across all classes, we focus on the simple accuracy as the performance measure. For the channel-wise evaluation, we combined the EMG feature values specific to the corrugator, zygomaticus and trapezius muscles to one feature vector (for each stimulus). For each temporal window and each channel, we analyse the binary class label combinations B/H1, B/H3 and H1/H3, as well as B/E1, B/E3 and E1/E3. In addition, we apply a feature fusion, which combines the features from all channels. Subsequently, we evaluate both of the four-class classification tasks, which are represented by the class label sets {B, H1, H2, H3} and {B, E1, E2, E3}, respectively. Heat and electric pain stimuli are treated separately, in all cases. To test for statistically significant differences, we apply the two-sided Wilcoxon signed-rank test [15] (discarding zero-differences). The significance level is set to 5%.

In Table 2, we present the LOSO-CV mean accuracy values, specific to the heat pain stimuli, in combination with all binary classification tasks. Table 2 includes the results for each channel, as well as the feature fusion approach. The mean accuracy for the ECG signal increases in general, when the temporal feature extraction window is shifted. In a few cases, the mean accuracy slightly drops. For the EMG signal, we observe minimal fluctuations, specific to the different temporal feature extraction windows. With an advanced shift of the temporal feature extraction window, the mean accuracy for the EDA signal increases as well (for each label combination). This observation differs for the outcomes based on the ECG signal. Regarding the feature fusion, an increase of the mean accuracy is observed in all binary tasks, except for B vs. H1.

In Table 3, we present the LOSO-CV mean accuracy values, specific to the electric pain stimuli, in combination with all binary classification tasks. Table 3 includes the results for each channel, as well as the feature fusion approach. For all channels, there is a tendency for a decrease, concerning the mean accuracy (with a few exceptions). The tendency is associated with the shift of the temporal feature extraction window.

In Table 4, we present the LOSO-CV mean accuracy values, specific to the heat and electric four-class tasks. The classification results are listed for each

Table 2. Binary heat pain classification tasks: single modalities and feature fusion. The LOSO-CV mean accuracy and standard deviation values are presented in %. The best performing temporal window approach is highlighted in bold. An asterisk (*) indicates a significant change over the *shift0s* approach, according to a two-sided Wilcoxon signed-rank test, at a significance level of 5%. The different temporal window approaches (column labels) are defined in Sect. 3.4.

		shift0s	shift1s	shift2s	shift2.5s
B/H1	ECG	50.86 ± 06.61	51.87 ± 06.13	51.27 ± 07.20	50.79 ± 06.33
	EMG	51.07 ± 07.94	50.98 ± 07.40	**52.13 ± 06.64**	50.59 ± 06.19
	EDA	56.29 ± 08.07	56.49 ± 08.78	56.53 ± 09.01	57.26 ± 08.92
	Fusion	55.82 ± 07.18	55.60 ± 08.07	56.15 ± 09.36	56.85 ± 09.00
B/H3	ECG	57.39 ± 10.19	61.61 ± 13.03*	64.92 ± 14.15*	65.67 ± 15.55*
	EMG	75.90 ± 16.89	78.34 ± 17.43*	78.22 ± 17.24*	78.69 ± 16.75*
	EDA	67.42 ± 13.53	74.05 ± 14.97*	79.45 ± 14.55*	81.55 ± 14.29*
	Fusion	77.08 ± 16.35	81.25 ± 16.05*	83.25 ± 14.89*	84.56 ± 14.02*
H1/H3	ECG	56.21 ± 10.34	61.45 ± 12.88*	64.24 ± 15.04*	65.57 ± 15.73*
	EMG	74.87 ± 17.15	78.33 ± 16.79*	77.95 ± 17.27*	77.95 ± 17.07*
	EDA	62.83 ± 12.27	68.88 ± 13.59*	74.12 ± 14.15*	76.06 ± 14.71*
	Fusion	75.86 ± 16.57	79.18 ± 16.12*	80.15 ± 15.78*	80.82 ± 15.82*

		shift0s	shift3s	shift1s+.5s	shift2s+.5s
B/H1	ECG	50.86 ± 06.61	51.51 ± 06.89	52.13 ± 06.34	**52.48 ± 06.69**
	EMG	51.07 ± 07.94	51.86 ± 05.98	51.84 ± 06.78	51.84 ± 06.94
	EDA	56.29 ± 08.07	57.01 ± 08.87	57.29 ± 09.28	**58.30 ± 09.07***
	Fusion	55.82 ± 07.18	57.18 ± 08.93	57.45 ± 08.50*	**57.68 ± 08.88***
B/H3	ECG	57.39 ± 10.19	**66.93 ± 15.19***	63.00 ± 13.74*	65.69 ± 14.86*
	EMG	75.90 ± 16.89	78.69 ± 16.89*	**78.93 ± 16.99***	78.74 ± 16.66*
	EDA	67.42 ± 13.53	**82.21 ± 14.54***	77.15 ± 14.52*	81.57 ± 14.33*
	Fusion	77.08 ± 16.35	**85.19 ± 13.63***	82.28 ± 15.51*	84.63 ± 13.93*
H1/H3	ECG	56.21 ± 10.34	**66.33 ± 15.71***	62.94 ± 13.97*	65.33 ± 15.28*
	EMG	74.87 ± 17.15	**78.43 ± 17.10***	78.34 ± 17.15*	78.09 ± 17.25*
	EDA	62.83 ± 12.27	**76.86 ± 14.74***	71.02 ± 14.08*	75.69 ± 14.37*
	Fusion	75.86 ± 16.57	**81.56 ± 15.63***	79.44 ± 15.72*	80.63 ± 15.54*

channel and the feature fusion. We can observe similar characteristics, as we discussed above for the binary classification tasks.

The significant changes between *shift0s* and all other temporal feature extraction windows, are computed with the Wilcoxon signed-rank test [15] (significance level of 5%). In Tables 2, 3 and 4, the significant changes are marked with an asterisk (*).

Table 3. Binary electric pain classification tasks: single modalities and feature fusion. The LOSO-CV mean accuracy and standard deviation values are presented in %. The best performing temporal window approach is highlighted in bold. An asterisk (*) indicates a significant change over the *shift0s* approach, according to a two-sided Wilcoxon signed-rank test, at a significance level of 5%. The different temporal window approaches (column labels) are defined in Sect. 3.4.

		shift0s	shift1s	shift2s	shift2.5s
B/E1	ECG	57.41 ± 11.40	57.42 ± 11.76	56.66 ± 10.51	57.21 ± 10.19
	EMG	**68.20 ± 17.38**	64.71 ± 16.56*	63.02 ± 15.83*	62.12 ± 15.64*
	EDA	**66.38 ± 14.87**	63.79 ± 15.00*	60.64 ± 13.26*	59.78 ± 12.10*
	Fusion	**71.13 ± 16.67**	69.08 ± 15.45*	65.87 ± 15.22*	65.03 ± 15.75*
B/E3	ECG	**79.79 ± 15.61**	79.61 ± 15.72	78.16 ± 15.80*	77.30 ± 15.58*
	EMG	**88.56 ± 12.05**	86.98 ± 12.27*	87.05 ± 12.24*	86.67 ± 12.96*
	EDA	**90.10 ± 11.86**	90.02 ± 11.57	86.25 ± 12.68*	85.30 ± 12.81*
	Fusion	**92.85 ± 09.56**	92.80 ± 09.59	91.27 ± 09.79*	90.72 ± 10.24*
E1/E3	ECG	**75.81 ± 14.07**	75.36 ± 13.87	74.03 ± 13.78*	72.69 ± 13.86*
	EMG	**82.86 ± 14.47**	81.64 ± 14.60*	81.76 ± 14.76	80.78 ± 15.56*
	EDA	84.62 ± 10.72	**84.65 ± 11.20**	82.46 ± 12.54*	81.70 ± 12.77*
	Fusion	**88.58 ± 10.61**	87.84 ± 11.02*	86.56 ± 11.86*	86.13 ± 12.41*

		shift0s	shift3s	shift1s+.5s	shift2s+.5s
B/E1	ECG	57.41 ± 11.40	56.76 ± 10.04	57.45 ± 11.58	**57.75 ± 10.65**
	EMG	**68.20 ± 17.38**	62.45 ± 15.53*	64.31 ± 15.78*	63.38 ± 15.10*
	EDA	**66.38 ± 14.87**	58.90 ± 11.80*	64.29 ± 14.76*	61.48 ± 13.24*
	Fusion	**71.13 ± 16.67**	64.58 ± 14.59*	69.22 ± 14.85*	66.12 ± 14.48*
B/E3	ECG	**79.79 ± 15.61**	76.00 ± 15.55*	79.71 ± 15.49	78.19 ± 15.70*
	EMG	**88.56 ± 12.05**	85.90 ± 13.11*	86.91 ± 12.33*	86.81 ± 12.44*
	EDA	**90.10 ± 11.86**	83.67 ± 13.44*	89.99 ± 11.65	87.08 ± 11.63*
	Fusion	92.85 ± 09.56	90.13 ± 10.86*	**93.05 ± 09.46**	91.65 ± 09.77*
E1/E3	ECG	**75.81 ± 14.07**	71.52 ± 13.48*	75.05 ± 13.92	73.75 ± 13.50*
	EMG	**82.86 ± 14.47**	80.28 ± 15.06*	82.13 ± 14.33	81.65 ± 14.79*
	EDA	84.62 ± 10.72	80.67 ± 12.98*	84.59 ± 11.94	82.36 ± 12.51*
	Fusion	**88.58 ± 10.61**	85.52 ± 12.31*	87.77 ± 11.26*	86.39 ± 12.05*

Table 4. Four-class heat and electric pain classification tasks: single modalities and feature fusion. The LOSO-CV mean accuracy and standard deviation values are presented in %. The best performing temporal window approach is highlighted in bold. An asterisk (*) indicates a significant change over the *shift0s* approach, according to a two-sided Wilcoxon signed-rank test, at a significance level of 5%. The different temporal window approaches (column labels) are defined in Sect. 3.4.

		shift0s	shift1s	shift2s	shift2.5s
Heat	ECG	28.52 ± 05.72	30.91 ± 07.18*	32.27 ± 08.30*	32.81 ± 08.01*
	EMG	38.59 ± 10.03	40.01 ± 10.39*	40.04 ± 10.02*	40.38 ± 09.41*
	EDA	34.51 ± 08.00	38.50 ± 09.19*	41.27 ± 09.76*	42.52 ± 10.16*
	Fusion	40.56 ± 09.78	42.95 ± 10.64*	43.80 ± 10.42*	44.31 ± 10.57*
Electric	ECG	**43.07 ± 11.90**	42.97 ± 12.03	41.55 ± 11.30*	40.39 ± 10.78*
	EMG	**53.47 ± 15.15**	51.02 ± 14.16*	49.56 ± 13.68*	48.89 ± 13.50*
	EDA	**52.76 ± 12.96**	52.25 ± 13.31	48.76 ± 12.21*	47.68 ± 11.79*
	Fusion	**59.84 ± 14.33**	58.43 ± 13.63*	55.51 ± 12.82*	54.22 ± 13.07*

		shift0s	shift3s	shift1s+.5s	shift2s+.5s
Heat	ECG	28.52 ± 05.72	**33.47 ± 08.34***	31.73 ± 07.55*	33.27 ± 08.38*
	EMG	38.59 ± 10.03	**40.89 ± 09.24***	40.86 ± 09.84*	39.84 ± 09.90*
	EDA	34.51 ± 08.00	**43.27 ± 09.53***	40.07 ± 10.27*	42.97 ± 10.08*
	Fusion	40.56 ± 09.78	**45.12 ± 10.83***	43.57 ± 10.55*	44.55 ± 10.53*
Electric	ECG	**43.07 ± 11.90**	39.98 ± 11.13*	42.80 ± 11.63	41.52 ± 11.24*
	EMG	**53.47 ± 15.15**	49.51 ± 12.99*	50.76 ± 14.18*	49.63 ± 13.57*
	EDA	**52.76 ± 12.96**	46.45 ± 11.29*	52.69 ± 13.16	49.50 ± 11.83*
	Fusion	**59.84 ± 14.33**	53.79 ± 12.93*	58.68 ± 13.49*	55.47 ± 13.11*

5   Discussion and Future Work

In this work, we evaluated different temporal feature extraction windows for heat and electric pain stimuli using the random forest method [1], based on the X-ITE Pain Database [3]. Higher accuracy values, based on LOSO-CV, were achieved for the classification tasks, in combination with electric pain stimuli.

For the shifted temporal windows (see Sect. 3.4), the mean accuracy increased, in combination with the heat pain stimuli. Especially, the features of the EDA signal benefited from a temporal shift. Due to the fact that heat needs time to decrease, valuable information can be found by applying a shift. This was also shown by Thiam et al. [8]. For the B vs. H3 setting, the EDA-based mean accuracy value significantly increased by 14.79%, in combination with the *shift3s* approach (with respect to *shift0s*). The mean accuracy increased also for the ECG and EMG signals. The lowest improvement was reported for the EMG signal (2.32%). Temporal shifts led to significant improvements in most of the cases, with respect to *shift0s*. Similar findings were observed for the heat pain four-class task.

The classification performance, in combination with electric pain stimuli, decreased when a temporal shift was applied. In contrast to the heat pain classification, ECG-based and EMG-based mean accuracy values decreased, in combination with the *shift3s* approach (with respect to *shift0s*). For the B vs. E3 setting, the mean accuracy decreased by 3.79%, in combination with the ECG signal and the *shift3s* approach. The EMG-based mean accuracy decreased by 2.66%. The temporal shift denoted by *shift3s* led to the most significant decrease of the mean accuracy (6.43%), with respect to the EDA signal. Similar findings were observed for the electric pain four-class task.

The temporal shift denoted by *shift3s* led to a significant reduction of the mean accuracy values, with respect to *shift0s*. This observation was also made, in combination with the other defined shifts. A reason for the observed differences between heat and electric pain classification, with respect to the described shifts, lies in the nature of the stimuli. Electrical pain starts and stops instantly. In contrast, the rise and fall in temperature for induced heat stimuli takes time. Applying temporal shifts seems to include important information, which is obtained during the decrease of the temperature [8]. Werner et al. [13] showed that classification based on phasic electric pain stimuli results in higher accuracy values, in comparison to classification on phasic heat pain stimuli. This is due to the fact, that electric pain stimuli are felt instantly. Our outcomes strengthen this assumption. Regarding the different temporal feature extraction windows, the application of *shift0s* resulted in higher performance values for phasic electric pain classification. Furthermore, we believe that because an electric pain stimulus ends instantly, the body of a test subject is instantly relieved as well. This indicates that less information is found when a temporal shift is applied, which is also reflected by the reported mean accuracy values (see Table 3). We propose extracting features for heat pain stimuli with an application of a temporal shift. Contrary, features from electric pain stimuli should be extracted without an application of a temporal shift.

For future work, we aim to examine the detailed effects of temporal feature extraction windows, with respect to the separate EMG signals. We also want to focus on the feature importance and its correlation, with respect to different temporal shifts. Additional work on window length variations should be carried out as well.

Acknowledgments. The work of Peter Bellmann and Friedhelm Schwenker is supported by the project *Multimodal recognition of affect over the course of a tutorial learning experiment* (SCHW623/7-1) funded by the German Research Foundation (DFG). We gratefully acknowledge the support of NVIDIA Corporation with the donation of the Tesla K40 GPU used for this research.

References

1. Breiman, L.: Random forests. Mach. Learn. **45**(1), 5–32 (2001)
2. Cao, C., Slobounov, S.: Application of a novel measure of EEG non-stationarity as 'Shannon-entropy of the peak frequency shifting' for detecting residual abnormalities in concussed individuals. Clinical Neurophysiology **122**(7), 1314–1321 (2011). https://doi.org/10.1016/j.clinph.2010.12.042
3. Gruss, S., et al.: Multi-modal signals for analyzing pain responses to thermal and electrical stimuli. JoVE (J. Vis. Exp.) (146), e59057 (2019)
4. Gruss, S., et al.: Pain intensity recognition rates via biopotential feature patterns with support vector machines. PLoS ONE **10**(10), e0140330 (2015)
5. Hamilton, P.: Open source ECG analysis. In: Computers in Cardiology, pp. 101–104. IEEE (2002)
6. Howell, L., Porr, B.: Popular ECG R peak detectors written in Python (2019). https://doi.org/10.5281/zenodo.3588108
7. Thiam, P., Bellmann, P., Kestler, H.A., Schwenker, F.: Exploring deep physiological models for nociceptive pain recognition. Sensors **19**(20), 4503 (2019)
8. Thiam, P., et al.: Multi-modal pain intensity recognition based on the SenseEmotion database. IEEE Trans. Affect. Comput., 1 (2019). https://doi.org/10.1109/taffc.2019.2892090
9. Thiam, P., Kestler, H.A., Schwenker, F.: Two-stream attention network for pain recognition from video sequences. Sensors **20**(3), 839 (2020)
10. Velana, M., et al.: The SenseEmotion database: a multimodal database for the development and systematic validation of an automatic pain- and emotion-recognition system. In: Schwenker, F., Scherer, S. (eds.) MPRSS 2016. LNCS (LNAI), vol. 10183, pp. 127–139. Springer, Cham (2017). https://doi.org/10.1007/978-3-319-59259-6_11
11. Walter, S., et al.: The BioVid heat pain database: data for the advancement and systematic validation of an automated pain recognition system, pp. 128–131, June 2013. https://doi.org/10.1109/CYBConf.2013.6617456
12. Werner, P., Al-Hamadi, A., Limbrecht-Ecklundt, K., Walter, S., Gruss, S., Traue, H.C.: Automatic pain assessment with facial activity descriptors. IEEE Trans. Affect. Comput. **8**(3), 286–299 (2017). https://doi.org/10.1109/TAFFC.2016.2537327
13. Werner, P., Al-hamadi, A., Werner, P., De, A.A.h., Gruss, S., Walter, S.: Twofold-multimodal pain recognition with the X-ITE pain database (2019)
14. Werner, P., Lopez-Martinez, D., Walter, S., Al-Hamadi, A., Gruss, S., Picard,R.: Automatic recognition methods supporting pain assessment: a survey. IEEE Trans. Affect. Comput. (2019)
15. Wilcoxon, F.: Individual comparisons by ranking methods. Biom. Bull. **1**(6), 80–83 (1945)

Pain Intensity Recognition - An Analysis of Short-Time Sequences in a Real-World Scenario

Peter Bellmann[1]([✉]), Patrick Thiam[1,2][iD], and Friedhelm Schwenker[1][iD]

[1] Institute of Neural Information Processing, Ulm University,
James-Franck-Ring, 89081 Ulm, Germany
{peter.bellmann,patrick.thiam,friedhelm.schwenker}@uni-ulm.de
[2] Institute of Medical Systems Biology, Ulm University,
Albert-Einstein-Allee 11, 89081 Ulm, Germany

Abstract. Pain intensity recognition still constitutes a challenging classification task. In this work, we focus on the physiological signals of the publicly available BioVid Heat Pain Database, which was collected at Ulm University. The BioVid Heat Pain Database consists of different recordings of healthy test subjects that were exposed to various short-time heat stimuli. The results reported in the literature, which are based on those short-time sequences do not justify the implementation of automated pain detection systems, due to unsatisfactory accuracy rates. In the current study, we show that the outcomes, which are stated in the literature, most likely represent lower bound estimations. For this purpose, we transfer the classification task, which is provided by the BioVid Heat Pain Database, to a real-world scenario. More precise, according to an expected hospital setting, we analyse the automated pain intensity recognition approach in combination with different sets of short-time sequences. Our outcomes indicate, that in real-world applications, where the detection of pain intensity is based on more than one single short-time sequence, the accuracy values can be significantly improved. In the current study, the classification performance of bagged decision tree ensembles is evaluated, based on a person-independent scenario.

Keywords: Pain intensity recognition · Physiological signals

1 Introduction

Pain detection, or more precise pain intensity recognition, constitutes an important up-to-date topic of machine learning based medical applications. The aim of implementing automated pain assessment devices is to have the ability to correctly classify different levels of pain, of patients that are not able to communicate the current state of their medical condition. This applies, for example, to newborns or people suffering from dementia. Automated pain detection systems would not only help the affected patients, but also facilitate the work of doctors and nurses.

© Springer Nature Switzerland AG 2020
F.-P. Schilling and T. Stadelmann (Eds.): ANNPR 2020, LNAI 12294, pp. 149–161, 2020.
https://doi.org/10.1007/978-3-030-58309-5_12

Some researchers collected pain-related data sets, such as the publicly available UNBC-McMaster Shoulder Pain Expression Archive Database (SPDB) [12], or the SenseEmotion Database (SEDB) [21]. While the SPDB includes different recordings specific to participants that were suffering from shoulder pain, the SEDB includes recordings specific to healthy test subjects that participated in strictly controlled pain elicitation experiments.

In the current study, we focus on the evaluation of data specific to healthy participants that were exposed to thermal stimuli. Usually, such kind of pain elicitations are applied in form of short-time sequences, mostly only covering a temporal range of a couple of seconds. However, identifying different levels of pain, based on short-time sequences, does not reflect a hospital-related real-world scenario. Therefore, in this study, we show that short-time recordings might be used to implement automated pain intensity recognition systems, which are based on *long-time* recordings (with lengths of up to 80 s).

Note that there are real-world application scenarios, such as examination-induced pain, for which short time sequences play an essential role. Thus, reliable short time sequences based classification models still cover an important research field of machine learning based pain assessment.

The remainder of this work is organised as follows. For a better understanding of the related work part, we first shortly describe the BioVid Heat Pain Database, in Sect. 2. We then provide some related work on pain intensity recognition, in Sect. 3. Subsequently, in Sect. 4, we present our approach for a real-world setting and define our evaluation protocol. The experimental results are presented and discussed in Sect. 5. Finally, in Sect. 6, we conclude the current work.

2   The BioVid Heat Pain Database

The BioVid Heat Pain Database (BVDB) [22] was recorded at Ulm University, to support the research fields of automated pain detection and emotion recognition. In this work, we focus on Part A of the BVDB[1], which constitutes a pain intensity recognition task. Interestingly, such kind of databases can also be used to evaluate the effectiveness of physiological signals, in different person identification scenarios [3].

2.1   Description

Healthy test subjects were recruited to participate in the multi-modal data acquisition experiments. For Part A of the BVDB, the recordings specific to 87 participants are available. Pain was induced in form of heat with the use of a Medoc thermode[2], under strictly controlled conditions.

In the first step of the experiments, each participant had to undergo an individual calibration phase, which led to four equidistant participant-specific

---

[1] Publicly available at http://www.iikt.ovgu.de/BioVid.print.

[2] More information at https://www.medoc-web.com/pathway.

levels of pain, denoted by T_1, T_2, T_3 and T_4. The global pain-free level was set to 32 °C (T_0), for each participant. Figure 1 provides an illustration of an exemplary recording sequence.

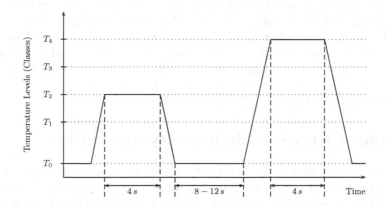

Fig. 1. An exemplary pain elicitation sequence.

Subsequently, each participant was stimulated 20 times with each of the predefined pain levels, in randomised order, for a duration of four seconds. In between two pain-related sequences, each participant was stimulated with the pain-free level, for a random duration of 8−12 s.

During the pain elicitation experiments, videos from three different angles, as well as different physiological signals, were recorded. In the current study, we focus on the biopotentials. These include the electrocardiogram (ECG), which measures a person's heart activity, and the electromyogram (EMG), which measures the muscle activity. In part A of the BVDB, the activity of the trapezius muscle was recorded, which is located in the shoulder area of a human torso. Moreover, sensors that were attached to the participants' ring and index fingers led to recordings of electrodermal (EDA) signals, which measure the skin conductance. For a full data set description, we refer the reader to [22].

Table 1. Properties of the BVDB. Total number of samples: 8700 ($87 \times 5 \times 20$). Total number of features: 194 ($68 + 56 + 70$).

Number of participants	87 (43 f, 44 m)
Number of classes	5 ($T_0, \ldots, T_4$)
Number of samples per class per participant	20
Feature dimension of ECG	68
Feature dimension of EMG	56
Feature dimension of EDA	70

2.2 Feature Extraction

Feature extraction is not part of our current contribution. Therefore, we refer the reader to [8] and [9], for the analysis of feature extraction and normalisation of the physiological signals of the BVDB. Note that in the current study, we are using exactly the same features. All features were extracted from windows of 5.5 s. The extraction of features included the computation of statistical descriptors, such as mean and extreme values from the temporal and frequency domains. Moreover, also signal-specific descriptors, such as the ECG-related P, Q, R, S, and T wavelets based features, were implemented.

The properties of the BVDB are summarised in Table 1.

3   Related Work on Pain Intensity Recognition

In the current section, we shortly present some of the many existing approaches, which were analysed in combination with the pain intensity recognition task.

One basic step in building a robust classification model is the choice of an appropriate **fusion approach** [15]. In cases, where the data is available specific to only a *few* recording signals/channels/modalities, one should focus on an *early fusion* (EF) architecture [2]. In the EF approach, the data from all recording sources is concatenated to one feature vector and fed to the classification model. In cases, where the data specific to *several* modalities (e.g. audio, video and different biopotentials) is available, one can make use of many existing intelligent *late fusion* architectures. As an example, in [17], the authors evaluate different pseudo inverse [13] based classification architectures [14], for different pain intensity recognition tasks, based on the SenseEmotion Database.

One effective way to improve the classification performance in a pain intensity classification task, is the application of so-called **personalisation techniques** [8,9,18]. Thereby, one makes use of the known (and unlabelled) data of the current test subject. Based on its data distribution and a predefined *similarity* or *distance* measure (e.g. Hausdorff distance [7]), one can determine a subset of training subjects, which are used to train the classification model, similar to an on-line setting. Usually those classification models outperform the models, which are trained on all available training data samples.

Kessler et al. propose to include **remote photoplethysmography** (rPPG), for the pain intensity recognition task [10,11]. The rPPG signal can be used as an additional physiological input modality for the detection of a participant's heart and respiration rates. The main idea is to filter the rPPG signal in multiple frequency ranges. Thereby, the authors make use of the fact that the rPPG signal consists of three colour channels, i.e. red, green and blue.

Moreover, one can define different classifier-specific data **transformation mappings** to improve the accuracy of the current classification model. Thereby, the transformation mappings are applied to the extracted features. As an example based on the SenseEmotion Database, in [1], the authors propose a so-called unsupervised *quartile-based data transformation* (QBDT), which maps each feature value to a corresponding element of the set $\{-2, -1, 1, 2\}$. The outcomes in [1] show that using the corresponding transformed feature values significantly improved the classification performance of basic *nearest neighbour* classifiers [6].

Following the current trend in machine learning, one can also implement **deep models** (DMs), for the pain intensity classification task. Thiam et al. analysed different DMs, in combination with physiological signals [16,19] and video sequences [20]. Thereby, the authors followed the so-called *end-to-end* approach, which constitutes an alternative to hand-crafted features based approaches. An end-to-end architecture is trained on (filtered) raw signals. Thus, there is no need for manual feature extraction, which requires modality-specific expert knowledge, in general.

We refer the reader to [23], for a recently published **survey article** on automated pain assessment methods, presented by Werner et al.

4 Evaluation Protocol

In the current section, we first present our approach for using short-time stimuli in real-world applications. Subsequently, we provide a list of settings, which will be included in the upcoming experimental evaluations.

4.1 Real-World Scenario

It is legitimate to assume that in a real-world application, i.e. in a hospital setting, automated pain detection devices will not evaluate a patient's condition based on one single short-time sequence. It is most likely that a pain detection system will provide its final decisions based on long-time sequences or on sets of short-time sequences. The BVDB consists of short-time sequences, therefore, we will evaluate the latter approach.

Note that in the BVDB, each participant constitutes a data subset, with 20 samples per class (see Sect. 2.1). Each of the samples reflects a heat stimulus of a duration of four seconds. Moreover, all of the samples were obtained during one single recording session, for each participant. Therefore, according to a real-world scenario, in this study, we evaluate the automated pain assessment approach based on different-sized sets of those samples. Note that since all recordings were collected in the course of a single experiment, the combinations of short-time sequences are expected to represent *good approximations* for consecutive long-time sequences.

By $n \in \mathbb{N}$, we denote the number of samples, which are combined to predict a participant's current pain level. Therefore, only samples specific to the test participant and to one single class are combined. Moreover, to include all of the initial samples from the test set, we restrict n to the set $N := \{1, 2, 4, 5, 10, 20\}$. Note that each participant's data subset contains 20 samples per class. Moreover, 20 is divisible without remainder by each element of N.

For each test subject, and for each $n \in N$, we randomly draw n samples without replacement, combine them to one set and define the corresponding set of samples as a new *data point*. This procedure is repeated until the set of initial samples is empty, for each class of the current participant. For $n = 1$, the data specific to the test subject remains unchanged. For $n = 2$, two randomly combined initial samples, specific to the current participant's class, define a new data point. Therefore, for $n = 2$, each test set participant constitutes a data subset consisting of 50 samples, including 10 samples per class. For $n = 20$, each participant constitutes a data subset consisting of solely 5 samples (1 per class).

Note that only the samples from the test set are combined to new data points. The implemented classification model is always trained on the initial, i.e. non-combined, data samples.

4.2   Experimental Settings

In our experiments, we apply the following settings.

Evaluation Approach. We apply the leave-one-person-out (LOPO) cross validation. Thereby, in iteration i, the data specific to participant i is used as the test, whereas the data specific to all remaining participants is used as the training set.

Classification Model. We apply the bagging method [4], in combination with 200 unpruned decision tree classifiers [5], with the Gini index as impurity measure, using the MATLAB[3] software. Note that in this study, we focus on the early fusion approach. Thereby, the data specific to ECG, EMG and EDA recordings is concatenated to one feature vector, for each sample.

Performance Measure. Since the BVDB constitutes a balanced data set, i.e. the amount of samples is equal for each class, we focus on the unweighted accuracy (acc), i.e.

$$acc = \frac{\text{number of correctly classified test samples}}{\text{total number of test samples}}.$$

We also provide class-specific accuracy values, in form of confusion matrices.

Significance Tests. We apply the two-sided Wilcoxon signed-rank test [24], at a significance level of 5%.

Definition of Tasks. Moreover, we define the following two classification tasks. An automated pain intensity recognition system should, at least, be able to correctly differentiate between the no-pain level and the highest level of pain. Therefore, we will first evaluate the binary task, $T_0 \, vs. \, T_4$. In the multi-class task, we include all available classes to evaluate the class-specific accuracy values. Note that for this task, the chance level accuracy is equal to 20%.

---

[3] https://www.mathworks.com/products/matlab.html.

5   Results

In the current section, we first provide the results for the binary task. Subsequently, we present the results for the multi-class task, including discussions of the obtained outcomes. Note that we evaluate the BVDB in combination with the set N, which we defined as $N = \{1, 2, 4, 5, 10, 20\}$. Thereby, $n \in N$ denotes the number of initial samples (from the same class) from the current test set, which are combined to represent a corresponding new data point.

The classification model is always trained on the initial samples. We classify a set of samples (new data point) by classifying each of the samples from the current set separately, and applying the simple *mean-rule* to the obtained class-specific scores. The final prediction of the classification model corresponds to the class with the highest score. Note that a more straightforward approach would be to predict the labels for each of the samples, and to apply the majority vote. However, since most elements from the set N are even numbers, we might obtain many indecisive votes in the binary tasks, leading to many random guesses. Therefore, we chose to implement the class-specific scores based mean-rule approach.

By C_n, we denote the confusion matrix specific to $n \in N$. The rows of C_n represent the true labels, whereas the columns of C_n represent the predicted labels. Moreover, we focus on the percentage values of each C_n, for a better comparison. Thus, the row elements of each C_n sum up to 100 (The sum is not always exactly equal to 100, since we state the corresponding rounded values). For the confusion matrices, we present the natural order of the rows and columns, i.e. the first row and column correspond to T_0, the second to T_1, etc.

5.1   Evaluation of the Binary Task

Table 2 states the averaged LOPO accuracy and standard deviation values, for the defined binary task. From Table 2, we obtain the following observations. Except for $n = 5$, the averaged accuracy values are monotonously increasing, in combination with an ascending number of combined test samples n. The difference between the obtained mean accuracy values, for $n = 4$ and $n = 5$, is not significant. For $n = 20$, we note an improvement in accuracy of approximately 14%, with comparison to $n = 1$. Moreover, for each $n \neq 1$, the improvement is statistically significant, in comparison to $n = 1$, according to a two-sided Wilcoxon signed-rank test, at a significance level of 5%.

Figure 2 provides the obtained box plots for all $n \in N$. From Fig. 2, we can observe that not only the mean accuracy values improved, but also the corresponding median values, with respect to $n = 1$. For $n \in \{4, 5, 10, 20\}$, the median accuracy values are equal to 100%.

Table 2. T_0 vs. T_4 Task: Averaged Leave-One-Person-Out accuracy and standard deviation values in %. The figures in the upper row (n) denote the number of stimuli that are defined as one sample. For all $n \neq 1$, the accuracy values improved significantly over $n = 1$, according to a two-sided Wilcoxon signed-rank test, with $p < 0.05$.

$n = 1$	$n = 2$	$n = 4$	$n = 5$	$n = 10$	$n = 20$
82.27 ± 14.5	87.13 ± 14.9	91.84 ± 13.1	91.67 ± 14.6	94.54 ± 12.9	96.56 ± 12.7

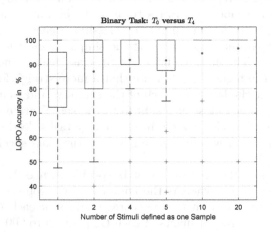

Fig. 2. T_0 vs. T_4 Task: LOPO accuracy values. LOPO: Leave-One-Person-Out. The mean and median values are denoted by a dot and a horizontal line, respectively. For $n \in \{4, 5\}$, the median values, as well as the upper quartile values, are all equal to 100%. For the two box plots on the right-hand side, the median values, the lower and upper quartiles, and both whiskers are all equal to 100%.

The corresponding confusion matrices, for the LOPO cross validation evaluation approach, are defined as follows,

$$C_1 = \begin{pmatrix} 85.5 & 14.5 \\ 20.9 & 79.1 \end{pmatrix}, \quad C_2 = \begin{pmatrix} 89.4 & 10.6 \\ 15.2 & 84.8 \end{pmatrix}, \quad C_4 = \begin{pmatrix} 93.1 & 6.90 \\ 9.40 & 90.6 \end{pmatrix},$$

$$C_5 = \begin{pmatrix} 92.5 & 7.50 \\ 9.20 & 90.8 \end{pmatrix}, \quad C_{10} = \begin{pmatrix} 95.4 & 4.60 \\ 6.30 & 93.7 \end{pmatrix}, \quad C_{20} = \begin{pmatrix} 97.7 & 2.30 \\ 4.60 & 95.4 \end{pmatrix}.$$

For $n = 1$, the class-specific accuracy values are equal to 85.5% and 79.1%, for T_0 and T_4, respectively. Whereas for $n = 20$, the accuracy values for the classes T_0 and T_4 are equal to 97.7% and 95.4%, respectively. These values denote the maxima across all computed confusion matrices.

5.2   Evaluation of the Multi-Class Task

Table 3 states the averaged LOPO accuracy and standard deviation values, for the multi-class task. From Table 3, we obtain the following observations.

The accuracy values are monotonously increasing, in combination with an ascending number of combined test samples n. For $n = 20$, we note an improvement in accuracy of approximately 14%, with comparison to $n = 1$. Moreover, for each $n \neq 1$, the improvement is statistically significant, in comparison to $n = 1$, according to a two-sided Wilcoxon signed-rank test, at a significance level of 5%. The smallest improvement is observed by changing the value of n from 4 to 5. This corresponds to the smallest change of n, in our experiments. In contrast to the binary task, the averaged standard deviation accuracy values also increased monotonously.

Table 3. Multi-Class Task: Averaged Leave-One-Person-Out accuracy and standard deviation values in %. The figures in the upper row (n) denote the number of stimuli that are defined as one sample. For all $n \neq 1$, the accuracy values improved significantly over $n = 1$, according to a two-sided Wilcoxon signed-rank test, with $p < 0.05$. Chance level accuracy: 20%.

$n = 1$	$n = 2$	$n = 4$	$n = 5$	$n = 10$	$n = 20$
37.60 ± 10.7	41.40 ± 12.7	44.60 ± 15.0	45.98 ± 15.4	49.31 ± 17.0	51.49 ± 23.7

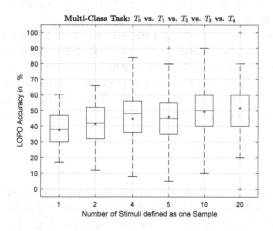

Fig. 3. Multi-Class Task: LOPO accuracy values. LOPO: Leave-One-Person-Out. The mean and median values are denoted by a dot and a horizontal line, respectively.

Figure 3 provides the obtained box plots for $n \in N$. From Fig. 3, we can observe that not only the mean accuracy values improved, but also the corresponding median values, with respect to $n = 1$. For $n = 20$, the accuracy values range from 0% to 100%, in the current task. Note that for $n = 20$, each test subject (participant) is represented by solely five data samples (one sample per class).

For the multi-class scenario, we focus on the confusion matrices C_1 (reflects testing on the initial data set) and C_{20}, which contain the following values:

$$C_1 = \begin{pmatrix} \mathbf{48.8} & 23.3 & 13.6 & 9.30 & 5.10 \\ 27.3 & \mathbf{31.1} & 19.4 & 15.1 & 7.10 \\ 19.6 & 26.2 & \mathbf{21.0} & 20.3 & 12.9 \\ 11.3 & 15.7 & 18.9 & \mathbf{27.8} & 26.3 \\ 4.10 & 6.60 & 11.0 & 19.0 & \mathbf{59.3} \end{pmatrix}, \quad C_{20} = \begin{pmatrix} \mathbf{59.8} & 19.5 & 11.5 & 9.20 & 0.00 \\ 20.7 & \mathbf{48.3} & 19.5 & 11.5 & 0.00 \\ 6.90 & 34.5 & \mathbf{27.6} & 24.1 & 6.90 \\ 2.30 & 6.90 & 16.1 & \mathbf{43.7} & 31.0 \\ 0.00 & 0.00 & 2.30 & 19.5 & \mathbf{78.2} \end{pmatrix}$$

By comparing C_1 to C_{20}, we can make the following observations. The averaged accuracy improved for each of the five classes (see the diagonal elements, denoted in bold for better readability). The lowest improvement was obtained for the class T_2 (third row and column, of the matrices C_1 and C_{20}). For $n = 20$ and the class T_2, the misclassification with *neighbouring* classes increased (denoted by red (C_{20}) and blue (C_1), in the colour print). We can clearly see that the class T_2 is more often misclassified as T_1 and T_3, for $n = 20$. However, the *crucial* classification errors decreased, for $n = 20$. As an example, for $n = 1$, the class T_0 (no pain at all) was misclassified as T_4 (highest level of pain), with a rate of 5.10%. Class T_4 was misclassified as T_0, with a rate of 4.10%. On the other hand, for $n = 20$, the highest level of pain (T_4) was **never** classified as the pain-free (T_0) level or the lowest level (T_1) of pain (see last row of matrix C_{20}). Moreover, the classes T_0 and T_1 were both never misclassified as the class T_4 (see last column of matrix C_{20}).

Note that the classes are clearly ordered, according to their natural occurrence, i.e. $T_0 < \ldots < T_4$. Therefore, the 5-class classification task can also be formulated as a regression task, including other performance metrics, such as the mean absolute error or the root mean squared error, which is applied for example in [9]. In the current study, we focused on the simple accuracy to keep the performance measure consistent for both tasks, i.e. the binary and multi-class task.

5.3   Discussion

For both, the binary as well as the multi-class classification tasks, the averaged LOPO accuracy improved from 82.27% to 96.56% and from 37.60% to 51.49% respectively, when using a set of 20 short-time stimuli sequences instead of one single sequence (see Tables 2 and 3). The median LOPO accuracy values improved also significantly, even more than the mean values (see Figs. 2 and 3). The results reported in the literature, which are based on the initial data samples, i.e. on single short-time sequences, could indicate that we are still *far away* from reliable automated pain intensity detection systems. However, combining random sets of participant- and class-specific samples to new data points (from the test set), significantly improves the classification performance. Note that for $n = 20$, for the binary task, we reached an outstanding averaged accuracy value of 96.56% (see Table 2). Therefore, the reported values in the literature might represent *lower estimation bounds*, for pain intensity recognition tasks, in real-world applications.

Note that the averaged accuracy increases with an ascending number of combined samples n, because the classification performance is significantly above chance level for $n = 1$.

6 Conclusion

In the current study, based on the physiological signals of the publicly available BioVid Heat Pain Database, we evaluated the pain intensity recognition task, in a simulated real-world scenario. For this purpose, we analysed the classification transfer from short-time (4 s) stimuli sequences to artificially generated long-time ($4 \times n$ s) sequences. Thereby, we showed that the classification accuracy significantly improved when the classification model's final decision was based on sets of twenty short-time sequences (STS). Note that a set of 20 STS reflects a recording duration of 80 s.

In general, it is safe to assume that in a real-world application, each prediction provided by an automated pain intensity recognition system is based on recordings (of e.g. physiological signals) that exceed the duration of four seconds. Therefore, we showed that in a hospital setting, one might use long-time sequences, randomly divide them into (non-overlapping) STS, and feed the resulting sets of samples to a classification architecture, which is trained on STS.

We believe that the results, which are reported in the literature, represent lower bound accuracy estimations for the pain intensity recognition task, because they are based on the samples specific to STS. For the person-independent differentiation of the neutral state and the highest pain level, we obtained a very impressive accuracy value of 96.56%. On the other hand, this value might represent an upper bound accuracy estimation. In this work, we artificially combined sets of STS to long-time sequences. In non-simulated real-world scenarios, long-time sequences might lead to significantly different characteristics of a person's physiology. For example, a patient's body functions might adapt to the persistent pain, leading to changes in physiological signals, after a certain period of time. Therefore, to be able to draw convincing conclusions, one has to collect data in hospitals.

Acknowledgments. The work of Peter Bellmann and Friedhelm Schwenker is supported by the project *Multimodal recognition of affect over the course of a tutorial learning experiment* (SCHW623/7-1) funded by the German Research Foundation (DFG). We gratefully acknowledge the support of NVIDIA Corporation with the donation of the Tesla K40 GPU used for this research. Moreover, we thank the reviewers for their constructive feedback, which helped to improve the current work.

References

1. Bellmann, P., Thiam, P., Schwenker, F.: Using a quartile-based data transformation for pain intensity classification based on the SenseEmotion database. In: 2019 8th International Conference on Affective Computing and Intelligent Interaction Workshops and Demos (ACIIW), pp. 310–316 (2019)

2. Bellmann, P., Thiam, P., Schwenker, F.: Multi-classifier-systems: architectures, algorithms and applications. In: Pedrycz, W., Chen, S.-M. (eds.) Computational Intelligence for Pattern Recognition. SCI, vol. 777, pp. 83–113. Springer, Cham (2018). https://doi.org/10.1007/978-3-319-89629-8_4

3. Bellmann, P., Thiam, P., Schwenker, F.: Person identification based on physiological signals: Conditions and risks. In: ICPRAM, pp. 373–380. Scitepress (2020)

4. Breiman, L.: Bagging predictors. Mach. Learn. **24**(2), 123–140 (1996)

5. Breiman, L., Friedman, J.H., Olshen, R.A., Stone, C.J.: Classification and regression trees. Wadsworth (1984)

6. Cover, T.M., Hart, P.E.: Nearest neighbor pattern classification. IEEE Trans. Inf. Theory **13**(1), 21–27 (1967)

7. Huttenlocher, D.P., Klanderman, G.A., Rucklidge, W.J.: Comparing images using the Hausdorff distance. IEEE Trans. Pattern Anal. Mach. Intell. **15**, 850–863 (1993)

8. Kächele, M., et al.: Adaptive confidence learning for the personalization of pain intensity estimation systems. Evol. Syst. **8**(1), 71–83 (2016). https://doi.org/10.1007/s12530-016-9158-4

9. Kächele, M., Thiam, P., Amirian, M., Schwenker, F., Palm, G.: Methods for person-centered continuous pain intensity assessment from bio-physiological channels. J. Sel. Top. Signal Process. **10**(5), 854–864 (2016)

10. Kessler, V., Thiam, P., Amirian, M., Schwenker, F.: Multimodal fusion including camera photoplethysmography for pain recognition. In: ICCT, pp. 1–4. IEEE (2017)

11. Kessler, V., Thiam, P., Amirian, M., Schwenker, F.: Pain recognition with camera photoplethysmography. In: IPTA, pp. 1–5. IEEE (2017)

12. Lucey, P., Cohn, J.F., Prkachin, K.M., Solomon, P.E., Matthews, I.A.: Painful data: The UNBC-McMaster shoulder pain expression archive database. In: FG, pp. 57–64. IEEE (2011)

13. Penrose, R.: A generalized inverse for matrices. Proc. Camb. Philos. Soc. **51**, 406–413 (1955)

14. Schwenker, F., Dietrich, C., Thiel, C., Palm, G.: Learning of decision fusion mappings for pattern recognition. Int. J. Artif. Intell. Mach. Learn. (AIML) **6**, 17–21 (2006)

15. Snoek, C., Worring, M., Smeulders, A.W.M.: Early versus late fusion in semantic video analysis. In: ACM Multimedia, pp. 399–402. ACM (2005)

16. Thiam, P., Bellmann, P., Kestler, H.A., Schwenker, F.: Exploring deep physiological models for nociceptive pain recognition. Sensors **19**(20), 4503 (2019)

17. Thiam, P., et al.: Multi-modal pain intensity recognition based on the SenseEmotion database. IEEE Trans. Affect. Comput., 1 (2019)

18. Thiam, P., Kessler, V., Schwenker, F.: Hierarchical combination of video features for personalised pain level recognition. In: ESANN, pp. 465–470 (2017)

19. Thiam, P., Kestler, H.A., Schwenker, F.: Multimodal deep denoising convolutional autoencoders for pain intensity classification based on physiological signals. In: ICPRAM, pp. 289–296. Scitepress (2020)

20. Thiam, P., Kestler, H.A., Schwenker, F.: Two-stream attention network for pain recognition from video sequences. Sensors **20**(3), 839 (2020)

21. Velana, M., et al.: The SenseEmotion database: a multimodal database for the development and systematic validation of an automatic pain- and emotion-recognition system. In: Schwenker, F., Scherer, S. (eds.) MPRSS 2016. LNCS (LNAI), vol. 10183, pp. 127–139. Springer, Cham (2017). https://doi.org/10.1007/978-3-319-59259-6_11

22. Walter, S., et al.: The BioVid heat pain database data for the advancement and systematic validation of an automated pain recognition system. In: CYBCONF, pp. 128–131. IEEE (2013)
23. Werner, P., Lopez-Martinez, D., Walter, S., Al-Hamadi, A., Gruss, S., Picard, R.: Automatic recognition methods supporting pain assessment: a survey. IEEE Trans. Affect. Comput., 1 (2019)
24. Wilcoxon, F.: Individual comparisons by ranking methods. Biom. Bull. 1(6), 80–83 (1945)

A Deep Learning Approach for Efficient Registration of Dual View Mammography

Sina Famouri(✉) , Lia Morra , and Fabrizio Lamberti

Dipartimento di Automatica e Informatica, Politecnico di Torino, Turin, Italy
{sina.famouri,lia.morra,fabrizio.lamberti}@polito.it

Abstract. In a standard mammography study, two views are acquired per breast, the Cranio-Caudal (CC) and Mediolateral-Oblique (MLO). Due to the projective nature of 2D mammography, tissue superposition may both mask or mimic the presence of lesions. Therefore, integrating information from both views is paramount to increase diagnostic confidence for both radiologists and computer-aided detection systems. This emphasizes the importance of automatically matching regions from the two views. We here propose a deep convolutional neural network for the registration of mammography images. The network is trained to predict the affine transformation that minimizes the mean squared error between the MLO and the registered CC view. However, due to the complex nature of the breast glandular pattern, deformations due to compression and the paucity of natural anatomic landmarks, optimizing the mean squared error alone yields suboptimal results. Hence, we propose a weakly supervised approach in which existing annotated lesions are used as landmarks to further optimize the registration. To this aim, the recently proposed Generalized Intersection over Union (GIoU) is exploited as loss. Experiments on the public CBIS-DDSM dataset show that the network was able to correctly realign the images in most cases; corresponding bounding boxes were spatially matched in 68% of the cases. Further improvements can be expected by incorporating an elastic deformation field in the registration network. Results are promising and support the feasibility of our approach.

Keywords: Mammography · Image registration · Spatial transformer · Convolutional neural networks

1 Introduction

Population screening by means of digital mammography was shown to reduce mortality associated to breast cancer. However, the 2D projective nature of mammography results in tissue superposition that may both mask and simulate the presence of lesions [13, 20]. This is especially true when breast tissue is very dense, [18], as the fibrous and glandular components have higher attenuation than fatty tissue, and more similar to that of potential lesions, especially masses.

© Springer Nature Switzerland AG 2020
F.-P. Schilling and T. Stadelmann (Eds.): ANNPR 2020, LNAI 12294, pp. 162–172, 2020.
https://doi.org/10.1007/978-3-030-58309-5_13

In a standard screening examination, two projection views are acquired for each breast, named craniocaudal (CC) and mediolateral oblique (MLO) [3]. The breast is positioned between two compression plates; in the MLO view, the compression plates are rotated by 45°–50°, towards the axilla. The radiologist is thus able to locate suspicious areas on both views by triangulating from these projections. This increases the diagnostic confidence as false positives due to tissue superposition are likely to disappear in the contralateral view. Computer Aided Detection (CAD) algorithms have also shown reduced false positive rates when the two views are taken into account [3,14,19].

The objective of our research is to design and evaluate a registration network for CC-MLO registration based on emerging deep learning technologies. Applications range from enhancing image presentation to the radiologist, to improving the performance of lesion detection algorithms that operate on single-view images [12,14,19]. Unfortunately, registration of the breast is considerably more challenging than other imaging modalities as the soft tissues in the breast are compressed and distorted during the acquisition [4]. To the best of our knowledge, few authors have explored the registration of CC and MLO views, and no established deep learning approach exists for this task [4,5].

Given the difficulty of estimating the deformation field between the CC and MLO views, many works in literature have resorted to matching Regions of Interest instead. The goal is not necessarily to establish the exact correspondence between lesions, but to minimize the chance that true positives are matched with false positive detections. This technique has largely been explored in combination with CAD algorithms that detect candidate lesions, which are then matched based on a combination of position and visual similarity. Visual similarity can be estimated based on hand-crafted features such as texture, size, intensity, etc. [19] or, with the advent of deep learning, by training a Siamese Convolutional Neural Network (CNN) [14]. Compared to this standard candidate-matching approach, our proposed registration technique works directly on the input image, and can be applied before, after or independently of other lesion detection or classification networks. At the same time, it is a flexible and versatile module that can be incorporated and jointly trained in more complex pipelines.

Successfully training a registration CNN requires defining a robust loss while reducing the cost of annotation [5]. To this aim, we augment the standard Mean Squared Error (MSE) loss exploiting available lesion annotations in the form of bounding boxes. The Generalized Intersection over Union (GIoU) forces the registration to match true lesions across both views. Preliminary experiments on the CBIS-DDSM dataset (presented in Sect. 5) with an affine transformation support the feasibility of our approach.

2   Background and Related Work

2.1   Deep Learning for Medical Image Registration

Registration requires estimating the spatial coordinate transformation that maximizes some measure of similarity between two images, usually denoted as the *fixed* and *moving* images [4,12]. Conventional registration methods are based on numerical optimization techniques, and may differ based on the domain of the transformation (global, local), its nature (rigid, affine, or elastic) and the optimization procedure [4,22].

Recently, CNN-based techniques have been proposed to regress the registration transformation from pairs of unregistered images [5]. Available solutions include fully convolutional networks or encoder-decoder architectures for elastic transformations [2,8,11,15] and Spatial Transformer Networks for affine transformations [23]. For a comprehensive review on the topic, the reader is referred to a recent survey by Haskins and colleagues [5].

Compared to traditional optimization approaches, CNN-based approaches are poised to have a substantial advantage: even if the training process is slower and requires hundreds or thousands of image pairs, at inference time it is usually much faster than optimizing the transformation on each image pair.

One of the main obstacles to efficient CNN-based registration is defining a suitable loss. In principle, the registration can be trained from image pairs, without additional annotations, by defining a similarity metric, such as the MSE, and a regularization term (registration is a generally ill-posed inverse problem). This approach forms the basis of unsupervised approaches, such as Voxelmorph [2], which has been applied to the registration of several imaging modalities, such as brain, breast and cardiac magnetic resonance imaging [1]. However, defining a robust image similarity measurement is notoriously challenging, especially in the presence of different source modalities, anatomical deformations or temporal changes [5,8]. Unlike common registration tasks in brain, cardiac or abdominal images, mammography images are characterized by stronger changes in viewpoint and high tissue deformation induced by organ compression; this fact makes the task more complex and, to the best of our knowledge, the feasibility of registering mammographic images has yet to be established.

An alternative strategy is supervised training, which however requires marking an appropriate number of manually matched points. Such ground truth is usually difficult and expensive to obtain in the medical domain. In our case, the breast is highly compressible and lacks rigid structures, and hence very few anatomical landmarks can be accurately matched. Large calcifications have been used as landmarks for validating registration algorithms as their location and correspondence can be determined very precisely [21]. However, collecting a large number of such annotations would be time consuming, and such benign structures are usually disregarded in radiological reports.

Our methodology falls into the semi-supervised domain, exploiting existing partial annotations. A similar strategy was successfully applied to train prostate MR registration from organ segmentation maps [8]. Our setting is more challenging as we assume that only coarse bounding boxes are available for training.

Finally, our work shares some similarities with multi-task learning settings in which the registration task is jointly learned with another task. For instance, Qin *et al.* combined estimation of cardiac motion and segmentation for cardiac MRI in a single network with shared weights [15]. Our approach is complementary since the bounding boxes, which are in any case an approximate ground truth, are used to supervise directly the registration task.

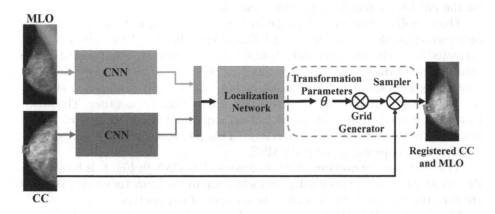

Fig. 1. Overall architecture of the registration network. From left to right: the CC and MLO views are passed through the shared convolutional layers; the feature map is concatenated and passed as input to the localization network; the CC image is registered by applying the estimated affine transformation parameters.

2.2   Spatial Transformer Networks

A Spatial Transformer network is a lightweight block which predicts and applies a spatial transformation to an input feature map during a single forward pass. It was proposed as a way to enhance an image classification network by allowing the network to transform feature maps to a canonical, expected pose to simplify inference in the subsequent layers [9]. The spatial transformer is composed of a localization network, which predicts the parameters of an affine transformation, which only requires six output parameters. Then, a sampling grid is created, that is a set of points where the input map should be sampled to produce the transformed output. Finally, the input feature map is resampled and interpolated to produce the output image (see Fig. 1). Spatial Transformers include a differentiable implementation of the sampling grid and resampling layer, allowing for end-to-end training, with standard back-propagation, of the models they are injected in. The network learns how to actively transform the feature maps to help minimise the overall cost function of the network during training.

3   Methodology

The proposed registration network is an end-to-end architecture which accepts as input a pair of unregistered CC and MLO images, and outputs the resampled CC image. We chose the MLO as fixed image and the CC as moving image since the former includes also the pectoral muscle, which is outside of the CC field of view. Registering the MLO to the CC would push the pectoral muscle out of the image pixels grid, and it would be impossible to estimate the correct deformation for the pixel belonging to the pectoral muscle.

The overall architecture, depicted in Fig. 1, is divided in two parts: the feature extraction block, and the Spatial Transformer block. The feature maps are extracted for each view separately, before being concatenated and passed to the Spatial Transformer network (introduced in Sect. 2.2). The proposed architecture implements an affine transformation, but can be easily extended to support other types of deformations by substituting the localization network. The architecture is trained in an end-to-end fashion exploiting the ground truth lesion bounding boxes as additional supervision. This provides cues for higher quality registration compared to the plain MSE.

The feature extraction backbone, marked as CNN in Fig. 1, is based on a ResNet50 network [6]. Specifically, we include up to the Conv4_x blocks. Weights are shared between views to reduce the number of parameters.

The Spatial Transformer is formed by a localization network and a resampling module. The localization network is made of a residual block (corresponding to the Conv5_x block of the ResNet50) followed by a dense layer to predict the parameters of the affine transformation:

$$\theta = \begin{bmatrix} a_{1,1} & a_{1,2} & t_1 \\ a_{2,1} & a_{2,2} & t_2 \\ 0 & 0 & 1 \end{bmatrix} \tag{1}$$

In the case of image registration, the sampling grid is simply the pixel grid of the fixed image, which greatly simplifies the implementation of the *grid generator* [9]. The output warped CC image is obtained by applying the affine transformation to this sampling grid using a bi-linear interpolation scheme.

The above resampling scheme can be applied indifferently to the original images (as done here), as well as to the feature maps (which could be useful if the feature maps were used for other tasks). Bounding boxes are converted by applying the inverse affine transformation and then rectifying the results. All layers including the bounding box registration are differentiable and, hence, can be trained end-to-end.

3.1   Loss

We argue that the MSE cannot by itself achieve successful registration. One of the underlying reasons is that the pectoral muscle is visible only in the MLO view. Experimentally, we observe that the CC may be overstretched over the pectoral muscle to achieve lower loss. If the registration is correct, the border of the CC should align to that of the pectoral muscle (see Fig. 2(a)).

To counterbalance this fact, we include in the loss only the region in which the moving CC image and the fixed MLO overlap (see Fig. 2(c)). The effect of the pectoral muscle, as well as of external air, is thus minimized. The resulting loss is defined as:

$$L_{MSE}(X^{mlo}, X^{cc}) = \left\| (X^{mlo} - X^{cc_{reg}})M \right\|^2 \tag{2}$$

where X^{mlo} is the MLO image, $X^{cc_{reg}}$ is the CC view after registration and M is the binary overlap mask.

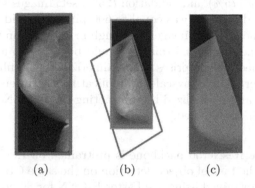

(a)   (b)   (c)

Fig. 2. Calculation of the overlap mask for the MSE loss. Unregistered (red box) and registered (green box) CC views are shown in (a) and (b). The shaded blue area is included in the calculation of the loss (b). In (c) the registered CC, fixed MLO and overlap mask are shown superimposed. It can be noticed how the margin of the CC view aligns with the pectoral muscle, outside of the overlap area. (Color figure online)

In order to exploit the lesion bounding boxes, we need a loss which reflects to which extent corresponding views are matched by the registration. The Intersection over Union (IoU) is a widely used measure to compare bounding boxes, but when the two bounding boxes do not overlap, the IoU is undefined. The recently proposed GIoU overcomes this limitation [16]. Given a pair of bounding boxes, it is defined as:

$$GIoU(B_i^{mlo}, B_i^{cc_{reg}}) = IoU(B_i^{mlo}, B_i^{cc_{reg}}) - \frac{A_c - U}{A_c} \tag{3}$$

where B_i^{mlo} and $B_i^{cc_{reg}}$ are the two bounding boxes, A_c is the area of the smallest enclosing box that includes both and U is their union. In short, when the bounding boxes don't overlap significantly, the GIoU takes their relative distance into account.

The GIoU loss ($L_{GIoU} = 1 - GIoU$) was initially proposed as a regression loss to train object detection networks. To the best of our knowledge, this is the first time it is used for the purpose of registration. To conclude, for each pair of mammographic views the total loss is calculated as

$$L_{total} = L_{MSE} + \lambda L_{GIoU} \tag{4}$$

where λ is a rescaling parameter.

4   Experimental Setup

Dataset. Our experiments were performed on the curated CBIS-DDSM collection [7, 10]. Each study comprises up to 4 images including both CC and MLO orientation. We selected cases with benign and malignant lesions visible on both views. Based on the standard training/test split, we obtained 985 cases for the development set and 122 for the test set. The training set was further split into a training (75%) and validation (25%) set. Images were downsampled so that the largest dimension was equal to 600 pixels. We did not exploit metadata available in the DICOM images; although in digital mammography patient positioning and other useful information would be available in the image headers, the DDSM collection comprises only scannerized screen-film mammography. Images were converted to grayscale by replicating the intensity values across the RGB channels and normalized by subtracting the ImageNet mean. No other pixel normalization was applied.

Pretraining. The ResNet50 backbone is pretrained on the ImageNet dataset and finetuned for the task of object detection on the same CBIS-DDSM dataset. Specifically, it is pretrained using the Faster R-CNN for 80 epochs before transferring to the registration [17]. This allows faster convergence than transferring directly from ImageNet (results not reported due to space limitations). This observation opens interesting prospects for feature sharing across multiple tasks, which however are outside of the scope of these experiments.

Hyperparameter Setup. Hyperparameters were experimentally finetuned on a smaller dataset. For the final training, we used the Adam optimizer (learning rate 10^{-4}, batch size 1). The network was trained for 300 epochs, each comprising 500 batches. The λ parameter (see Eq. 4) is set to 1000. The output dense layer of the Spatial Transformer is randomly initialized using Glorot initialization. The affine transformation parameters bias parameters are initialized to a 45 degree counterclockwise rotation, which is based on prior knowledge of the acquisition process. The network was implemented in Keras 2.2 with Tensorflow 1.13.1. All experiments were conducted on an AWS px2.large GPU instance.

Evaluation. Evaluation is not straightforward given the absence of a ground truth. Since the GIoU takes into account both the intersection and the distance of each pair of bounding boxes, we consider it as a viable evaluation metric. In addition, we visually inspected the registration results for the test set.

5   Results

The network was trained for 300 epochs without showing signs of overfitting (see Fig. 3). Both MSE and GIoU decreased indicating a synergistic behaviour of the two losses. When two bounding boxes do not overlap ($IoU = 0$), the

GIoU loss simplifies to $L_{GIoU} = 2 - \frac{U}{A_c} \geq 1$ [16]. In order to minimize $\frac{U}{A_c}$, the distance between the two bounding boxes must be reduced to the point where they eventually overlap.

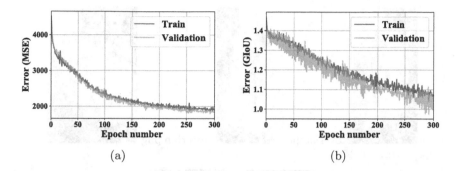

(a)   (b)

Fig. 3. Evolution of the loss during training: MSE (a) and GIoU (b)

The distribution of the L_{GIoU} on the test set is shown in Fig. 4. The bounding boxes for the registered CC and MLO overlap in 66.7% of the cases, which is an encouraging result. When L_{GIoU} approaches 0, the bounding boxes tend to perfectly overlap. In practice, due to the rectification process, the bounding boxes are unlikely to achieve perfect overlap, and lower IoU values are to be expected. Visually, in the large majority of cases the registration was successful in aligning the two views in terms of shape and global features, although an evaluation by a trained radiologist would be needed for confirmation.

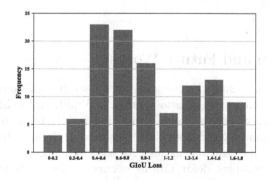

Fig. 4. Histogram of the GIoU loss for the test set

Examples of successful and unsuccessful registration results are shown in Fig. 5. In roughly 10% of the cases, the CC is still slightly overstretched to cover the pectoral muscle (Fig. 5a). It can be shown that in two cases, even if global alignment is successful, the bounding boxes do not overlap, sometimes by a large amount (Fig. 5c): this indicates that certain deformations cannot be recovered with the proposed affine transformation.

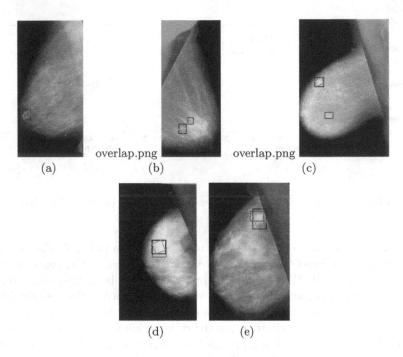

Fig. 5. Registration examples: the MLO and registered CC views are shown overlapped. The MLO bounding box is shown in red, the CC in blue, before and after rectification. (Color figure online)

6   Conclusion and Future Works

The presented work tackles the challenge of registering CC and MLO views by designing a fully trainable registration network. Weakly supervision that exploits available lesion annotations achieves promising results both in terms of visual alignment and lesion registration. The proposed technique has been demonstrated using an affine transformation. As a consequence, the network cannot fully capture the complex deformations occurring due to breast compression. Further improvements can be expected by substituting the Spatial Transformer with a different module to estimate a pixel-wise deformation field. This work lays the basis for several future developments. We will investigate how to combine the proposed network with other architectures, e.g., for object detection, to achieve multi-view analysis of mammographic images. The proposed technique could also be adapted to related tasks, such as the temporal registration of images from subsequent screening rounds.

Acknowledgements. The authors wish to thank Antonio Gaetano Barletta for his contributions in the implementation of the registration module.

References

1. Alfano, F., et al.: Prone to supine surface based registration workflow for breast tumor localization in surgical planning. In: 2019 IEEE 16th International Symposium on Biomedical Imaging (ISBI 2019), pp. 1150–1153. IEEE (2019)
2. Balakrishnan, G., Zhao, A., Sabuncu, M.R., Guttag, J., Dalca, A.V.: VoxelMorph: a learning framework for deformable medical image registration. IEEE Trans. Med. Imaging **38**(8), 1788–1800 (2019)
3. van Engeland, S., Snoeren, P., Hendriks, J., Karssemeijer, N.: A comparison of methods for mammogram registration. IEEE Trans. Med. Imaging **22**(11), 1436–1444 (2003)
4. Guo, Y., Sivaramakrishna, R., Lu, C.C., Suri, J.S., Laxminarayan, S.: Breast image registration techniques: a survey. Med. Biol. Eng. Comput. **44**(1–2), 15–26 (2006). https://doi.org/10.1007/s11517-005-0016-y
5. Haskins, G., Kruger, U., Yan, P.: Deep learning in medical image registration: a survey. Mach. Vis. Appl. **31**(1), 1–18 (2020). https://doi.org/10.1007/s00138-020-01060-x
6. He, K., Zhang, X., Ren, S., Sun, J.: Deep residual learning for image recognition. In: Proceedings of the IEEE Conference on Computer Vision and Pattern Recognition, pp. 770–778 (2016)
7. Heath, M., Bowyer, K., Kopans, D., Moore, R., Kegelmeyer, W.P.: The digital database for screening mammography. In: Proceedings of the 5th International Workshop on Digital Mammography, pp. 212–218. Medical Physics Publishing (2000)
8. Hu, Y., et al.: Weakly supervised convolutional neural networks for multimodal image registration. Med. Image Anal. **49**, 1–13 (2018)
9. Jaderberg, M., Simonyan, K., Zisserman, A., Kavukcuoglu, K.: Spatial transformer networks. In: Cortes, C., Lawrence, N.D., Lee, D.D., Sugiyama, M., Garnett, R. (eds.) Advances in Neural Information Processing Systems, vol. 28. pp. 2017–2025. Curran Associates, Inc. (2015). http://papers.nips.cc/paper/5854-spatial-transformer-networks.pdf
10. Lee, R.S., Gimenez, F., Hoogi, A., Miyake, K.K., Gorovoy, M., Rubin, D.L.: A Curated Mammography Data Set for Use in Computer-Aided Detection and Diagnosis Research, vol. 4, p. 170177. Nature Publishing Group, Berlin (2017)
11. Li, H., Fan, Y.: Non-rigid image registration using self-supervised fully convolutional networks without training data. In: 2018 IEEE 15th International Symposium on Biomedical Imaging (ISBI 2018), pp. 1075–1078. IEEE (2018)
12. Morra, L., Delsanto, S., Correale, L.: Artificial Intelligence in Medical Imaging: From Theory to Clinical Practice. CRC Press, Boca Raton (2019)
13. Morra, L., et al.: Breast cancer: computer-aided detection with digital breast tomosynthesis. Radiology **277**(1), 56–63 (2015)
14. Perek, S., Hazan, A., Barkan, E., Akselrod-Ballin, A.: Siamese network for dual-view mammography mass matching. In: Stoyanov, D., et al. (eds.) RAMBO/BIA/TIA -2018. LNCS, vol. 11040, pp. 55–63. Springer, Cham (2018). https://doi.org/10.1007/978-3-030-00946-5_6

15. Qin, C.: Joint learning of motion estimation and segmentation for cardiac MR image sequences. In: Frangi, A.F., Schnabel, J.A., Davatzikos, C., Alberola-López, C., Fichtinger, G. (eds.) MICCAI 2018. LNCS, vol. 11071, pp. 472–480. Springer, Cham (2018). https://doi.org/10.1007/978-3-030-00934-2_53

16. Rezatofighi, H., Tsoi, N., Gwak, J., Sadeghian, A., Reid, I., Savarese, S.: Generalized intersection over union: a metric and a loss for bounding box regression. In: Proceedings of the IEEE Conference on Computer Vision and Pattern Recognition, pp. 658–666 (2019)

17. Ribli, D., Horváth, A., Unger, Z., Pollner, P., Csabai, I.: Detecting and classifying lesions in mammograms with deep learning. Sci. Rep. 8(1), 4165 (2018)

18. Sacchetto, D., et al.: Mammographic density: comparison of visual assessment with fully automatic calculation on a multivendor dataset. Eur. Radiol. 26(1), 175–183 (2016). https://doi.org/10.1007/s00330-015-3784-2

19. Samulski, M., Karssemeijer, N.: Optimizing case-based detection performance in a multiview CAD system for mammography. IEEE Trans. Med. Imaging 30(4), 1001–1009 (2011)

20. Sechopoulos, I.: A review of breast tomosynthesis. Part I. The image acquisition process. Med. Phys. 40(1), 014301 (2013)

21. Van Schie, G.: Correlating locations in ipsilateral breast tomosynthesis views using an analytical hemispherical compression model. Phys. Med. Biol. 56(15), 4715 (2011)

22. Viergever, M.A., Maintz, J.A., Klein, S., Murphy, K., Staring, M., Pluim, J.P.: A survey of medical image registration-under review. Med. Image Anal. 33, 140–144 (2016)

23. de Vos, B.D., Berendsen, F.F., Viergever, M.A., Staring, M., Išgum, I.: End-to-end unsupervised deformable image registration with a convolutional neural network. In: Cardoso, M.J., et al. (eds.) DLMIA/ML-CDS -2017. LNCS, vol. 10553, pp. 204–212. Springer, Cham (2017). https://doi.org/10.1007/978-3-319-67558-9_24

Deep Transfer Learning for Texture Classification in Colorectal Cancer Histology

Srinath Jayachandran$^{(\boxtimes)}$ (iD) and Ashlin Ghosh (iD)

Robert Bosch Engineering and Business Solutions, Bangalore, India
{jayachandran.srinath,ashlin.ghosh}@in.bosch.com

Abstract. Microscopic examination of tissues or histopathology is one of the diagnostic procedures for detecting colorectal cancer. The pathologist involved in such an examination usually identifies tissue type based on texture analysis, especially focusing on tumour-stroma ratio. In this work, we automate the task of tissue classification within colorectal cancer histology samples using deep transfer learning. We use discriminative fine-tuning with one-cycle-policy and apply structure-preserving colour normalization to boost our results. We also provide visual explanations of the deep neural network's decision on texture classification. With achieving state-of-the-art test accuracy of 96.2% we also embark on using a deployment friendly architecture called SqueezeNet for memory-limited hardware.

Keywords: Histology · Colorectal cancer · Transfer learning · Texture classification

1 Introduction

According to the statistics provided by the American Cancer Society, colorectal cancer (CRC) is the third and second most commonly occurring cancer in men and women, respectively [3]. Histopathology provides one of the diagnosis procedures wherein, suspicious tissue is sampled by biopsy and examined under a microscope. A typical pathology report consists of tissue cell structural information which is used by the pathologist to decide upon any presence of malignant tumours. Such histological samples may typically contain more than two tissue types. Automating texture classification in CRC histology images will aid pathologists in making informed clinical decisions. Figure 1 represents randomly sampled histological images of eight tissue types in CRC. Histology image analysis for cancer diagnosis can be extremely challenging because of issues with slide preparation, variations in staining and inherent biological structure [14]. This makes the need for domain-specific input very important for feature generation. Deep learning being domain agnostic can make use of rich information present in histology images.

© Springer Nature Switzerland AG 2020
F.-P. Schilling and T. Stadelmann (Eds.): ANNPR 2020, LNAI 12294, pp. 173–186, 2020.
https://doi.org/10.1007/978-3-030-58309-5_14

2 Related Work

Studies relating to different texture analysis in human CRC have been lacking, although [14] has introduced a dataset containing 8 different classes of textures in CRC. They used traditional machine learning with hand-engineered texture descriptor features like histogram, local binary patterns, gray level co-occurrence matrix, gabor filters and perception like features with a reported accuracy of 87.4%. Work involving the use of deep neural networks on this dataset is very limited. Authors in [4] have designed a fully convolutional neural network (CNN) of 11 layers which could only classify with 75.5% accuracy. In another work [30], stain decomposition has boosted the performance on this dataset. They have derived hematoxylin and eosin (H&E) image components using an orthonormal transformation of the original RGB images and fed it into a bilinear convolutional neural network giving an accuracy of 92.6%. More recent works have explored the use of transfer learning. [28] has made use of a weakly labelled dataset and transferred features to retrain the model to reach an accuracy level of 92.70%. Another work experimented with different transfer learning approaches, generating features from pre-trained models without fine-tuning to train a classifier and fine-tuning pre-trained models, with the former method achieving the best accuracy of 95.40% [22].

Recent advancements in deep learning has led to the development of intelligent microscope-based slide scanners which have embedded software solutions for a specific task, like cell counting, tumour segmentation etc. [2]. Moreover, telepathology is the future of digital pathology where tissue images can be acquired and transmitted to pathology experts to facilitate rapid diagnostics using a smartphone. These technologies have limited computing resources and require specialized software and highly accurate small-sized CNN architecture [7, 31]. Much of the work does not explain in detail the fine-tuning method in transfer learning and has only focused on achieving good accuracy but not resource efficiency. In the rest of the paper, we demonstrate and achieve state-of-the-art results on CRC dataset [14] using transfer learning with discriminative fine-tuning and one cycle policy on a lightweight CNN architecture.

3 Methodology

3.1 Dataset and Preprocessing

The CRC dataset contains 5000 RGB histological images of 150-by-150 pixels each belonging to one of the 8 tissue categories. Each category has 625 images of H&E stained tissue samples digitized with an Aperio ScanScope [14]. We follow two steps of image preprocessing for this dataset. Firstly, the H&E staining process enables a clear view of morphological changes within a tissue [6]. But this process is prone to undesirable colour variations across tissue types because of differences in tissue preparation, staining protocols, the colour response of scanners, and raw materials used in stain manufacturing.

Any learning algorithm weighing-in more on the colour variation will lead to error in classification [15]. We use a structure-preserving colour normalization technique with sparse stain separation on these images given by [29]. Keeping one target image, other images are normalized by combining their respective stain density maps with a stain colour basis of the target image thus preserving the morphology. Figure 2 illustrates

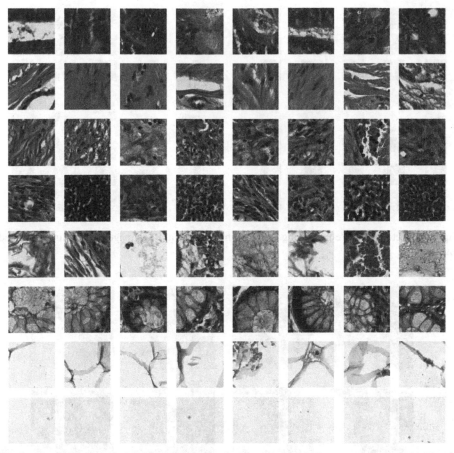

Fig. 1. Randomly sampled 8 images from each class (row) - a) Tumour epithelium b) Simple stroma c) Complex stroma d) Immune cell e) Debris f) Mucosal glands g) Adipose tissue h) Background (no tissue) in row-wise order starting from the top.

the effect of colour normalization. Second, the colour normalized dataset is then scaled between 0 and 1 followed by normalizing each channel to the ImageNet dataset [5].

3.2 Data Augmentation

With limited data, convolutional neural networks may overfit. Data augmentation improves the generalization capability of these networks by transforming images such that the network becomes robust to unseen data [20].

Random zoom crops were applied, as image patches in histopathology are invariant to translation in the input space. Tissue diagnosis is rotation invariant, which means that the pathologists can study histopathological images from different orientations. We introduce vertical and horizontal flips, and rotations, restricted to 90, 180 and 270° because of interpolation issues. The other augmentation techniques used were lighting, warps, gaussian blur, and elastic deformation. We applied in-memory dynamic data

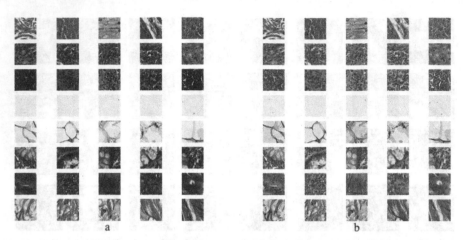

Fig. 2. Effect of structure-preserving colour normalization: a) Raw images b) Colour normalized images.

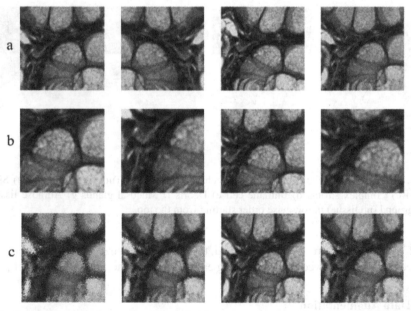

Fig. 3. Examples of image transformations for data augmentation using a) Rotations b) Random zoom crops and c) Jitter.

augmentation that applies random transformations on a batch of images during training. Figure 3 shows examples of a few transformations.

3.3   Transfer Learning Using SqueezeNet

Advancements in Deep Learning have led to super-human level performance on the ImageNet large scale visual recognition challenge. State-of-the-art deep neural networks (DNN) trained on the ImageNet dataset possess generic feature computation capabilities like gabor filters and colour blobs in their first layer that are very generic to any dataset or task. On the other hand, the final layer of these architectures become task-specific [32]. Given a new target visual dataset with a limited number of training examples, features from the pre-trained neural networks can be repurposed to adapt to this new dataset, called transfer learning. Since the AlexNet [18] breakthrough in ImageNet classification, many variants of convolutional neural networks have been submitted to the ImageNet challenge achieving state-of-the-art results. There is a high correlation between top-1 accuracy ImageNet architectures and their transfer learning capabilities [17], which makes it obvious to pick an architecture that performs the best on ImageNet. The focus of the majority of the models has not been on resource utilization, hence they are not practically deployable on resource-limited hardware. There has been an emergence of lightweight models like MobileNets, EfficientNets and ShuffleNet, but for this work, we choose an architecture called SqueezeNet [12] which has fewer parameters among the recent architectures and that performs reasonably well on ImageNet. With only 1,267,400 parameters and a model size of 4.85 MB, SqueezeNet proves to be a very lightweight model.

SqueezeNet is a convolutional neural network that is carefully designed such that it has few parameters but with competitive accuracy on ImageNet. For this, they follow strategies like replacing 3 × 3 filters with 1 × 1 filters, reducing the number of input channels to 3 × 3 filters and maintaining large activation maps for the convolutional layers by downsampling late in the network. These strategies are bundled into a module called Fire module (Fig. 4) which consists of a set of 1 × 1 filters in the squeeze convolutional layer, and a mix of 1 × 1 and 3 × 3 filters in the expand layer. The network macro architecture and architectural dimensions are presented in Fig. 5 and Table 1, respectively.

Pre-trained SqueezeNet was used as a backbone network and its penultimate layer is used as a feature extractor. The final output layer is replaced with a series of fully connected layers with Kaiming initialization [8], coupled with BatchNorm [13], and Dropout [27] layers which we call the head (layer group 4, Table 1). The rectified linear unit (ReLU) was used as the activation function.

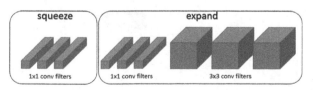

Fig. 4. Fire module of SqueezeNet

Fig. 5. SqueezeNet macro architecture

3.4 Finding the Optimal Learning Rate for Super-Convergence

The ability of an architecture to converge towards global minima on the loss function topology is an active area of research and is guided by hyperparameters like learning rate, batch size, momentum, and weight decay. Optimizers like Adam, AdaGrad, AdaDelta, and Nesterov momentum use a piecewise constant learning rate, starting with a global learning rate while carefully reducing it on the test set reaching a plateau [23]. Such strategies do not have a mechanism to automatically choose large learning rates that may help the network converge faster. We unfreeze the weights of the head (layer group 4, Table 1) and make them learnable while freezing the rest of the layers and fine-tune for

Table 1. SqueezeNet architectural dimensions

Learning rate	Layer group	Layer	#1 × 1	#1 × 1	#3 × 3	Filter size/Stride	Output shape
		Input image					224 × 224 × 3
0.0001	1	Conv1				7 × 7/2	96 × 109 × 109
		Maxpool1				3 × 3/2	96 × 54 × 54
		Fire1	16	64	64		64 × 54 × 54
		Fire2	16	64	64		64 × 54 × 54
0.003	2	Fire3	32	128	128		128 × 54 × 54
		Maxpool2				3 × 3/2	256 × 27 × 27
		Fire4	32	128	128		128 × 27 × 27
0.006	3	Fire5	48	192	192		192 × 27 × 27
		Fire6	48	192	192		192 × 27 × 27
		Fire7	64	256	256		256 × 27 × 27
		Maxpool3				3 × 3/2	512 × 13 × 13
		Fire8	64	256	256		256 × 13 × 13

(continued)

Table 1. (*continued*)

Learning rate	Layer group	Layer	#1 × 1	#1 × 1	#3 × 3	Filter size/Stride	Output shape
0.01	4	Adaptiveavgpool2d					512 × 1 × 1
		Adaptivemaxpool2d					512 × 1 × 1
		Flatten					1024
		BatchNorm1d					1024
		Dropout					1024
		Linear					512
		Batchnorm1d					512
		Dropout					512
		Softmax					8

two epochs. Next, we unfreeze the entire network and test its ability to super-converge [24].

We run a learning rate (LR) range test [25], a mock training on the network on a large range of learning rates for 100 batches and generate a loss vs learning curve as given in Fig. 6. This gives us an idea of the maximum learning rate up to which the model converges, beyond that the test or validation loss starts increasing leading to overfitting and poor accuracy. Learning rates between 0.0001 and 0.01 prove to reduce the loss, whereas beyond 0.01 the network starts to unlearn.

3.5 Discriminative Fine-Tuning with One-Cycle-Policy

Instead of using a global learning rate and monotonically decreasing it, we implement one cycle policy (cyclical learning rate) [26] with decoupled weight decay (AdamW: beta1 = 0.9, beta2 = 0.99) [19] that lets the learning rate cyclically vary between reasonable boundary values. This leads the network to converge faster and attain improved accuracy. Setting the learning rate to slightly below the maximum learning rate as the maximum bound, and to 10 times less this value as the lower bound, we train the network by starting at the lower bound and linearly increasing the learning rate up to maximum bound. At the same time momentum is decreased from 0.95 to 0.85 linearly. Then we perform cosine-annealing on the learning rate down to zero, while applying symmetric cosine annealing on momentum from 0.85 to 0.95 as shown in Fig. 7 [9].

Pretrained architectures exhibit different levels of information in their layers, starting from initial layers learning generic features to the final layer learning task-specific high-level features. Hence, different layers require different learning rates when being fine-tuned for a new task [11]. As shown in Table 1, the network is divided into 4 layer groups. The maximum bound learning rate discovered using the LR range test is assigned to the final (4th) layer group and the preceding layers are assigned with evenly

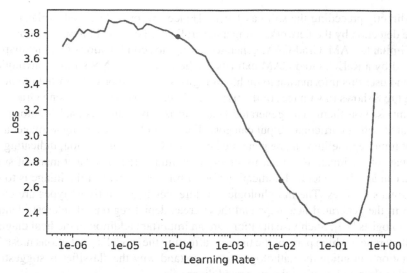

Fig. 6. Learning rate range test – training the entire network over a range of learning rates. The red markers define boundary values for a learning rate range where the network is still learning. (Color figure online)

spaced decreasing learning rates up to the boundary value marked in red. Each layer then undergoes one cycle policy with new maximum and minimum bounds.

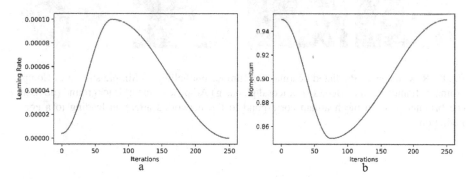

Fig. 7. Progression of a) learning rate and b) momentum during the one cycle training policy.

4   Visual Explanation Using Gradient-Based Localization

Making deep learning models transparent and explainable helps in understanding the failure modes as well as establishing trust and confidence by its users. However, decomposing deep neural networks into intuitive and interpretable components is difficult. A technique known as Class Activation Map (CAM) is very popular for interpreting the decisions made by deep learning models [33] but is limited to architectures with feature maps directly preceding the softmax layers. Hence, we provide visual explanations of texture detection by the networks using a generalization of CAM known as Grad-CAM [21]. Similar to CAM, Grad-CAM generates a weighted combination of feature maps but followed by a ReLU. Grad-CAM extracts gradients from a CNN's final convolutional layer and uses this information to highlight regions most responsible for the prediction.

Figure 8 shows raw images of the tissues and heatmaps generated using Grad-CAM superimposed on them. The generated heat map is a two-dimensional fractional grid associated with a particular output category. For example, for each input image, a heat map of tumor epithelium can be generated by Grad-CAM visualization, indicating how every part of the image is similar to tumor epithelium's features. Heat maps of simple stroma can also be generated, indicating how similar every part of the image is to simple stroma's features. The morphological differences between tissue types are clearly visible in the generated heatmaps and these areas depict regions of interest identified by pathologists [16]. Such visualizations are an important addition in medical diagnosis since they reflect which parts of the tissue is affecting the model's predictions most. This information can guide the pathologist to understand why the classifier is suggesting a particular class and confirm the suspected diagnosis.

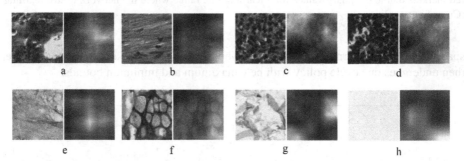

Fig. 8. Raw image vs Predicted heatmap. a) Tumour epithelium b) Simple stroma c) Complex stroma d) Immune cell e) Debris f) Mucosal glands g) Adipose tissue h) Background (no tissue). The brighter parts in the heatmap correspond to the network's attention leading to a correct prediction.

5   Results

The dataset is randomly shuffled and stratified, and divided into 3 sets: training (60%), validation (20%) and test (20%). We conducted different experiments related to the use of the pre-trained SqueezeNet network, and the use of the optimizer with one cycle policy. In Table 2 we compare the results of SqueezeNet architecture trained with and without the pre-trained weights. Also, the difference in the results is analyzed when a traditional piecewise constant learning rate scheduler optimizer like Adam is used.

It is realized that while training the network with Adam [23] and pre-trained weights set to true, the network achieves below par accuracy. One cycle policy and AdamW without using the pre-trained weights performs better than Adam, and with weights,

Table 2. Comparison of one cycle policy with Adam optimizer

Squeezenet pretrained weights	Optimizer	Validation accuracy (%)	Test accuracy (%)
True	One cycle policy with AdamW	97.4	96.2
False	One cycle policy with AdamW	80.1	75.6
True	Adam	71.5	64.8

we achieve the state of the art results of accuracies 97.4% and 96.2% on the validation and test set respectively. These experiments were carried out using the fast.ai [10] and Tensorflow [1] frameworks. We computed the receiver operating characteristic (ROC) curves (Fig. 9) for each of the classes based on different threshold settings and generated the area under the curve (AUC) plots. The ROC curve is a plot of the true positive rate (TPR) against the false positive rate (FPR) at various threshold settings. An AUC of 1 is a perfect scenario of the model predicting every class correctly in the test set. With the SqueezeNet architecture, by computing average ROC the overall sensitivity and specificity is approximately 99%.

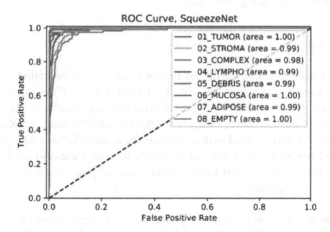

Fig. 9. ROC curves for the test set using SqueezeNet's fine-tuned architecture.

Figure 10(a) and (b) shows the training progress for 14 epochs of training, specifically the training and validation loss, as well as the accuracy, for training with a batch size of 32 and one threshold setting. The network saturated at an accuracy of 97.4% on the validation set, yielding 96.2% on the test set. The final model is serialized and made ready for deployment. We tested the model's performance in terms of inference time vs batch size (Fig. 10c) on a 640 cores Quadro M2000M GPU which is a mid-range

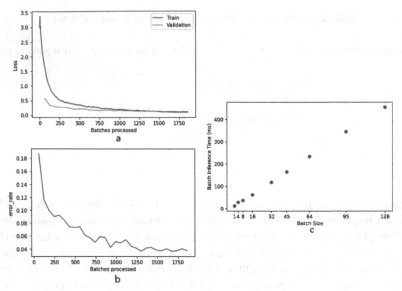

Fig. 10. a) Train and validation loss, b) error rate, and c) batch inference time vs batch size

graphics card, keeping in mind that the technologies where our models could be used would have limited computing resources.

6   Conclusion

In this paper, we achieve state-of-the-art results in texture classification in human colorectal cancer using transfer learning with super-convergence. The work also takes into consideration a deployment friendly DL model and network visualization to make neural networks decision making more transparent and explainable. SqueezeNet, a model of very small size (4.8 MB) is used to demonstrate the results. Various data augmentation techniques and structure preserving colour normalization were also used to boost the results. For future work, we aim to investigate tumour progression from the learned network using a similar dataset.

Acknowledgement. We wish to acknowledge the help and support provided by Sujit Kumar Ojha from the Engineering Data Science (EDS) department and Dr Kumar Rajamani from the Engineering Technology & Strategy (ETS) department of Robert Bosch Engineering and Business Solutions Private Limited.

References

1. Abadi, M., et al. TensorFlow: a system for large-scale machine learning. In: Proceedings of the 12th USENIX Symposium on Operating Systems Design and Implementation, OSDI, Savannah, GA, USA, 2–4 November 2016 (2016)

2. Bera, K., Schalper, K.A., Rimm, D.L., Velcheti, V., Madabhushi, A.: Artificial intelligence in digital pathology - new tools for diagnosis and precision oncology. Nat. Rev. Clin. Oncol. **16**(11), 703–715 (2019). https://doi.org/10.1038/s41571-019-0252-y

3. Bray, F., Ferlay, J., Soerjomataram, I., Siegel, R.L., Torre, L.A., Jemal, A.: Global cancer statistics 2018: GLOBOCAN estimates of incidence and mortality worldwide for 36 cancers in 185 countries. CA Cancer J. Clin. **68**(6), 394–424 (2018). https://doi.org/10.3322/caac. 21492

4. Ciompi, F., et al.: The importance of stain normalization in colorectal tissue classification with convolutional networks. In: 14th International Symposium on Biomedical Imaging, Melbourne, VIC, pp. 160–163 (2017). https://doi.org/10.1109/ISBI.2017.7950492

5. Deng, J., Dong, W., Socher, R., Li, L.J., Li, K., Fei-Fei, L.: ImageNet: a large-scale hierarchical image database. In: IEEE Conference on Computer Vision and Pattern Recognition, Miami, FL, pp. 248–255 (2010). https://doi.org/10.1109/cvpr.2009.5206848

6. Fischer, A.H., Jacobson, K.A., Rose, J., Zeller, R.: Hematoxylin and eosin staining of tissue and cell sections. Cold Spring Harb. Protoc. (2008). https://doi.org/10.1101/pdb.prot4986

7. Hartman, D.J., Parwani, A.V., Cable, B., et al.: Pocket pathologist: a mobile application for rapid diagnostic surgical pathology consultation. J. Pathol. Inform. 5–10 (2014). https://doi. org/10.4103/2153-3539.129443

8. He, K., Zhang, X., Ren, S., Sun, J.: Delving deep into rectifiers: surpassing human-level performance on imagenet classification. In: Proceedings of the IEEE International Conference on Computer Vision, Santiago, pp. 1026–1034 (2015). https://doi.org/10.1109/ICCV.2015.123

9. Howard, J., Gugger, S.: callbacks.one_cycle, fastai. https://docs.fast.ai/callbacks.one_cycle. html. Accessed 17 July 2020

10. Howard, J., Gugger, S.: Fastai: a layered API for deep learning. Information (Switzerland) **11**(2), 108 (2020). https://doi.org/10.3390/info11020108

11. Howard, J., Ruder, S.: Universal language model fine-tuning for text classification. In: Proceedings of the 56th Annual Meeting of the Association for Computational Linguistics, Melbourne, Australia, vol. 1, pp. 328–339 (2018). https://doi.org/10.18653/v1/p18-1031

12. Iandola, F.N., Moskewicz, M.W., Ashraf, K., Han, S., Dally, W.J., Keutzer, K.: SqueezeNet: AlexNet-level accuracy with 50x fewer parameters and < 0.5 MB model size. arXiv preprint arXiv:1602.07360 (2016)

13. Ioffe, S., Szegedy, C.: Batch normalization: accelerating deep network training by reducing internal covariate shift. In: Proceedings of the 32nd International Conference on Machine Learning, vol. 37, pp. 448–456 (2015)

14. Kather, J.N., et al.: Multi-class texture analysis in colorectal cancer histology. Sci. Rep. **6**, 27988 (2016). https://doi.org/10.1038/srep27988

15. Khan, A.M., Rajpoot, N., Treanor, D., Magee, D.: A nonlinear mapping approach to stain normalization in digital histopathology images using image-specific color deconvolution. IEEE Trans. Biomed. Eng. **61**, 1729–1738 (2014). https://doi.org/10.1109/TBME.2014.230 3294

16. Korbar, B., et al.: Looking under the hood: deep neural network visualization to interpret whole-slide image analysis outcomes for colorectal polyps. In: IEEE Computer Society Conference on Computer Vision and Pattern Recognition Workshops, pp. 821–827. IEEE (2017). https://doi.org/10.1109/CVPRW.2017.114

17. Kornblith, S., Shlens, J., Le, Q.V.: Do better imagenet models transfer better? arXiv preprint arXiv:1805.08974 (2018)

18. Krizhevsky, A., Sutskever, I., Hinton, G.E.: ImageNet classification with deep convolutional neural networks. Commun. ACM **60**(6), 84–90 (2017). https://doi.org/10.1145/3065386

19. Loschilov, I., Hutter, F.: Decoupled weight decay regularization. In: 7th International Conference on Learning Representations, ICLR, New Orleans (2019)

20. Perez, L., Wang, J.: The effectiveness of data augmentation in image classification using deep learning. arXiv preprint arXiv:1712.04621 (2017)
21. Selvaraju, R.R., Cogswell, M., Das, A., Vedantam, R., Parikh, D., Batra, D.: Grad-CAM: visual explanations from deep networks via gradient-based localization. arXiv preprint arXiv: 1610.02391 (2016)
22. Alinsaif, S., Lang, J.: Histological image classification using deep features and transfer learning. In: 17th Conference on Computer and Robot Vision (CRV), Ottawa, ON, Canada, 13–15 May 2020, pp. 101–108 (2020). https://doi.org/10.1109/CRV50864.2020.00022
23. Ruder, S.: An overview of gradient descent optimization algorithms. https://ruder.io/optimizing-gradient-descent/. Accessed 17 Jul 2020
24. Smith, L.N., Topin, N.: Super-convergence: very fast training of neural networks using large learning rates. arXiv preprint arXiv:1708.07120 (2019)
25. Smith, L.N.: Cyclical learning rates for training neural networks. In: IEEE Winter Conference on Applications of Computer Vision, Santa Rosa, CA, pp. 464–472 (2017). https://doi.org/10.1109/WACV.2017.58
26. Smith, L.N.: Disciplined approach to neural network. arXiv preprint arXiv:1803.09820v2 (2018)
27. Srivastava, N., Hinton, G., Krizhevsky, A., Sutskever, I., Salakhutdinov, R.: Dropout: a simple way to prevent neural networks from overfitting. J. Mach. Learn. Res. 15, 1929–1958 (2014)
28. Teh, E.W., Taylor, G.W.: Learning with less data via weakly labeled patch classification in digital pathology. In: Proceedings - International Symposium on Biomedical Imaging, Iowa City, IA, USA, pp. 471–475 (2020). https://doi.org/10.1109/ISBI45749.2020.9098533
29. Vahadane, A., et al.: Structure-preserving color normalization and sparse stain separation for histological images. IEEE Trans. Med. Imaging 35, 1962–1971 (2016). https://doi.org/10.1109/TMI.2016.2529665
30. Wang, C., Shi, J., Zhang, Q., Ying, S.: Histopathological image classification with bilinear convolutional neural networks. In: Proceedings of the Annual International Conference of the IEEE Engineering in Medicine and Biology Society, EMBS, Seogwipo, pp. 4050–4053 (2017). https://doi.org/10.1109/EMBC.2017.8037745
31. Weinstein, R.S., Descour, M.R., Liang, C., et al.: Telepathology overview: from concept to implementation. Hum. Pathol. 32(12), 1283–1299 (2001). https://doi.org/10.1053/hupa.2001.29643
32. Yosinski, J., Clune, J., Bengio, Y., Lipson, H.: How transferable are features in deep neural networks? In: Proceedings of the 27th International Conference on Neural Information Processing Systems, vol. 2, pp. 3320–3328 (2014)
33. Zhou, B., Khosla, A., Lapedriza, A., Oliva, A., Torralba, A.: Learning deep features for discriminative localization. In: Proceedings of the IEEE Computer Society Conference on Computer Vision and Pattern Recognition, Las Vegas, NV, pp. 2921–2929 (2016). https://doi.org/10.1109/CVPR.2016.319

Applications of Generative Adversarial Networks to Dermatologic Imaging

Fabian Furger[1], Ludovic Amruthalingam[2]($\boxtimes$), Alexander Navarini[2,3], and Marc Pouly[1]

[1] Department of Information Technology, Lucerne University of Applied Sciences and Arts, Lucerne, Switzerland
furgerfabian@hotmail.com, marc.pouly@hslu.ch
[2] Department of Biomedial Engineering, University of Basel, Basel, Switzerland
ludovic.amruthalingam@unibas.ch
[3] Department of Dermatology, University Hospital of Basel, Basel, Switzerland
alexander.navarini@usb.ch

Abstract. While standard dermatological images are relatively easy to take, the availability and public release of such data sets for machine learning is notoriously limited due to medical data legal constraints, availability of field experts for annotation, numerous and sometimes rare diseases, large variance of skin pigmentation or the presence of identifying factors such as fingerprints or tattoos. With these generic issues in mind, we explore the application of Generative Adversarial Networks (GANs) to three different types of images showing full hands, skin lesions, and varying degrees of eczema. A first model generates realistic images of all three types with a focus on the technical application of data augmentation. A perceptual study conducted with laypeople confirms that generated skin images cannot be distinguished from real data. Next, we propose models to add eczema lesions to healthy skin, respectively to remove eczema from patient skin using segmentation masks in a supervised learning setting. Such models allow to leverage existing unrelated skin pictures and enable non-technical applications, e.g. in aesthetic dermatology. Finally, we combine both models for eczema addition and removal in an entirely unsupervised process based on CycleGAN without relying on ground truth annotations anymore. The source code of our experiments is available on https://github.com/furgerf/GAN-for-dermatologic-imaging.

Keywords: Generative Adversarial Networks · Dermatology

1 Introduction

Generative Adversarial Networks (GANs), initially proposed in [8] have since then produced impressive results in a variety of synthetic data generation tasks. In contrast to other deep learning methods, which are notoriously data-intensive, GANs achieve good results even with relatively small data sets [2,7]. This makes GANs attractive for domains where training data is difficult or expensive to obtain. A standard example is the medical field, where specialized machinery

© Springer Nature Switzerland AG 2020
F.-P. Schilling and T. Stadelmann (Eds.): ANNPR 2020, LNAI 12294, pp. 187–199, 2020.
https://doi.org/10.1007/978-3-030-58309-5_15

may be needed or occurrences of pathologies may be hard to find. Using data sets augmented with GAN-generated synthetic data to train machine learning models has improved performance in a variety of medical domains [3,9,12].

Dermatology is one domain particularly suited for the application of deep learning models, but with far too few publicly-available data sets compared to the diversity of the cases encountered in clinical practice. Therefore, the idea to leverage the GAN framework to generate new samples is very promising. However, applications in dermatology are to this date still rare. One example is *MelanoGAN* [2], which generates images of skin lesions from ISIC 2017 [5]. The authors compare the results of different GAN models by training a lesion classifier on synthetic data only. In another work, [3] generate skin lesions from ISIC 2018 by translating lesion segmentation masks to images. The resulting images are thus directly associated with ground truth segmentations, which can be leveraged for further applications.

In this paper we present our results for two different types of skin lesions: eczema and moles. For eczema we use a private data set (due to identifying patient information) but for moles we use an established public data set for reproducibility and as an example of the generality of our approach.

Besides technical applications such as data augmentation or the creation of paired data, image transformation also enables domain-specific use cases such as prediction of a skin lesion evolution or the evaluation of aesthetic effects of treatment. With this in mind, we train our GAN models to add or remove eczema from skin pictures pursuing two different strategies: a supervised approach where we use ground truth lesion segmentation masks to target modifications to precisely defined areas as well as an unsupervised process entirely freed from the availability of training data.

2   Materials and Methods

2.1   Data Sets

We conduct experiments on 3 different types of dermatologic images:

Sets of Hands. The first set of experiments is conducted on photos of hands. Each of the 246 individual pairs of hands was photographed from the front and the back side, for a total of 492 photos. They were taken under uniform condition with green background and downscaled to 640×480 pixels.

Patches of Skin. Most of the remaining experiments leverage high-resolution photos (3456×2304 pixels) of the back side of hands from the EUSZ2 data set collected in the SkinApp project [17]. There are 79 photos available for training and we use a test set of 52 photos to analyze the overfitting of the discriminator. The photos are annotated with segmentations marking the contour of the hands and eczema lesions. From these photos, we extract patches of skin fulfilling the following criteria: a patch consists of skin only (no background) with a specified amount of skin being afflicted with eczema. We create a data set with *healthy skin* patches and a data set with *skin with eczema* patches, where 10–80% of the skin pixels are annotated as eczema. For these experiments, patches of 128×128 pixels are used. This procedure yields 51023 patches of healthy skin and 2872 patches of skin with eczema. Larger patch sizes yield smaller data sets and significantly increase overfitting, especially in the case of skin with eczema.

Skin Lesions. The final data sets consist of dermoscopic images of skin lesions from the ISIC archive 2018 [5,22]. In particular, we generate new lesion images of *Dermatofibroma* (DF) and *Melanoma* (MEL) with 115 and 1113 samples available for training, respectively. These different data set sizes allow to analyze the effects on GAN performance. The original images have varying sizes and are resized to a common resolution of 256×256 pixels.

2.2 Model Architecture

This section describes the architecture of the generator and discriminator models for the experiments. Our models are based on the architecture of DCGAN [19] with the changes described in the following paragraphs. All models are optimized using Adam [16] with a learning rate of $5 \cdot 10^{-5}$ and default moment decays $\beta_1 = 0.9$, $\beta_2 = 0.999$ (values determined experimentally for model convergence). The training was organized in batches of varying size depending on the image resolution and was stopped when the training metrics converged.

Unconditional Generator. The generator for unconditional image synthesis receives a 100-dimensional input vector (drawn independently from a standard Gaussian), which is first passed through a dense layer to produce 64 initial feature maps. The layer's output is reshaped based on the desired aspect ratio of the generated images with lower resolution. Then, a sequence of fractionally-strided convolutions (deconvolutions) increases the image size until the desired output resolution is achieved.

Following common practice, the number of feature maps per convolution are halved at each resolution stage. After each convolution, the output is passed through batch normalization [13] and activated with LeakyReLU [18]. Finally, a regular convolution with 3 output feature maps is activated with tanh to produce the RGB-channels of the generated image.

The hand images generator benefits from unstrided convolutions after each deconvolution to refine the intermediate representations. This is attributed to the comparatively large complexity of these images and does not help with the generation of patches of skin and skin lesions. The size of the initial dense layer and the number of deconvolutions determine the image resolution. Table 1 summarizes the model parametrizations.

Table 1. Unconditional generator: image resolution overview.

Experiment	Dense layers	Deconvolution	Resolution
Full hands (Sect. 3.1)	$20 \times 15 \times 64$	5	640×480
Skin patches (Sect. 3.1)	$8 \times 8 \times 64$	4	128×128
Skin lesions (Sect. 3.1)	$8 \times 8 \times 64$	5	256×256

Image Translation Generator. The image translation model is based on the U-Net architecture [20]: an encoder with increasing number of features, which reduces the image resolution, and a decoder to reverse the process. Additionally, the encoded representation is translated with a sequence of residual blocks [10]. We find experimentally (with the FID score and a qualitative review of the results) that 2 strided convolutions in the encoder and 2 deconvolutions in the decoder yield the best results. Consequently, the residual blocks translate features with a resolution of 32 × 32 pixels. We find that 4 residual blocks are ideal, which is surprisingly low but can be attributed to the fact that the skin images are small and relatively simple. Skip connections between the encoder and the corresponding decoder stages are used as suggested by [14]. These connections forward intermediate features from the encoder that are combined with the decoder features by concatenation.

Finally, we task the image translation generator with *image modification*. To that end, the input image is added to the 3 output channels of the generator, so that it is essentially tasked with generating an image *residual*. The generated residual contains the information to modify the input photo in the desired way.

Discriminator. All experiments leverage the same *multi-scale discriminator* architecture [23]: two individual discriminators process an input image and a downscaled version of the image. Afterwards, their outputs are averaged. This improves the sensibility to low-level details and high-level structures. We observed that more than two discriminators do not improve results, which can be explained by our images' lower resolution when compared with [23].

Both discriminators have the same architecture: a sequence of strided convolutions with batch normalization and LeakyReLU activation, followed by a dense layer with one output neuron to produce the prediction. The features are doubled after each convolution and the number of convolution layers matches the deconvolution layers of the corresponding generators, as summarized in Table 1. All the image translation experiments operate on patches of skin image with 4-convolution discriminators. As the generators produce normalized images, the channels of the real images are also normalized before discrimination.

Model Balance and Selection. The balance between the generator and discriminator is difficult to maintain, as neither should overpower the other [25]. Model balance is adjusted by selecting the number of *initial features* of the generator and discriminator. Table 2 summarizes the initial features of all models in this work's experiments. The ideal numbers of features are determined empirically with the restriction of the available GPU memory.

Besides visual inspection, we minimize the Fréchet Inception Distance (FID) [11] to select the best model. The FID measures the dissimilarity between real and generated images, it is commonly used to quantitatively compare the results of GAN models. In our experiments, this metric works well with unconditional generation, but not with image translations showing that the generator's secondary objective of retaining certain image regions penalizes image realism. Furthermore, we observe that FID scores computed on different data sets should not be compared as the data set's inherent statistics and variability greatly influence the FID scores.

Model selection is additionally guided by the discriminator's predictions confidence and consistency, which indicate whether the discriminator requires additional capacity to adequately distinguish real and generated samples, and thus, to better guide generator learning.

3 Experiments

3.1 Unconditional Dermatology Data Synthesis

The first experiments concern the unconditional generation of dermatology data. The objective is to explore the quality of generated images for different target data sets. The findings indicate the expected performance when the GAN task is not restricted and serves as a baseline for later comparisons with the results of restricted tasks.

Table 2. Initial features for the generator and discriminator models.

Experiment	Generator	Discriminator
Full hands (Sect. 3.1)	512	32
Healthy patches (Sect. 3.1)	1024	128
Eczema patches (Sect. 3.1)	1024	256
Skin lesions (Sect. 3.1)	512	64
Targeted eczema (Sect. 3.2)	1024	256
Untargeted eczema (Sect. 3.3)	1024	256

Sets of Hands. There are two central aspects to the quality of the generated images: high-level structures like anatomy and low-level details like textures. Here, the multi-scale discriminator architecture proves useful, as the two discriminators each focus on one of these aspects. However, many of the generated images still contain visible defects such as hands with more than 5 fingers. These issues are linked to unlikely generator input vectors and can be mitigated using the *truncation trick* [23] to improve the quality of the generated images.

The truncation technique includes the truncation of the input below some a priori defined threshold. Every exceeding component of the input vector is re-sampled. Truncation trades sample variability for quality: aggressive truncation significantly reduces variability, while sample quality increases. We determine empirically that a threshold of 0.1 is suitable for the generation of hands, based on the generated samples and FID scores. These scores are summarized in Table 3. Figure 1 shows the results with a truncation threshold of 0.1.

While the samples do not show great variability, their quality is generally high. The hands' textures look realistic, the side (front or back) of most pairs of hands can be determined in most samples and most hands consist of four fingers and a thumb.

Table 3. Truncation threshold selection with FID score.

Threshold	0.01	0.02	0.05	**0.1**	0.2	0.5	1	None
FID	111.4	94.5	75.0	**69.5**	69.5	70.3	74.1	74.2

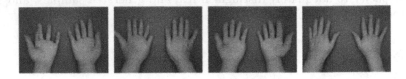

Fig. 1. Samples of the unconditional generation of hands.

This application shows that high-resolution dermatology images can be generated with a relatively small data set. These images could be mistaken for real photos at short glance. The model obtains a FID score of 74.2 without truncation, a significantly lower value than in all other experiments. This indicates that FID scores on different data sets should not be compared.

Patches of Skin. We further experiment with the unconditional generation of images of healthy skin and of skin that contains eczema. These experiments are a prerequisite for later eczema modification experiments.

Healthy Skin. With the large data set of 51023 patches of skin that do not contain any eczema, our GAN is able to generate high-quality images. Samples are shown in Fig. 2. The generated samples look very realistic and are also very diverse. Different types of skin, as well as creases and wrinkles are generated. The selected model achieves a FID score of 538.7.

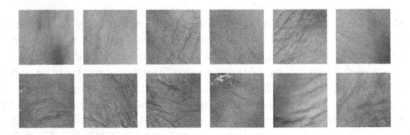

Fig. 2. Samples of the unconditional generation of healthy skin (first line) and skin with eczema (second line).

Skin with Eczema. We observe that the discriminator's task becomes more difficult when classifying patches of skin with eczema, so that the best results are achieved when the discriminator contains more feature maps. Sample results are shown in Fig. 2. The quality of the generated images is comparable with the synthetic healthy skin. The skin is detailed and contains different kinds of wrinkles and eczema. Overall, there are more creases than in the patches of healthy skin, which is attributed to the increased prevalence of eczema in such areas of the hand. The model achieves a FID score of 599.6 for this task.

Perceptual Study. We further evaluate the generated images quantitatively in a perceptual study. The results are presented in Sect. 3.1 along with the analysis of synthetic skin lesion images.

Overfitting. Finally, we analyze the models' overfitting, quantitatively for the discriminator and qualitatively for the generator. For patches of skin with eczema, the discriminator increasingly overfits over the course of the training. Samples from the training set are predicted as real with high likelihood, while testing samples are increasingly being rejected as generated. We observe that this is not the case for the discriminator of healthy skin. As the discriminator for skin with eczema has greater capacity, it is more prone to overfitting. However, we find that overfitting is mainly linked to the data set size. Low-capacity discriminators also overfit to the set of 2872 images, while high-capacity discriminator do not overfit on larger data sets.

We further investigate how the overfitting of the discriminator for patches of skin with eczema impacts the generator. We perform a qualitative assessment of the generator overfitting with the common method of comparing generated samples with their nearest training samples [4, 6, 15]. In our experiments, the *structural similarity index* [24] yields more similar samples than the *mean squared error*. We find that the generated samples do not contain memorized parts of the training set, so we can conclude that the discriminator's overfitting is not leading the generator to overfit as well.

Skin Lesions. Finally, we generate images of skin lesions. Samples of generated DF and MEL lesions are shown in Fig. 3.

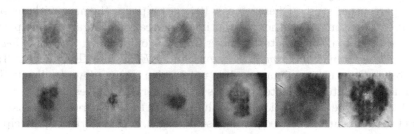

Fig. 3. Samples of the unconditional generation of DF (first line) and MEL (second line) lesions.

Dermatofibroma. While these images resemble the samples of the training set, they lack variability. Furthermore, they show clear tiling artifacts, i.e. patterns that are repeated within a generated image. In this case, the discriminator is trained with only 115 real samples and overfits severely. This visibly impacts the generator: we observe structures, such as lesion shapes or the hairs in the bottom left corners across different samples. With these negative aspects, the generator achieves a FID score of 822.9.

Melanoma. The generated images of MEL lesions contain far greater variability but also suffer from significant tiling. In this case, the generator's FID is 607.8. There is significantly less overfitting, as this data set contains 1113 samples. However, some of the hairs are still repeated. We hypothesize that such specific and distinctive hairs are prone to be copied, as they are rare among the real samples.

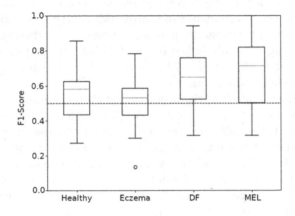

Fig. 4. Perceptual study: the box plots show the three quartiles of the obtained F1-scores for each data set.

Perceptual Study. We assess the realism of the generated patches of skin lesions with a perceptual study, where we ask 104 participants (laymen without prior training) to determine whether a given image is real or generated. The participants are asked to discriminate 20 images from one of four sets: *patches of healthy skin, patches of skin with eczema, DF lesions,* and *MEL lesions.* They have 2–3 seconds observation time per image and do not receive intermediate feedback. Such experiments are often conducted to assess if the generated images are easily identified [14, 21, 23]. The classifications are evaluated with the F1-score and the distribution of the results are visualized per data set in Fig. 4. The majority of participants are unable to distinguish real and generated patches of skin, regardless of the presence of eczema: the mean F1-scores are just above random guessing, with 0.58 and 0.53. The third quartiles are also very low, with 0.63 and 0.59. This result confirms that the models are able to generate realistic skin patches. On the other hand, skin lesions are simpler to distinguish, with a mean F1-scores of 0.65 and 0.71. This reflects the observations of the qualitative analysis, where generated lesions look less realistic than synthetic patches of skin. Interestingly, DF lesions are perceived as slightly more realistic than MEL lesions.

3.2   Targeted Eczema Modification

We formulate eczema addition and removal as an image translation task: the generator receives a skin photo and an eczema segmentation mask as input and should either remove or add eczema within the indicated areas. This is performed by generating a residual, which is added to the input image. To encourage pairing between the generator's input and output, its adversarial objective is combined with the *relevancy loss* [1].

The translations are performed between the data sets of skin with and without eczema, two data sets with very different sample sizes. Thus, the set of patches of healthy skin is truncated to 2872 samples, to match the smaller data set. We use additional healthy skin images to train the discriminator for eczema removal, which effectively prevents overfitting. Furthermore, we use the same segmentation with multiple photos of healthy skin. This also helps with generalization, though the effects of this technique are less pronounced.

Eczema Removal. In Fig. 5 we show the translation results of removing eczema from afflicted skin. Columns 3 and 6 still show the same parts of hands as the input photos in columns 1 and 4, but they no longer contain the structures and skin disruptions associated with eczema. However, the generated patches generally lose some fine details such as creases, which are often less visible, compared to the inputs. We observe that the FID score applies poorly to the results of image translation. For these experiments, the FID is often oscillating, in this case between 600 and 1100. Thus, we rely on the visual qualitative evaluation of the generated samples.

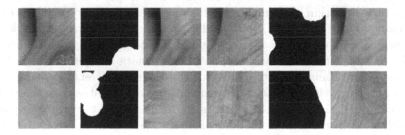

Fig. 5. Eczema removal (first line) and addition (second line) from afflicted skin: columns 1 and 4 show the input photos, columns 2 and 5 the input segmentations and columns 3 and 6 the generation results.

Eczema Addition. We modify photos of healthy skin by adding eczema to specified areas. Figure 5 shows sample results of this translation. The generator again produces realistic images, as we show in columns 3 and 6. Generally, the structures of the skin are retained and fewer details are lost, compared to eczema removal. Further, realistic-looking eczema is placed in the desired parts of the images. These results show that convincing eczema can be in-painted accurately in the indicated locations, which enables applications such as simulating the progression of untreated eczema.

3.3 Untargeted Eczema Modification

We experiment the cyclic translation between patches of skin with and without eczema. No segmentation masks are used and the translations are learned with the completely unsupervised CycleGAN framework [26]. The pairing between generator input and output is achieved with the *cycle consistency loss* [26], which penalizes differences between a generator's input and its reconstruction. While placing a greater emphasis on cycle consistency does increase the pairing, this benefit comes at the cost of reduced sample quality. Sample results of unsupervised eczema modification are shown in Fig. 6.

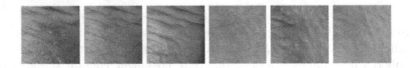

Fig. 6. Unsupervised cyclic eczema transformation: columns 1 and 4 show the sick and healthy input photos, columns 2 and 5 the generated translations without and with eczema and columns 3 and 6 the input reconstructions.

The results are realistic and the original inputs are reasonably reconstructed although some details are missing. This is to be expected, as the generated patches of healthy skin in column 2 should not contain any hints on where or how to in-paint specific eczema. Eczema addition produces realistic-looking lesions, however, it is no longer targeted and can not always be clearly determined.

The loss of details observed in previous translation experiments is barely noticeable here, likely a positive effect of the cycle consistency objective. The metrics of these cyclic translation experiments are more stable than those of the individual translations. For completeness, we mention that the synthetic patches of healthy skin have a FID of 654.7 to the real data, while the synthetic patches of skin with eczema have a FID of 690.2. These scores are reasonably similar to the scores of unconditional generation, with 538.7 and 599.6, respectively.

4   Conclusion

We present different applications of GANs on dermatologic images. First, unconditional image generation is performed successfully with photos of hands and patches of skin in particular. This is also shown for skin patches in the perceptual study. The validity of our approach is therefore confirmed and our initial objective to create realistic synthetic data achieved.

In the case of generated skin lesions, the results do not look as realistic. This could be corrected by further filtering of the images with rare features (such as hair in our particular case), when compared to the other images in the data set. Our analysis shows that the discriminator already overfits with data sets of several thousand images. On the other hand, we only notice overfitting in the generator when using smaller data sets of merely hundreds of samples. Thus, we conclude that the discriminator complexity should be especially controlled when working with small data sets

In the second part of this work, we explore the task of image modification, with eczema addition or removal within a specified area. The obtained results are again visually appealing but we observe that the FID score may be unsuitable to assess the quality of image translation experiments. In particular, we demonstrate the precise addition of eczema to the areas indicated by the segmentation mask. These results open the door for new applications in dermatology such as anomaly detection in a disease appearance or the visualization of the long term aesthetic effects of a disease.

Finally, we also perform domain translation between healthy skin and skin with eczema lesions in an entirely unsupervised experiment. In particular, the eczema removal results may be interesting for future applications, such as weakly-supervised eczema segmentation similar to [1]. This is certainly the most probable case that researchers will encounter as labeling is a costly step. In practice, before labeling is even considered, it is often necessary to first get prototyping results which could be achieved following this approach.

References

1. Andermatt, S., Horváth, A., Pezold, S., Cattin, P.C.: Pathology segmentation using distributional differences to images of healthy origin. CoRR abs/1805.10344 (2018). http://arxiv.org/abs/1805.10344
2. Baur, C., Albarqouni, S., Navab, N.: Melanogans: high resolution skin lesion synthesis with GANs. CoRR abs/1804.04338 (2018). http://arxiv.org/abs/1804.04338
3. Bissoto, A., Perez, F., Valle, E., Avila, S.: Skin lesion synthesis with generative adversarial networks. CoRR abs/1902.03253 (2019). http://arxiv.org/abs/1902.03253
4. Brock, A., Donahue, J., Simonyan, K.: Large scale GAN training for high fidelity natural image synthesis. CoRR abs/1809.11096 (2018). http://arxiv.org/abs/1809.11096
5. Codella, N.C., et al.: Skin lesion analysis toward melanoma detection: a challenge at the 2017 international symposium on biomedical imaging (ISBI), hosted by the international skin imaging collaboration (ISIC). In: 2018 IEEE 15th International Symposium on Biomedical Imaging (ISBI 2018), pp. 168–172. IEEE (2018)
6. Denton, E.L., Chintala, S., Szlam, A., Fergus, R.: Deep generative image models using a Laplacian pyramid of adversarial networks. In: Cortes, C., Lawrence, N.D., Lee, D.D., Sugiyama, M., Garnett, R. (eds.) Advances in Neural Information Processing Systems 28, pp. 1486–1494. Curran Associates, Inc. (2015). http://papers.nips.cc/paper/5773-deep-generative-image-models-using-a-laplacian-pyramid-of-adversarial-networks.pdf

7. Frid-Adar, M., Diamant, I., Klang, E., Amitai, M., Goldberger, J., Greenspan, H.: GAN-based synthetic medical image augmentation for increased CNN performance in liver lesion classification. CoRR abs/1803.01229 (2018). http://arxiv.org/abs/1803.01229

8. Goodfellow, I., et al.: Generative adversarial nets. In: Ghahramani, Z., Welling, M., Cortes, C., Lawrence, N.D., Weinberger, K.Q. (eds.) Advances in Neural Information Processing Systems 27, pp. 2672–2680. Curran Associates, Inc. (2014). http://papers.nips.cc/paper/5423-generative-adversarial-nets.pdf

9. Guibas, J.T., Virdi, T.S., Li, P.S.: Synthetic medical images from dual generative adversarial networks. CoRR abs/1709.01872 (2017). http://arxiv.org/abs/1709.01872

10. He, K., Zhang, X., Ren, S., Sun, J.: Deep residual learning for image recognition. CoRR abs/1512.03385 (2015). http://arxiv.org/abs/1512.03385

11. Heusel, M., Ramsauer, H., Unterthiner, T., Nessler, B., Klambauer, G., Hochreiter, S.: GANs trained by a two time-scale update rule converge to a nash equilibrium. CoRR abs/1706.08500 (2017). http://arxiv.org/abs/1706.08500

12. Hiasa, Y., et al.: Cross-modality image synthesis from unpaired data using cyclegan: effects of gradient consistency loss and training data size. CoRR abs/1803.06629 (2018). http://arxiv.org/abs/1803.06629

13. Ioffe, S., Szegedy, C.: Batch normalization: accelerating deep network training by reducing internal covariate shift. CoRR abs/1502.03167 (2015). http://arxiv.org/abs/1502.03167

14. Isola, P., Zhu, J.Y., Zhou, T., Efros, A.A.: Image-to-image translation with conditional adversarial networks. In: The IEEE Conference on Computer Vision and Pattern Recognition (CVPR), pp. 1125–1134 (2017)

15. Karras, T., Aila, T., Laine, S., Lehtinen, J.: Progressive growing of GANs for improved quality, stability, and variation. CoRR abs/1710.10196 (2017). http://arxiv.org/abs/1710.10196

16. Kingma, D.P., Ba, J.: Adam: a method for stochastic optimization. CoRR abs/1412.6980 (2014). http://arxiv.org/abs/1412.6980

17. Koller, T., Schnürle, S., vor der Brück, T., Christen, R., Pouly, M.: Skinapp deeplearning. Technical report, Lucerne University of Applied Sciences, September 2018

18. Maas, A.L., Hannun, A.Y., Ng, A.Y.: Rectifier nonlinearities improve neural network acoustic models. In: ICML Workshop on Deep Learning for Audio, Speech and Language Processing, p. 3 (2013)

19. Radford, A., Metz, L., Chintala, S.: Unsupervised representation learning with deep convolutional generative adversarial networks. CoRR abs/1511.06434 (2015). http://arxiv.org/abs/1511.06434

20. Ronneberger, O., Fischer, P., Brox, T.: U-net: Convolutional networks for biomedical image segmentation. CoRR abs/1505.04597 (2015). http://arxiv.org/abs/1505.04597

21. Salimans, T., Goodfellow, I.J., Zaremba, W., Cheung, V., Radford, A., Chen, X.: Improved techniques for training GANs. CoRR abs/1606.03498 (2016). http://arxiv.org/abs/1606.03498

22. Tschandl, P., Rosendahl, C., Kittler, H.: The ham10000 dataset, a large collection of multi-source dermatoscopic images of common pigmented skin lesions. Sci. Data 5, 180161 (2018)

23. Wang, T., Liu, M., Zhu, J., Tao, A., Kautz, J., Catanzaro, B.: High-resolution image synthesis and semantic manipulation with conditional GANs. CoRR abs/1711.11585 (2017). http://arxiv.org/abs/1711.11585

24. Wang, Z., Bovik, A.C., Sheikh, H.R., Simoncelli, E.P.: Image quality assessment: from error visibility to structural similarity. IEEE Trans. Image Process. **13**(4), 600–612 (2004). https://doi.org/10.1109/TIP.2003.819861
25. Yi, X., Walia, E., Babyn, P.: Generative adversarial network in medical imaging: a review. ArXiv e-prints (2018)
26. Zhu, J., Park, T., Isola, P., Efros, A.A.: Unpaired image-to-image translation using cycle-consistent adversarial networks. CoRR abs/1703.10593 (2017). http://arxiv.org/abs/1703.10593

Typing Plasmids with Distributed Sequence Representation

Moritz Kaufmann[1], Martin Schüle[2], Theo H. M. Smits[1] (iD), and Joël F. Pothier[1](✉) (iD)

[1] Environmental Genomics and Systems Biology Research Group, Institute of Natural Resource Sciences, Zurich University of Applied Sciences (ZHAW), Einsiedlerstr. 31, 8820 Wädenswil, Switzerland
{moritz.kaufmann,theo.smits,joel.pothier}@zhaw.ch
[2] Bio-Inspired Modeling and Learning Systems, Institute of Applied Simulation, Zurich University of Applied Sciences (ZHAW), Schloss 1, 8820 Wädenswil, Switzerland
martin.schuele@zhaw.ch

Abstract. Multidrug resistant bacteria represent an increasing challenge for medicine. In bacteria, most antibiotic resistances are transmitted by plasmids. Therefore, it is important to study the spread of plasmids in detail in order to initiate possible countermeasures. The classification of plasmids can provide insights into the epidemiology and transmission of plasmid-mediated antibiotic resistance. The previous methods to classify plasmids are replicon typing and MOB typing. Both methods are time consuming and labor-intensive. Therefore, a new approach to plasmid typing was developed, which uses word embeddings and support vector machines (SVM) to simplify plasmid typing. Visualizing the word embeddings with *t*-distributed stochastic neighbor embedding (*t*-SNE) shows that the word embeddings finds distinct structure in the plasmid sequences. The SVM assigned the plasmids in the testing dataset with an average accuracy of 85.9% to the correct MOB type.

Keywords: Plasmid typing · Word embedding

1 Background

1.1 Plasmids

Plasmids are extrachromosomal DNA elements with a characteristic number of copies in the host. Plasmids are found in representatives of all three domains *Archaea, Bacteria* and *Eukarya* [1]. Plasmids encode nonessential but often valuable genes for their host [2]. The plasmids allow genes to be horizontally exchanged via recombination and transposition. Since plasmids can enter new hosts via a variety of mechanisms, they can be regarded as a pool of extrachromosomal DNA that is shared across populations. The acquisition of such genes on plasmids enables the bacteria to react quickly to changing environmental influences, e.g. the presence of antibiotics, which would not be the case if bacterial fitness were only dependent on *de novo* evolution [3]. Plasmids contain genes that are responsible for initiation and the control of replication. In addition they

© Springer Nature Switzerland AG 2020
F.-P. Schilling and T. Stadelmann (Eds.): ANNPR 2020, LNAI 12294, pp. 200–210, 2020.
https://doi.org/10.1007/978-3-030-58309-5_16

contain genes that encode a wide variety of phenotypes that help their bacterial hosts to exploit and adapt to their environments [4]. These properties are considered as additional functions and include antibiotic and heavy metal resistance, metabolic properties and pathogenicity factors. Such phenotypes have important consequences for human and animal health, environmental processes and microbial adaptation and evolution [5].

1.2 Plasmid Typing

The classification of plasmids can provide insights into the epidemiology and transmission of plasmid-mediated antibiotic resistance. The previous methods to classify plasmids are replicon typing and MOB typing which use variation in replication loci and relaxase proteins, respectively. Replicons include various loci, none of which are universally present in plasmids [6]. On the other hand, relaxases are thought to occur in all plasmids mobilized by the relaxase-*in-cis* mechanism [7, 8]. Nevertheless, the relaxase homology may be distant, even in plasmids of the same MOB type [9]. Recent studies show that the current typing schemes are not able to classify the complete diversity of plasmids [10]. As an example, 11% of the plasmids from the dataset ($n = 2097$) of Orlek et al. [10] could not be replicon-typed or MOB typed.

1.3 Word Embeddings

In natural language processing (NLP) a powerful method to represent language is by learning so-called embeddings. An embedding is a vector representation of a text data token. Commonly the tokens are words, and therefore we refer in our explanations to word embeddings, but the method is not restricted to words. In contrast to word vectors created by one-hot-encoding, which are binary, sparse (mostly made of zeros), high-dimensional (same dimensionality as vocabulary), word embeddings are low-dimensional floating-point vectors. In a good word embedding space synonyms have similar word vectors. Also, distance between word vectors reflect semantic and syntactic distances between those words [11]. A popular training technique to learn word embeddings is Word2Vec [12, 13]. Word2vec consists of a two-layer neural network that is trained on the current word and its surrounding context words. The use of context words is inspired by the linguistic concept of *distributional hypothesis*, which states that words that appear in the same context have a similar meaning [14].

1.4 Aim of Study

The aim of this study is to determine whether plasmids can be represented as word embeddings, a method normally used in natural language processing, and subsequently classified by machine learning methods.

2 Methods

2.1 Preparing the Dataset

In order to test the new classification method for plasmids, the database with the original queries of Orlek et al. [15] was downloaded [16]. The database consisted of 2097 fully typed, complete, clinically relevant *Enterobacteriaceae* plasmids from the NCBI database. The data of the nucleotide sequences were loaded with the Biostrings package version 2.36.4 [17] to R version 3.5.1 using RStudio version 1.3.959. The nucleotide sequences were translated to amino acid sequences using the Biostrings package [17] using the standard genetic code. All fuzzy and stop codons automatically translated to X and *, respectively, were removed. To remove outliers, which could influence the training behavior of the machine learning methods, a box plot of the plasmid length was created. All plasmids marked as outliers were removed.

2.2   Embedding Representation

Inspired by NLP word embeddings, we created an embedding representation for amino acid sequences. Following Asgari and Mofrad [18], all amino acid sequences were split into triplets. Then, from one sequence, three sequences were created (see Fig. 1). These triplets are the "words" for which the word embedding is constructed. This is done since the most common techniques to study sequences in bioinformatics involves fixed–length overlapping n-grams [19–21].

Original Sequence

$(1)\overrightarrow{M}\,(2)\,\overrightarrow{A}\,(3)\,\overrightarrow{F}\,SAEDVLKEYDRRRRMEAL..$

Splittings

1)   MAF, SAE, DVL, KEY, DRR, RRM, ..
2)   AFS, AED, VLK, EYD, RRR, RME, ..
3)   FSA ,EDV, LKE, YDR, RRR, MEA, ..

Fig. 1. Schematic illustration of the generated three sequences [18].

The word embeddings were trained using the Skip-Gram algorithm. To calculate the vectors for the embedding the R-package wordVectors version 2.0 was used [22]. The Skip-Gram model learns embeddings by trying to predict context words based on the given target word. Context words are words that occur in a defined window around the target word. Skip-Gram tries to find the corresponding *n*-dimensional vectors for a given training sequence of words, which maximize the log probability function. This gives similar words a similar representation in vector space

$$argmax_{v,y}\frac{1}{N}\sum_{i=1}^{N}\sum_{-c\le j\le c, j\neq 0}\log p\left(w_{i+j}|w_i\right)$$

$$p\left(w_{i+j}|w_i\right) = \frac{\exp\left(v'^T_{w_{i+j}} v_{wi}\right)}{\sum_{k=1}^{w}\exp(v'^T_{w_k} v_{wi})} \tag{1}$$

where N is the length of the training sequence, $2c$ is the considered window size for the context, w_i is the center of the window, W is the number of words in the dictionary and v_w and v'_w are input and output *n*-dimensional representations of word w, respectively. The probability $p(w_{i+j}|w_j)$ is defined by a softmax function. Hierarchical softmax or negative sampling are effective approximations of such a softmax function. The wordVectors package uses negative sampling to approximate the softmax function. Negative sampling uses the following objective function to calculate the word vectors

$$argmax_\theta \prod_{(w,c)\in D} p(D = 1|c, w; \theta) \prod_{(w,c)\in D'} p(D = 0|c, w; \theta) \tag{2}$$

where D is a set of word and context pairs (w, c) existing in the training data set (positive samples) and D' is a randomly generated set of false word and context pairs (w, c) (negative samples). $p(D = 1|w, c; \theta)$ is the probability that (w, c) comes from the training data. $p(D = 0|w, c; \theta)$ is the probability does not come from the training data. The term $p(D = 1|c, w, \theta)$ can be defined as a sigmoid function which can be used for the wordVectors

$$p(D = 1|w, c; \theta) = \frac{1}{1 + e^{-v_c v_w}} \tag{3}$$

Here, the parameters θ are the word vectors we train within the optimization framework v_c while $v_w \in R^d$ are vector representations for the context c and the word w, respectively [23]. In Eq. (2), the positive samples maximize the probabilities of the observed (w, c) pairs in the training data, while the negative samples prevent all vectors from having the same value by not allowing certain incorrect (w, c) pairs [18]. To train different embeddings different vector sizes and context sizes were chosen. The vocabulary to train the word embeddings were all 8000 possible amino acid triplets. We then represent the entire plasmid as a word embedding, where the amino acid triplets of each reading frame were added for each plasmid. This method follows Asgari and Mofrad [18].

2.3 t-Distributed Stochastic Neighbor Embedding (t-SNE)

High-dimensional word embeddings can be displayed and interpreted two-dimensionally with the t-SNE algorithm. We used the R-package Rtsne version 0.15 [24]. To evaluate whether the individual MOB types are grouped into clusters, the data points were colored according to the MOB types assigned by Orlek et al. [10]. The t-SNE algorithm works as follows: first the similarity score in the original space is calculated from a distance matrix (Euclidean distance) of the input objects

$$p_{j|i} = \frac{exp\left(\frac{-||D_{ij}||^2}{2\sigma_i^2}\right)}{\sum_{k \neq i} exp\left(\frac{-||D_{ij}||^2}{2\sigma_i^2}\right)} \tag{4}$$

which is then symmetrized using

$$p_{ij} = \frac{p_{j|i} + p_{i|j}}{2n} \tag{5}$$

The parameter σ of each object is selected so that the perplexity in the original space takes a value as close as possible to the defined perplexity. The perplexity is a parameter that controls how many nearest neighbors are considered when the embedding is generated in low dimensional space. For the low dimensional space, the Cauchy distribution (t-distribution with one degree of freedom where the degree of freedom is the number of parameters that may vary independently) is used to represent the distribution of the objects

$$q_{ij} = \frac{\left(1 + ||y_i - y_j||^2\right)^{-1}}{\sum_{k \neq l}\left(1 + ||y_k - y_l||^2\right)^{-1}} \tag{6}$$

The positions of the points in the low dimensional space are determined by minimizing the Kullback-Leiber divergence (KL) of the distribution Q to the distribution P. To minimize the KL-divergence a gradient descent algorithm is used. Since for large datasets a normal gradient descent algorithm would be very computational expensive $O(n^2)$, a Barnes-Hut implementation of the algorithm, is used which leads to a computational complexity of $n\log(n)$. The θ parameter was set to zero to perform an exact t-SNE. The *max_iter* parameter was set to 1000. The *PCA* parameter was set to TRUE to perform a PCA prior to the t-SNE. To find the best parameters for the perplexity, each model was iterated over 50 cycles. The perplexity parameter was adjusted from 1 to 50. The best fitting perplexity value was chosen according to the lowest KL-divergence.

2.4 Support Vector Machine Classification

To classify the plasmids based on the embedding representation support vector machines are used. SVM with a linear kernel was chosen and implemented with the caret package 6.0–81 [25]. The caret package uses the implementation of the SVM algorithm by kernlab [26]. The SVM algorithm of the kernlab package uses the Sequential Minimal Optimization (SMO) algorithm of Platt [27] to solve the quadratic programming (QP) optimization problem of the SVM. Training an SVM usually requires solving a very big QP optimization problem. The SMO algorithm breaks these big QP optimization problems into a series of smallest possible QP problems. The small QP problems can then be solved analytically, saving the time consuming numerical solving of a large QP problem [27]. To train the SVM, the data were first centered by subtracting the mean and then scaled by the division of the standard deviation. The partition of the data was 0.8/0.2 for training and test for each iteration. The method for optimizing the tuning parameters was random search.

2.5 BLAST

The BLAST searches to confirm the MOB types of before unclassified plasmids were carried out using the NCBI online tool tblastn version 2.8.1. The algorithm parameters were set to default. The search results were then filtered according to the used thresholds for original MOB type queries used by Orlek et al. [10].

2.6 System

The analyses were run on a PC equipped with an Intel Core i7-3930 K processor cadenced at 3.20 GHz (6 physical cores, 12 logical cores) and with 64 GB of physical memory.

3 Results

3.1 Data Exploration

Unknown plasmids account for around 700 occurrences in the data set with the original queries of Orlek et al. [15]. The types MOB_F and MOB_P occur about 450 times each. MOB_Q and MOB_H were already significantly less present with around 150 counts. The types MOB_C and in particular MOB_V were very limited represented, which could lead to classification problems. The dataset was analyzed to check the plasmid lengths and the MOB class distribution (Fig. 2). In total, 61 plasmids were marked as outliers. Even though all outliers are most likely plasmids, they were removed from the dataset, as outliers can have a negative effect on the training of the embedding. The average length is shown in the figure as a dashed line. MOB_V plasmids were the shortest in length, while the group of MOB_Q plasmids encompassed mainly short plasmids. The mean lengths of MOB_C, MOB_P and unclassified plasmids were almost comparable. The longest plasmids were in the MOB_F and MOB_H group, with MOB_H plasmids being longer than MOB_F plasmids. Except for the MOB_V plasmids, all plasmids had a variance of about 50 kb in length.

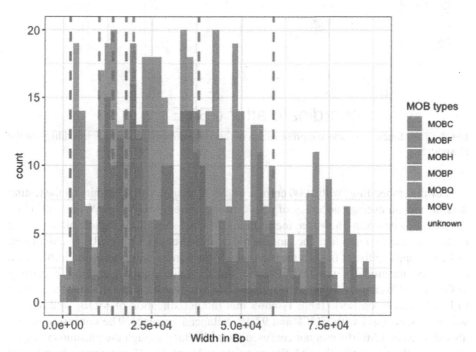

Fig. 2. Distribution of the length of the different MOB types of plasmids. Bp: base pairs.

3.2 *t*-Distributed Stochastic Neighbor Embedding (*t*-SNE)

The lowest KL-divergence was achieved with a perplexity of 49. Figure 3 shows the 1000-dimensional space of the embedding reduced to two dimensions. Each point is a vector representation of a plasmid. A clearly shaped structure was obtained.

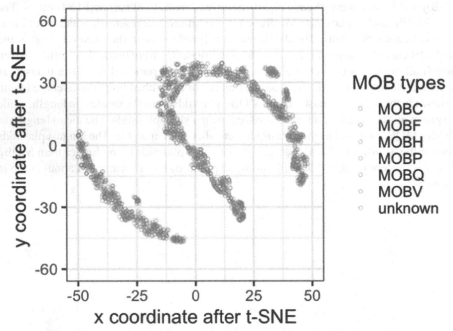

Fig. 3. Two-dimensional space representation of the 1000-dimensional word embeddings of the plasmids.

For the embeddings with 1000 entries, the SVM assigned the plasmids in the testing dataset with an average accuracy of 85.9% to the correct MOB type. With multi-class classification problems, however, the accuracy does not show the complete picture of the performance of the classifier. The same would apply to a data set with imbalanced classes. Cohen's kappa statistics (κ) is a measure which can handle multi-class and imbalanced classes. For the model, κ was 0.80, indicating that the value is good to excellent according to Greve and Wentura [28] and at the upper end with substantial agreement according to Landis and Koch [29]. Table 1 shows that the classification of MOB_F and MOB_H was very successful with 93.8% and 97.4% balanced accuracy. The confusion matrix also showed that MOB_H was not confused with MOB_F, although the plasmids in Fig. 3 were very close to each other. MOB_P was detected with 87.6% accuracy, which is still above the SVM average for all classes. However, MOB_P was confused with MOB_F in 10.7% of the cases. Furthermore, MOB_P was in 3.6% wrongly assigned to type MOB_Q. Orlek et al. [10] reported problems to distinguish between MOB_P and MOB_Q. In the prediction column of MOB_Q, 13 of the total of 51 MOB_Q plasmids were assigned to class MOB_P, which corresponded to 25.5% of all MOB_Q plasmids. However, this also meant that the representation of the plasmids with embeddings worked well, as the results were congruent with the previously obtained results from Orlek et al. [10]. MOB_Q and MOB_V showed the smallest accuracies, related to an underrepresentation of both classes in the training set. In the testing dataset, only very few plasmids with the respective classes were present and an inconsistent classification has a fatal effect on the accuracy.

Table 1. Confusion matrix SVM.

Predictions	Reference						Balanced accuracy
	MOB_C	MOB_F	MOB_H	MOB_P	MOB_Q	MOB_V	
MOB_C	14	2	1	0	2	0	0.812
MOB_F	2	135	0	15	2	0	0.938
MOB_H	0	0	37	0	0	0	0.974
MOB_P	6	5	1	120	13	2	0.876
MOB_Q	0	0	0	5	34	0	0.826
MOB_V	0	0	0	0	0	1	0.667

To investigate whether the SVM can be used to classify plasmids, which were previously unclassifiable by Orlek et al. [10], 939 unclassified plasmids from the data set were tested with the SVM. The plasmids that could get classified with the SVM were subsequently checked with BLAST against the corresponding proteins used by Orlek et al. [10] for testing. Table 2 shows that 96 plasmids were assigned to a MOB type. Taking into account the thresholds of Orlek et al. [10], 62 plasmids could, after checking with BLAST, still be assigned to a MOB type with relative security. This corresponds to a decrease of unclassifiable plasmids of 3.04%.

Table 2. BLAST verification of the SVM predicted MOB types.

MOB type	Total predicted	Verified with BLAST	Number of predictions < E-value threshold	E-value threshold
MOB_C	101	42	42	0.001
MOB_F	127	14	8	0.01
MOB_H	17	8	0	0.01
MOB_P	582	27	12	1
MOB_Q	92	2	0	0.0001
MOB_V	20	3	0	0.01
Total	939	96	62	–

4 Discussion

The aim of this work was to represent plasmids as word embeddings and to perform MOB typing using the word embeddings. Asgari and Mofrad [18] were able to classify proteins using word embeddings. However, the method has never been applied to whole plasmids. As could be shown in this work, the word embeddings of entire plasmids can be used to assign the correct MOB types to these plasmids. Based on the available data, MOB typing using SVM seems to be the most successful approach. On the other side, it is possible that with more plasmid sequences present, an approach based on a neural network outperforms the SVM.

By means of the *t*-SNE of the word embeddings, it became clear that the word embeddings represent an up to now not identified structure found in plasmids. The position on the Y-axis could correlate with the length of the plasmids. However, the factor that influenced the position on the X-axis could not be identified. The reconstruction of the plasmid typing, where only the word embeddings of the entire amino acid sequences were used, was functional. In the data set, the accuracy of the test data set was 85.9%, even though the whole plasmid sequences were only represented by a vector of 1000 entries. Nevertheless, the important factors to assign a MOB type seem to be precisely represented in the word embeddings. As the current version of the SVM was only trained on the known MOB types, one of the currently included MOB type is assigned to each plasmid, since the SVM does not know an unknown type. Nevertheless the MOB type could be set for 62 plasmids, which were before not assigned to any MOB type. These results were then confirmed with a BLAST search.

The classification of the word embeddings is currently based on the biological approach of MOB typing. As long as it is not clear for the already used biological method, which proteins have to be used as queries to get the best results or how many different MOB types exist, the word embedding classification cannot be improved. However, as soon as more biological information about MOB types is available, reconstructing typing with word embeddings offers an interesting alternative. The model only needs to be trained once and can then readily be used. An assignment to a MOB type only takes a fraction of a second and does not require any time-consuming BLAST analysis.

For the next steps it would be conceivable to create a new data set of plasmids. The GenBank database at NCBI continuously includes more plasmids from genome sequencing projects and probably contains a more balanced representation of all plasmid types than at the creation of the used dataset by Orlek et al. [10]. Furthermore, the scripts can be optimized to improve performance. Further tuning of the hyperparameters of the SVM will lead to even better results in the future. It is also conceivable that the MOB typing by word embeddings can be used to establish previously unknown MOB types. As shown in this paper, machine learning methods offer interesting alternatives for conventional bioinformatics approaches and will certainly make their way into biological research soon.

References

1. Woese, C.R., Kandler, O., Wheelis, M.L.: Towards a natural system of organisms: proposal for the domains Archaea, Bacteria, and Eucarya. Proc. Natl. Acad. Sci. **87**, 4576–4579 (1990). https://doi.org/10.1073/pnas.87.12.4576
2. Novick, R.P., Hoppensteadt, F.C.: On plasmid incompatibility. Plasmid **1**, 421–434 (1978). https://doi.org/10.1016/0147-619X(78)90001-X
3. Smets, B.F., Barkay, T.: Horizontal gene transfer: perspectives at a crossroads of scientific disciplines. Nat. Rev. Microbiol. **3**, 675–678 (2005). https://doi.org/10.1038/nrmicro1253
4. Frost, L.S., Leplae, R., Summers, A.O., Toussaint, A.: Mobile genetic elements: the agents of open source evolution. Nat. Rev. Microbiol. **3**, 722–732 (2005). https://doi.org/10.1038/nrmicro1235
5. Johnson, T.J., Nolan, L.K.: Plasmid replicon typing. In: Caugant, D.A. (ed.) CEUR Workshop Proceedings, vol. 551, pp. 27–35. Humana Press, Totowa (2009). https://doi.org/10.1007/978-1-60327-999-4_3
6. del Solar, G., Giraldo, R., Ruiz-Echevarría, M.J., Espinosa, M., Díaz-Orejas, R.: Replication and control of circular bacterial plasmids. Microbiol. Mol. Biol. Rev. **62**, 434–464 (1998). https://doi.org/10.1128/MMBR.62.2.434-464.1998
7. Garcillán-Barcia, M.P., Alvarado, A., de la Cruz, F.: Identification of bacterial plasmids based on mobility and plasmid population biology. FEMS Microbiol. Rev. **35**, 936–956 (2011). https://doi.org/10.1111/j.1574-6976.2011.00291.x
8. Ramsay, J.P., et al.: An updated view of plasmid conjugation and mobilization in *Staphylococcus*. Mob. Genet. Elements **6**, e1208317 (2016). https://doi.org/10.1080/2159256X.2016.1208317
9. Garcillán-Barcia, M.P., Francia, M.V., de La Cruz, F.: The diversity of conjugative relaxases and its application in plasmid classification. FEMS Microbiol. Rev. **33**, 657–687 (2009). https://doi.org/10.1111/j.1574-6976.2009.00168.x
10. Orlek, A., et al.: Ordering the mob: insights into replicon and MOB typing schemes from analysis of a curated dataset of publicly available plasmids. Plasmid **91**, 42–52 (2017). https://doi.org/10.1016/j.plasmid.2017.03.002
11. Chollet, F.F., Allaire, J.J.: Deep Learning with R. Manning Publications, Shelter Island (2018)
12. Mikolov, T., Yih, W., Zweig, G.: Linguistic regularities in continuous space word representations. In: Proceedings of the 2013 Conference of the North American Chapter of the Association for Computational Linguistics: Human Language Technologies (NAACL-HLT 2013), pp. 746–751. Association for Computational Linguistics, Atlanta (2013)
13. Brownlee, J.: Word embeddings. In: Deep Learning for Natural Language Processing, pp. 114–143. Machine Learning Mastery, Vermont Victoria (2017)
14. Harris, Z.S.: Distributional structure. Word **10**, 146–162 (1954). https://doi.org/10.1080/00437956.1954.11659520
15. Orlek, A., et al.: A curated dataset of complete Enterobacteriaceae plasmids compiled from the NCBI nucleotide database. Data Br. **12**, 423–426 (2017). https://doi.org/10.1016/j.dib.2017.04.024
16. Orlek, A., et al.: Figshare (2017). https://figshare.com/s/18de8bdcbba47dbaba41
17. Pagès, H., Abonyoun, P., Gentleman, R., DebRoy, S.: Biostrings: efficient manipulation of biological strings. R package version 2.56.0 (2018)
18. Asgari, E., Mofrad, M.R.K.: Continuous distributed representation of biological sequences for deep proteomics and genomics. PLoS ONE **10**, e0141287 (2015). https://doi.org/10.1371/journal.pone.0141287
19. Ganapathiraju, M., Weisser, D., Rosenfeld, R., Carbonell, P., Reddy, R., Klein-Seetharaman, J.: Comparative N-gram analysis of whole-genome protein sequences. In: Proceedings of

the Second International Conference on Human Language Technology Research, pp. 76–81. Morgan Kaufmann, San Francisco (2002)

20. Srinivasan, S.M., Vural, S., King, B.R., Guda, C.: Mining for class-specific motifs in protein sequence classification. BMC Bioinform. **14**, 96 (2013). https://doi.org/10.1186/1471-2105-14-96

21. Vries, J.K., Liu, X.: Subfamily specific conservation profiles for proteins based on n-gram patterns. BMC Bioinform. **9**, 72 (2008). https://doi.org/10.1186/1471-2105-9-72

22. Bmschmidt.: WordVectors. github (2017). https://github.com/bmschmidt/wordVectors

23. Goldenberg, Y., Levy, O.: word2vec explained: deriving Mikolov et al.'s negative-sampling word-embedding method. ArXiv 1402.3722 (2014)

24. Krijthe, J.H.: Rtsne: T-Distributed Stochastic Neighbor Embedding using Barnes-Hut Implementation (2015). https://github.com/jkrijthe/Rtsne

25. Kuhn, M.: Building predictive models in R using the caret package. J. Stat. Softw. **28**, 1–26 (2008). https://doi.org/10.18637/jss.v028.i05

26. Karatzoglou, A., Smola, A., Zeileis, A.: Kernlab – an S4 package for kernel methods in R. J. Stat. Softw. **11**, 1–20 (2004)

27. Platt, J.C.: Sequential Minimal Optimization : A Fast Algorithm for Training Support Vector Machines. MSR-TR-98-14 (1998)

28. Greve, W., Wentura, D.: Wissenschaftliche Beobachtung eine Einführung. Beltz, Weinheim (1997)

29. Landis, R., Koch, G.: The Measurement of Observer Agreement for Categorical Data. Biometrics **33**, 159–174 (1977)

KP-YOLO: A Modification of YOLO Algorithm for the Keypoint-Based Detection of QR Codes

Nouredine Hussain[iD] and Christopher Finelli[(⊠)][iD]

Icare Research Institute, Rue de Technopôle 10, 3960 Sierre, Switzerland
`info@icare.ch`

Abstract. The YOLO (v1/v2/v3) algorithm is a well-known, extensively used, and much studied detection algorithm. YOLO detects objects in an image by predicting their bounding boxes and class distributions. This paper studies a modified version of the YOLO model applied to a specific problem: the detection of a set of feature points in a one-class setting. We apply this approach to the particular case of detecting QR Codes. Instead of detecting rectangular, axis-aligned bounding boxes, our objective is to detect a set of keypoints—the QR Code's *finder patterns*—and hence the model's name KP-YOLO. Although QR Code detection was chosen to showcase this experiment, the KP-YOLO algorithm can be applied to other keypoint-based detection problems (under certain constraints).

Keywords: Object detection · Convolutional neural networks · QR Codes

1 Introduction

Detection algorithms have greatly improved in the last few years, thanks to the advent of deep learning and, more specifically, convolutional neural networks (CNNs). Many different CNN architectures have been proposed for the task of detecting objects in an image. The first generation of such algorithms proceeded in two steps: a first process proposed a number of regions of interest (ROIs) and a second one classified these regions [4,5,13]. This approach's biggest weakness is its relative slowness [11].

YOLO (You Only Look Once) [10] was one of the first models not to require this two-step approach. Instead, it divides the image into a grid (multiple grids in later versions). Each grid cell can detect a set of objects whose center lies within that particular cell. To allow for the detection of more than one object per grid cell, it uses several anchor boxes. Thus, every object is associated with a grid cell and with the anchor box that best fits its bounding box. After the inference step, a non-maximum suppression step is performed in order to keep only the best candidates and remove overlaps.

F.-P. Schilling and T. Stadelmann (Eds.): ANNPR 2020, LNAI 12294, pp. 211–222, 2020.
https://doi.org/10.1007/978-3-030-58309-5_17

Other models, using similar approaches, brought enhancements to the original YOLO algorithm. For example, the single-shot multibox detector [8] model uses multiple grids at different stages in a CNN in order to better detect objects at different scales. ResNets [6] introduces the novel idea of shortcuts that significantly enhances the performance of deep neural networks. The latest version of the YOLO algorithm (v3) integrates many of these enhancements, making it both more precise and faster than previous versions [12].

2   State of the Art

QR Code detection is traditionally performed using *classic* computer vision algorithms, i.e., hand-tuned algorithms rather than ones based on a machine learning process. This approach is generally reliable and fast; it also has the advantage of being easier to analyze when an error occurs. However, in recent years, deep learning has been investigated as a replacement for or a complement to *classic* algorithms. Deep-learning-based algorithms could potentially operate on very blurry and distorted images.

Leonardo Blanger et al. [3] used multiple variations of the popular Single Shot Detector architecture [8] on the task of QR Code (bounding boxes) detection. By incorporating the detection of the finder patterns to the task of QR Code detection, Blanger et al. observed enhanced detection performance. Their paper used axis-aligned bounding boxes for detection.

Instead of detecting axis-aligned bounding boxes, another approach is to work at the pixel level and try to assign a particular label to each pixel. Qijie Zhaoa et al. [15] used a dual pyramid structure-based segmentation network, BarcodeNet, in conjunction with a synthetic dataset, to detect QR Codes at the pixel level. Kazuya Nakamura et al. [9] studied a similar approach, using a CNN to decode QR Codes under non-uniform geometric distortion. They treated the problem of localizing the finder patterns as a binary classification problem: given a target pixel, it assigned one of two kinds of labels, either "a finder pattern" or "others". Baoxi Yuan et al. [14] used a modified version of Mask R-CNN [5], which they called MU R-CNN, for the segmentation of QR Codes, also at the pixel level.

To sum up, these approaches all fall into one of the following two categories:

- A classic approach, using regular axis-aligned bounding boxes.
- A pixel level (segmentation) approach, where a label is assigned to each pixel (whether the output image is the same size or smaller than the input image).

3   Experiment

This paper proposes a third approach: we detect keypoints that can be used to draw a non-rectangular boundary around an object of interest. This approach can be applied to an arbitrary number of keypoints if they respect certain constraints.

3.1 Datasets

The quality of the dataset is one of the most important factors for success when developing deep learning. We started our experimentation with a dataset of approximately 10 k video frames of real QR Codes taken from different angles and distances and using different light conditions. This dataset was then augmented (see Data augmentation) to produce a dataset of approximately 1.5M images.

We quickly realized that our models were overfitting to the QR Codes' content (although we had a great variety of shots, they were always shots of the same ten or so physical QR Codes). To solve this problem, we generated an additional dataset of 500k synthetic images, which we then also augmented, reaching a total of 2.5M annotated images.

This process was the first of three key steps used to reduce the problem of overfitting. The second step was a massive reduction in the number of parameters in our model, and the third was the use of dropout.

Data Augmentation. To help our models to generalize better, each image from our original dataset (real + synthetic images) was used to generate a number of additional images by applying random transformations such as cropping, rotation, blurring, and noise. Examples of the augmented images can be seen in Fig. 1.

Our experience has shown us that it is important for the augmentation to be representative of the transformations that a QR Code might experience under real-world conditions. Otherwise, the models will learn the *abnormal* aspects and perform poorly when tested on real images.

Synthetic Dataset. To create our synthetic dataset, we used an open dataset [1] composed of 100k random images as backgrounds. We then generated random QR Codes of different sizes and intensities and projected them randomly onto these backgrounds. The random projections were chosen so as to represent common, realistic optical effects. Examples of the synthetic images can be seen in Fig. 2.

Creating a dataset of photos of real QR Codes is a time-consuming and costly task. However, training models using entirely synthetic datasets is also

Fig. 1. Augmented training images

Fig. 2. Synthetic training images

ill-advised, as it would likely cause the models to perform poorly with real images. For an acceptable balance, we used the real images for the detection assessment and the synthetic ones for the localization assessment since we needed a ground truth.

3.2 Model

Our deep learning model, shown in Fig. 3, is a modification of the YOLO v3 model, which we chose because it is widely known (it is easy to find open-source implementations in different languages and frameworks) and yields good results in comparison to other models [12].

There are five differences between our model and the standard one:

- The input size is 256×256 (instead of 416×416 in the original model). This resolution is sufficient to clearly distinguish the finder patterns and optimize the model's training and execution time.
- The output grid sizes are 8, 16, and 32 (instead of 13, 26, and 52). This comes from the modification of the input size and fits better with the number of QR Codes present in the images.
- There are far fewer channels. One of the first problems encountered with KP-YOLO was overfitting. This was expected, however, since detecting finder patterns is *easier* than detecting thousands of different classes of objects. Reducing the number of channels was one of the elements that helped with this problem.
- The components of the output grid's channels have different meanings in KP-YOLO. The first component represents the probability that the center of gravity of a set of finder patterns is inside this specific cell. The other 6 components represent the (normalized) finder patterns' coordinates.
- KP-YOLO's loss function considers an arbitrary number of keypoints in a one-class setup under some constraints. More details are given in the subsection on the Loss function.

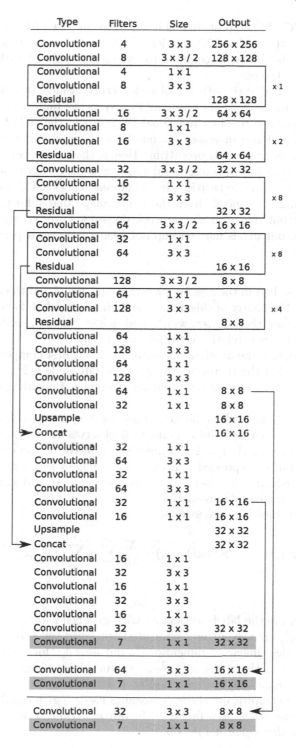

Type	Filters	Size	Output
Convolutional	4	3 x 3	256 x 256
Convolutional	8	3 x 3 / 2	128 x 128
Convolutional	4	1 x 1	
Convolutional	8	3 x 3	
Residual			128 x 128
Convolutional	16	3 x 3 / 2	64 x 64
Convolutional	8	1 x 1	
Convolutional	16	3 x 3	
Residual			64 x 64
Convolutional	32	3 x 3 / 2	32 x 32
Convolutional	16	1 x 1	
Convolutional	32	3 x 3	
Residual			32 x 32
Convolutional	64	3 x 3 / 2	16 x 16
Convolutional	32	1 x 1	
Convolutional	64	3 x 3	
Residual			16 x 16
Convolutional	128	3 x 3 / 2	8 x 8
Convolutional	64	1 x 1	
Convolutional	128	3 x 3	
Residual			8 x 8
Convolutional	64	1 x 1	
Convolutional	128	3 x 3	
Convolutional	64	1 x 1	
Convolutional	128	3 x 3	
Convolutional	64	1 x 1	8 x 8
Convolutional	32	1 x 1	8 x 8
Upsample			16 x 16
Concat			16 x 16
Convolutional	32	1 x 1	
Convolutional	64	3 x 3	
Convolutional	32	1 x 1	
Convolutional	64	3 x 3	
Convolutional	32	1 x 1	16 x 16
Convolutional	16	1 x 1	16 x 16
Upsample			32 x 32
Concat			32 x 32
Convolutional	16	1 x 1	
Convolutional	32	3 x 3	
Convolutional	16	1 x 1	
Convolutional	32	3 x 3	
Convolutional	16	1 x 1	
Convolutional	32	3 x 3	32 x 32
Convolutional	7	1 x 1	32 x 32
Convolutional	64	3 x 3	16 x 16
Convolutional	7	1 x 1	16 x 16
Convolutional	32	3 x 3	8 x 8
Convolutional	7	1 x 1	8 x 8

The blocks are repeated: x 1, x 2, x 8, x 8, x 4 respectively.

Fig. 3. Model architecture

FC vs. NO-FC. In the past, it was usual to include two fully connected (FC) layers at the bottom of a CNN model [7]. In recent years, this tendency has changed, and it is now more common to have a CNN model with no fully connected (NO-FC) layers.

We experimented with both model architectures. Since, by default, the standard YOLO v3 model does not have any FC layers, we built a model called KP-YOLO-FC, which features two additional FC layers at the end of each detection grid. We found that this increased the numbers of parameters considerably and, with it, the tendency towards overfitting. Hence, the removal of these FC layers was beneficial, both in terms of generalization and execution time performance.

One important thing to keep in mind when removing the FC layers is that this restricts the number of pixels from the input image that influence a particular pixel of the output image. If the network does not have enough convolutional layers, some output pixels may end up not seeing important parts of the input image.

Loss Function. It is in the loss function that our model and the original YOLO v3 differ most. The source of this difference is that instead of detecting bounding boxes by specifying their center, width, and height, we are specifying a number of keypoints with their relative coordinates.

More specifically, the depth of our output is 7: the first component represents the probability that the center of gravity of the QR Code is inside the grid cell, the six following components represent the normalized coordinates of the three patterns.

So, for a particular cell i, the first component p_i is set to 1 if the center of gravity of a QR Code falls into it and to 0 otherwise. The coordinates of the finder patterns (x_{ij}, y_{ij}), $j = 1, 2, 3$ are given relative to the center of the cell i, with the distances expressed relative to the image's width and height; these normalized coordinates are therefore always between -1 and 1. See Fig. 4 for a visualization of the coordinate system.

The loss can then be expressed as:

$$\frac{\lambda_1}{N} \sum_{i=1}^{N} -p_i \log(p_i') - (1 - p_i) \log(1 - p_i') + \frac{\lambda_2}{3N} \sum_{i=1}^{N} p_i \sum_{j=1}^{3} (x_{ij} - x_{ij}')^2 + (y_{ij} - y_{ij}')^2$$

where:

- p_i, x_i and y_i are the labels associated with grid cell i for $i = 1, 2, \ldots, N$.
- p_i', x_{ij}' and y_{ij}' are the outputs associated with grid cell i for $i = 1, 2, \ldots, N$.
- Assuming the linear 7-dimensional output o_i for cell i, we have $p_i' = \sigma(o_{i1})$ with $\sigma(x) = \frac{1}{1+e^{-x}} \in [0, 1]$ the sigmoid function, and $(x_{i1}', y_{i1}', x_{i2}', y_{i2}', x_{i3}', y_{i3}') = \tanh(o_{ik})$ with $k = 2, 3, \ldots, 7$.
- λ_1 and λ_2 are empirically chosen constants used to give equal importance to the two terms of the loss function. In our experiment, we set $\lambda_1 = 1$ and $\lambda_2 = grid\_width * grid\_height * 5$.

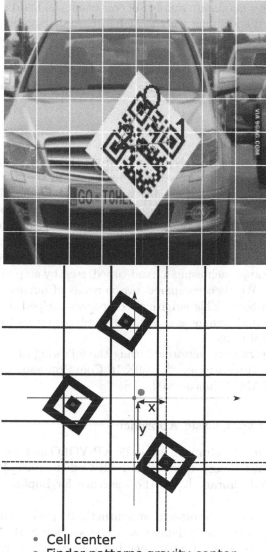

- Cell center
- Finder patterns gravity center
- Finder pattern center

Fig. 4. Coordinate system

3.3 Keypoint Constraints

Our algorithm imposes some constraints on the keypoints that it tries to detect:

1. The number of keypoints must be fixed. Our model does not yet support a variable number of keypoints since the number of coordinates is determined by the number of output channels. One way to mitigate this constraint would

be to determine a maximum number of constraints and add a third coordinate to each keypoint specifying whether it had been detected in the image.

2. Keypoints need to be distinguishable and intrinsically ordered. Otherwise, the model will mix them up and will not be able to learn anything. In our dataset, the finder patterns were ordered using the middle keypoint as the first one, and then moving clockwise to the other two keypoints (Fig. 4).

4   Results

Since our annotation method is different from that of our peers, comparing our results with theirs is not possible, as explained in the section on the State Of The Art. Indeed, approaches using rectangular bounding boxes succeed in detection but not in localization. On the other hand, pixel-based methods can localize finder patterns but cannot retrieve their intrinsic order. Furthermore, these methods could miss some finder patterns, forcing us to handle these cases separately. As an alternative, we first compared our model with the *classic approach*, that is, the approach using a hand-tuned, step-by-step algorithm, with no learning involved. We then compared two versions of our model: one with FC layers and one without. This empirical comparison helped us to highlight our model's strengths and weaknesses, with an emphasis on its future (potentially industrial) practical usage.

All measurements were performed using the following platform: AMD® Ryzen™ Threadripper™ 1950X 16-Core Processor x32, GeForce GTX 1080 Ti, 32 GB RAM, Ubuntu 18.04 LTS.

4.1   KP-YOLO vs. Classic Approach

We compared the results obtained using the KP-YOLO and KP-YOLO-FC models to those obtained using a conventional QR Code scanner. We used the Icare Institute's QR Code library [2] as the reference for implementing the classic algorithm.

To evaluate the classic approach, we synthetically generated a test dataset of 500 images. Using a synthetic dataset as a test set is not ideal, for the reasons mentioned above (see Synthetic dataset), but this was the only way to generate a ground truth independently of a classic algorithm (Fig. 5). As a metric, the Mean Squared Error (MSE) was defined as follows:

$$MSE = \frac{1}{N_{detected\ QR\ Codes}} \sum_{i=1}^{N_{detected\ QR\ Codes}} \frac{1}{6} \sum_{j=1}^{3} [(x_{ij} - x'_{ij})^2 + (y_{ij} - y'_{ij})^2]$$

Fig. 5. Ground truth vs. prediction

Fig. 6. Example of test (real) images

Table 1. KP-YOLO vs. Classic approach

Algorithm	Recall	MSE	MedSE	Execution time
Classic	0.476	**0.155**	**0.139**	**5.9 ms**
KP-YOLO	**0.968**	30.320	0.513	19 ms
KP-YOLO-FC	0.868	52.435	11.009	39 ms

The results, shown in Table 1, confirmed our initial intuition that the KP-YOLO model was capable of detecting QR Codes in images where the classic approach failed, but this was at the cost of a far worse performance in the prediction of the finder patterns' positions and slower execution time. This may be explained by the fact that neural networks can be unstable, depending on their inputs. In such a case, the predicted output is far from its expected value, and it yields outliers with large deviations. Since the median is far less affected by outliers than the mean is, a modified version of MSE (MedSE) is also shown to illustrate this phenomenon. The MedSE consists of computing the median over the Squared Errors of each detected sample. Based on this metric, KP-YOLO's loss in performance is much more acceptable.

A word of caution regarding the execution time: when it comes to deep learning models, accurately measuring the execution (inference) time is hard. This is because these algorithms are highly dependent on the particular framework, optimizations, and hardware used to run the model. Making one small optimization may make the same model run 10× faster. Since we did not spend much time trying to optimize our models' execution time, these figures should be taken as mere indications and not as definitive measurements.

4.2   KP-YOLO vs. KP-YOLO-FC

We also compared two versions of our model, one with two FC layers at the end of each output grid (KP-YOLO-FC) and one without (KP-YOLO). The aim was to see whether adding FC layers would bring an improvement that was worth the additional execution time and memory usage.

We tested our models on two test datasets: the synthetic dataset used in our first experiment (500 images) and a second one composed of 255 real images annotated using the *classic* algorithm (see Fig. 6). The results are shown in Table 2.

Table 2. KP-YOLO vs. KP-YOLO-FC (Learning Time: per batch of 16 images, and Inference Time: per image)

Model	Number of parameters	Learning time	Inference time	Error on real images	Error on synthetic images
KP-YOLO	**912,009**	**57 ms**	**19 ms**	**0.1706**	**0.0632**
KP-YOLO-FC	102,616,628	68 ms	39 ms	0.2624	0.7861

Our results showed that the FC layers provided no improvement at all. On the contrary, the two FC layers increased the number of parameters by a factor of 113, which not only slowed down training and inference but also accentuated the problem of overfitting.

This confirmed the state-of-the-art tendency on CNNs which is to remove FC layers altogether.

5   Conclusion

Our experiment showed that YOLO architecture could be modified to detect keypoints associated with a particular class of objects, provided that these keypoints respect a number of constraints. We showed how the KP-YOLO model compared to hand-tuned, classic approaches, as well as its advantages and disadvantages.

We believe our model could be used for the detection of objects in an industrial setting, and we think it would be interesting to extend this approach in two ways:

1. Alleviate the constraints on the keypoints: find a way to detect an arbitrary number of keypoints that are indistinguishable from each other.
2. Alleviate the one-class constraint: extend the KP-YOLO model to a multiple-class setup, with each class having its own set of keypoints.

References

1. https://storage.googleapis.com/openimages/web/index.html
2. https://icare.ch/
3. Blanger, L., Hirata, N.S.: An evaluation of deep learning techniques for QR code detection. In: 2019 IEEE International Conference on Image Processing (ICIP), pp. 1625–1629. IEEE (2019)
4. Girshick, R.: Fast R-CNN. In: Proceedings of the IEEE International Conference on Computer Vision, pp. 1440–1448 (2015)
5. Girshick, R., Donahue, J., Darrell, T., Malik, J.: Rich feature hierarchies for accurate object detection and semantic segmentation. In: Proceedings of the IEEE Conference on Computer Vision and Pattern Recognition, pp. 580–587 (2014)
6. He, K., Zhang, X., Ren, S., Sun, J.: Deep residual learning for image recognition. In: Proceedings of the IEEE Conference on Computer Vision and Pattern Recognition, pp. 770–778 (2016)

7. LeCun, Y., Bottou, L., Bengio, Y., Haffner, P.: Gradient-based learning applied to document recognition. Proc. IEEE **86**(11), 2278–2324 (1998)

8. Liu, W., et al.: SSD: single shot multibox detector. In: Leibe, B., Matas, J., Sebe, N., Welling, M. (eds.) ECCV 2016. LNCS, vol. 9905, pp. 21–37. Springer, Cham (2016). https://doi.org/10.1007/978-3-319-46448-0_2

9. Nakamura, K., Kamizuru, K., Kawasaki, H., Ono, S.: Multi-agent-based two-dimensional barcode decoding robust against non-uniform geometric distortion. Int. J. Comput. Inf. Syst. Ind. Manag. Appl. **9**(2017), 60–70 (2017)

10. Redmon, J., Divvala, S., Girshick, R., Farhadi, A.: You only look once: unified, real-time object detection. In: Proceedings of the IEEE Conference on Computer Vision and Pattern Recognition, pp. 779–788 (2016)

11. Redmon, J., Farhadi, A.: YOLO9000: better, faster, stronger. In: Proceedings of the IEEE Conference on Computer Vision and Pattern Recognition, pp. 7263–7271 (2017)

12. Redmon, J., Farhadi, A.: YOLOv3: an incremental improvement. arXiv preprint arXiv:1804.02767 (2018)

13. Ren, S., He, K., Girshick, R., Sun, J.: Faster R-CNN: towards real-time object detection with region proposal networks. In: Advances in Neural Information Processing Systems, pp. 91–99 (2015)

14. Yuan, B., et al.: MU R-CNN: a two-dimensional code instance segmentation network based on deep learning. Future Internet **11**(9), 197 (2019)

15. Zhao, Q., Ni, F., Song, Y., Wang, Y., Tang, Z.: Deep dual pyramid network for barcode segmentation using barcode-30k database. arXiv preprint arXiv:1807.11886 (2018)

Using Mask R-CNN for Image-Based Wear Classification of Solid Carbide Milling and Drilling Tools

Jasmin Dalferth[1]([✉]) [iD], Sven Winkelmann[1] [iD], and Friedhelm Schwenker[2] [iD]

[1] c-Com GmbH, Heinkelstraße 11, 73431 Aalen, Germany
{jasmin.dalferth,sven.winkelmann}@c-com.net
[2] Ulm University, Institute of Neural Information Processing,
James-Franck-Ring, 89081 Ulm, Germany
friedhelm.schwenker@uni-ulm.de
https://www.uni-ulm.de/index.php?id=11107, https://c-com.net/

Abstract. In order to ensure high productivity and quality in industrial production, early identification of tool wear is needed. Within the context of Industry 4.0, we integrate wear monitoring of solid carbide milling and drilling cutters automatically into the production process. Therefore, we propose to analyze wear types with image instance segmentation using Mask R-CNN with feature pyramid and bounding box regression. Our approach is able to recognize the five most important wear types: flank wear, crater wear, fracture, built-up edge and plastic deformation. While other methods use image classification and classify only one wear type for each image, our model is able to detect multiple wear types. Over 35 models with different hyperparameter settings were trained on 5,000 labeled images to establish a reliable classifier. The results show up to 82.03% accuracy and benefit for overlapping wear types, which is crucial for using the model in production.

Keywords: Tool wear detection · Machine learning · Convolutional Neural Network (CNN) · Mask R-CNN · Supervised learning · Multi-class classification · Image segmentation

1 Introduction

Cutting tools used in machining production wear out in daily operation after a certain period of use. Generally speaking, one could say: The tool becomes blunt. To prevent excessive or rapid wear, it is advisable to identify the type of wear, which can vary depending on the material and cutting parameters. Currently, wear types for different tools are not classified and evaluated automatically, instead they have to be examined manually by a technical expert or tools have to be used until they become broken. The goal of Industry 4.0 and digitization is to automate industrial processes to increase efficiency.

© Springer Nature Switzerland AG 2020
F.-P. Schilling and T. Stadelmann (Eds.): ANNPR 2020, LNAI 12294, pp. 223–234, 2020.
https://doi.org/10.1007/978-3-030-58309-5_18

Therefore, neural networks are a popular method for implementation in the field of machine learning algorithms. Our approach to detect wear is based on a special Convolutional Neural Network that combines image segmentation and classification. Therefore, standardized images taken from the experimental laboratory of the company MAPAL Dr. Kress KG[1] are used. For future practice, images taken from different camera systems (e.g. smartphones, adjustment tools) should reach a sufficient classification result. This offers various use cases for predictive maintenance, starting with a mobile application and ending with a completely automated process of quality assurance.

1.1   Related Work

To measure tool wear, methods such as real-time data monitoring [2,16,22] or direct measurement are often used [16,21]. These variants of measurement are suitable if the data is recorded continuously and the tool is tracked during its lifetime. In practice, however, tools are often used on different machines in different applications and it is not economically feasible to analyze how each tool is used over time. For this reason, it is important to estimate the wear status of a tool on a snapshot basis. Hence, the history and application of the tool does not matter when analyzing wear. Therefore, this work uses camera-based snapshots for wear detection and classification. In the field of image analysis various methods and approaches exist to classify wear. At first, a distinction is made between different classes: Papers like [3,8] distinguish two or three different categories of wear: Either broken and unbroken [8] or new, used and broken [3]. In comparison, this paper differs between five wear types: flank wear, crater wear, fracture, built-up edge and plastic deformation. The benefit of an accurate determination of wear is that causes can be identified with higher probability and cutting parameters or the use of coatings can be optimized. We propose using Mask R-CNN as a neural network to classify wear of solid carbide milling and drilling tools. To classify various wear types for each cutting edge, our method uses image segmentation and analyses each Region of Interest (RoI) individually. As wear types overlap to some extent, this enables a precise classification.

In terms of methodology, some of our approaches were influenced by other papers and led to a successful result: To record the images - as [3,20,24,25] - the position, exposure and background of the tool or cutting edge are mostly standardized. It is agreed that gray-scale images are adequate, since wear is determined by the reflection and the difference in brightness, among other things [3,8,9,20,24,25]. In addition, sources such as [8] and [24] are using image segmentation to analyze sections individually and identify the edges of the tool. If the image is divided in different regions, Support Vector Machines (SVMs) are a popular classifier [9]. Some other works, such as [3], [5] and [20] are also using neural networks of different kinds, e.g. Feed-Forward Neural Networks (FNNs [5]) or Single Category-Based Classifier (SCBC [20]).

---

[1] MAPAL Fabrik für Präzisionswerkzeuge Dr. Kress KG, URL: https://www.mapal.com/.

The paper is organized as follows: First we define different wear types and describe our data set in Sect. 2. In Sect. 3, we explain our approach to classify wear with Mask R-CNN. In an iterative implementation, we trained different models with various hyperparameter settings, which is described in Sect. 4. The results of the most promising models are presented in Sect. 5. Finally, we conclude in Sect. 6 that classifying wear of solid carbide milling and drilling tools with Mask R-CNN is possible and outperforms common classification methods.

2   Wear Classification of Milling and Drilling Tools

For the classification of the different wear types solid carbide milling and drilling shank tools are considered. These tools are frequently used, but are complex to analyze due to their multi-edged geometry. *Drills* create cylindrical holes in the workpiece by rotation. There are various drilling methods and countless variants of the tool. *Milling cutters* have several geometry-specific cutting edges and are used on milling machines. To remove material, the tool rotates around its own axis and the workpiece is moved over the machine table. For more information, see [6,7]. In the following, these two tool groups are summarized as *round tools*.

Round tools have a similar structure, consisting of flank face, guide chamfer and cutting face, see Fig. 1. Wear (see Sect. 2.1) occurs at different regions of a round tool, thus it is required to record and analyze images of at least three components to determine the wear status of the tool. Therefore, a detailed image is taken of each component and is analyzed separately. Then all results are combined to identify the wear type of the tool.

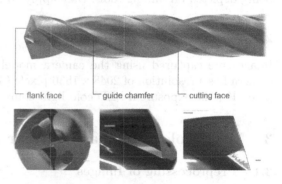

Fig. 1. Components of a drill: flank face (green), guide chamfer (red) and cutting face (blue) (cf. [19]). (Color figure online)

2.1   Wear Types

Tool wear is the abrasion of the tool cutting edge due to mechanical and thermal stress. Depending on the stress on the tool, there are different types of wear. Focusing on round tools we mainly considered [14] and [15] as a source of information in addition to the advice of technical experts. Considering these sources, the most important wear type classes are: flank wear, crater wear, fracture, built-up edge and plastic deformation. According to consultation with experts, these occur frequently on round tools and are relevant for industrial production.

Table 1. Wear types with sketch and sample images (figures are partly taken from [15])

Flank wear	Crater wear	Fracture	Built-up edge	Plastic deformation

The types of wear are illustrated in Table 1. Despite the following samples being depicted on milling tools, they apply to drilling tools, too.

2.2 Source Images to Classify Wear

Images are captured using the camera model UI-1460SE from iDS [13]. The camera has a resolution of 2048 × 1536 pixels (3.15 MP) and a size of 3.2 µm per pixel. For the exposure a LED cold light source (CV-KLQ-LED-9 [18]) is used.

3 Methodology to Classify Wear by Using Mask R-CNN

3.1 Preprocessing of Images

A total of 26,293 images of tool cutting edges are provided by the experimental laboratory of MAPAL Dr. Kress KG (see Footnote 1). The images are sorted according to their tool and wear types. For learning a neural network it is essential to consider an evenly distributed number of types and data that is as variable as possible. Finally, 4,307 images (augmented to 5,577) are considered. As augmentation methods, cropping and resizing were used to retain the tool cutting position, which is essential for later implementation.

The images always measure 2048 × 1536 pixels and contain three channels (RGB). Most of them are standardized images, i.e. the exposure is always kept to the cutting edge of the tool and the background is largely uniform. Depending on the tool there is a different number of images, which rely on the number of cutters. The cutting edges are not always in the same position on the images, which makes it difficult to detect wear. During preprocessing, the images are scaled, normalized and transformed to gray-scale, resulting in uniform images with the same size, the same value range and only one channel.

As wear is not identified by the color but by the reflection of the light, a gray-scale image is sufficient (cf. Sect. 1.1). The images are divided into a training (60%), validation (20%) and test dataset (20%). The training and validation sets are used during the learning process of the neural network, the test dataset is used separately for evaluation.

3.2 Labeling of Images

The images are labeled with the help of tool experts. The wear types are marked on the cutting edge of the tool. As a result, all shapes denoting wear are exported to a JSON file. To label the images, the VGG Image Annotator [4] is used, as proposed by the authors of Mask R-CNN [1]. An example of the labeling is shown in Fig. 2.

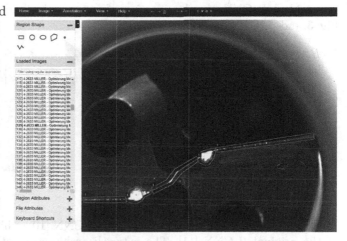

Fig. 2. Interface of the VIA tool [4] with an image of a flank face of a drill with fractures and flank wear

3.3 Model Selection

For classifying wear types, *Convolutional Neural Networks (CNN)* are considered. Convolution is an operation where an image is transformed by a kernel matrix, i.e. the image gets filtered. Since several network architectures of CNNs exist, we have to assess which one is able to classify wear types.

First the *VGG16* [23] is considered. It is a deep CNN, which is often used for image recognition and consists of Convolution, Max Pooling and fully-connected layers. We have chosen this network first, because it is clearly structured and can be used for many application scenarios. The most difficult parts of wear type classification are the proportion of small wear regions to the image size and overlapping wear types. VGG16 is able to detect one class per image and does not adapt its features to small objects, thus its results are insufficient for our requirements. The objective was to detect all occurring wear types and then to find the one with maximum size and expression.

Mask R-CNN [12] is used for image segmentation and object recognition, which can be used to classify multiple wear types for each image. R-CNN stands for region-based Convolutional Neural Network.

Mask means that the output of the model contains a binary mask over each segmented object. Therefore, Mask R-CNN is able to classify various objects in parallel but independent of each other. Nevertheless, this makes the network complex and difficult to implement, but it is flexible in its application and promises high accuracy as long as enough training data is available.

3.4   Network Architecture of Mask R-CNN

Using Mask R-CNN, the classification of an image is divided into three steps (cf. Fig. 3 and [12]):

1. *Bounding Boxes*, i.e. rectangular boxes, are placed around the objects to be detected.
2. *Classification with CNN*: Each box or RoI is classified individually with a CNN.
3. *Binary Mask*: The output is extended by a binary mask, which is placed over the object as an outline within a bounding box.

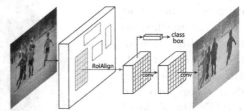

Fig. 3. Mask R-CNN [12]

Mask R-CNN was developed from Faster R-CNN [10], which is an improvement of the original R-CNN [11]. The network architecture of the Mask R-CNN is divided into two parts:

- The convolutional backbone architecture performs feature extraction over the entire image. This part consists of a *ResNet* of depth 50 or 101 (depending on the configuration) and a *Feature Pyramid Network (FPN)*.
- The head of the network performs bounding box detection, i.e. classification and regression, and creates a mask for each RoI.

Especially for the identification of small objects, the Pyramid Network feature is intended to improve the classification, which is an advantage for the detection of wear types in images of tool cutting edges.

3.5   Using Mask R-CNN for Model Training

The Mask R-CNN reference implementation based on Python 3, Keras and TensorFlow is provided on GitHub [1]. To apply *transfer learning*, pre-trained weights of MS COCO [17] are given. So first the code is adapted to our six-class segmentation problem and several parameters of the network are adjusted for better performance. The six-class problem distinguishes five wear types and the class new - technically speaking the class new stands for the *background* class. In total, 4,307 images are labeled and used for training and evaluation of the network. The dataset is split into 60% training set, 20% validation set and 20% test set. The images are passed to the network as input in JSON files.

The JSON files include image paths, the respective wear class and coordinates of the associated wear region. Important to mention is that the number of images per class is equally distributed or else the detection will not work properly.

After some experiments, it is concluded that pre-trained weights improve the flexibility of the network. In the beginning, *overfitting* has to be dealt with, which results that *Early Stopping* is added, so that the learning curve can be stopped at its optimum and the ideal weights can be saved. Starting with 500 images the dataset is slowly increased and improved to handle overfitting. On average, about 100 epochs are needed to reach optimal results using transfer learning. Due to the extensive amount of resources needed for model training, computing power optimized for machine learning and graphic-intensive applications are utilized from a public cloud provider. The training is conducted on a NVIDIA Tesla T4 graphics card with CUDA version 10.1 within a few hours. Two to four images are analyzed in a batch on a single GPU. Finally, a representative classification result can be achieved with a study of the network architecture, its parameters and the image data.

3.6   Evaluation of Mask R-CNN Model

In order to be able to evaluate the classification results, an evaluation method is defined. Depending on the confidence value, all masks of the detected RoIs that exceed the confidence value threshold are displayed after the classification. The default threshold of the Mask R-CNN is 0.7, that means all detected RoIs with at least 70% probability are considered. To determine the unique result, only the class with the highest probability is considered for evaluation. 30 images per class are selected from the validation dataset. After each classification, the selected class and its probability is saved. Finally, the average of all detected probabilities is calculated for each class. These values are visualized in a confusion matrix. On this basis all trained neural networks are analyzed in the same manner, allowing for a uniform evaluation and comparison of the models. During our study we observed stable classification results in this setting and therefore we do not expect different results when changing the testset size within a suitable range.

4   Implementation

In total 35 Mask R-CNN models are trained to compare different parameter settings. Besides parameter optimization, such as learning rate or momentum, the development of the final model is based on a step-by-step solution, which first solves a simpler partial problem and finally transfer its solution to the overall problem. Certain conclusions can be drawn from each training of a network, where positive changes are adopted and negative ones are rejected. The first step is the *reduction of the classes* from six to three classes: New (no wear), flank wear (mild wear) and general wear (consisting of crater wear, fracture, built-up edge and plastic deformation).

Therefore, transfer learning was applied on the pre-trained weights of the
COCO dataset [17]. This achieves a significant improvement in the learning curve
(see Fig. 4), but the classification results are not sufficient yet. The detection rate
is about 60% and some wear types overlap with the class new, which is contra-
dictory. That is why a second reduction step is required: The *concentration of the
classification on the cutting positions*, i.e. on three different components of the tool
(cf. Fig. 1). Three different networks are trained, one network per cutting position.
This shows improvements, as the classification is easier, because the wear is now
mainly visible at the same position. During the analysis of different image types a
few observations have been made: Depending on the cutting position, some types
of wear occur more often or less on a certain component. Additionally, crater wear
occurs only on the cutting face. To proceed strategically, one of the three networks
is first trained and optimized on a cutting position, in this case the guide chamfer.
Until approximately 750 images per class are labeled for the pictures of the guide
chamfer, the overfitting improves as expected (see Fig. 4).

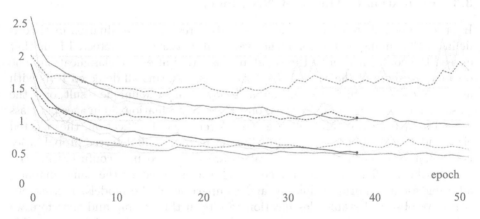

Fig. 4. Learning curve (loss = continuous line and validation loss = dashed line) of the
different reduction steps: First training (red), the reduction of classes (purple) and the
reduction of cutting positions (orange). (Color figure online)

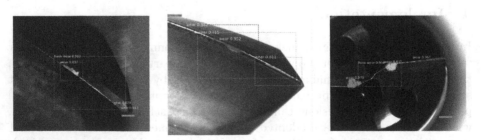

Fig. 5. Samples of the classification results of the reduced task with three classes and
different networks per cutting position

After tuning of the hyper parameters, the results of the classification are satisfying, with an average precision of 82.03%. In order not to repeat the training effort transfer learning is applied, from the neural network weights of the guide chamfer to those of the cutting face and the flank face, respectively. Therefore, fewer images and epochs are sufficient. For the cutting face, about 400 images per class and only 20 epochs are adequate for good results. For the flank face about 200 images and also 20 epochs are trained. In Fig. 5, some classification results are shown. As output masks and class assignment are promising, the *problem is extended* step-by-step to enable each network classifying all five wear types and new cutting edges (six classes). The three networks are trained in the same way as before, resulting in a similar learning curve (see Fig. 6).

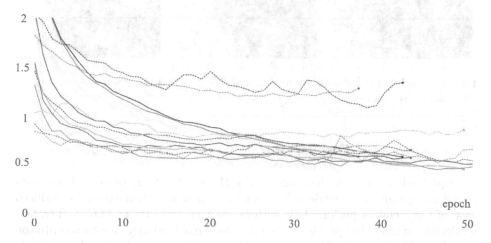

Fig. 6. Learning curve (loss = continuous line and validation loss = dashed line) of the three networks before and after the reduction: Guide chamfer (before: pink, after: purple), cutting face (before: light blue, after: dark blue) and flank face (before: green, after: dark green). (Color figure online)

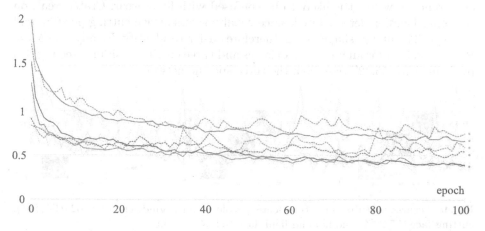

Fig. 7. Learning curve (loss = continuous line and validation loss = dashed line) in comparison: The reduction of classes on the guide chamfer (before: pink, and after extension: purple) and the extension to all images and classes (blue). (Color figure online)

The crucial point is to generalize the findings of the three single networks to one model, which is able to classify wear types for all cutting positions. Therefore, one network with all images and classes is trained. The remaining issue of unequal distribution of classes and cutting position explains the slightly increased loss, as depicted in Fig. 7. Some sample results are shown in Fig. 8.

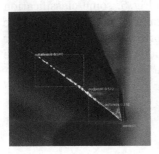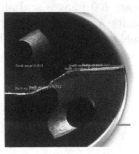

Fig. 8. Samples of the classification results of the extended task with one network for all images and classes

5   Results and Discussion

As explained in Sect. 3.6, the results of the Mask R-CNN are presented and compared using confusion matrices. The networks that were trained on the reduced task of three classes perform best with an average accuracy of 74.65%. The networks on our extended problem with six classes reach 60.41%, see each confusion matrix on Fig. 9 and 10. The confusion between classes can be mainly explained by the images used for training and evaluation: Generally, the classes built-up edge and plastic deformation rarely occur, therefore fewer images are available and thus the classification results are inferior to other classes. In case of a slight occurrence of wear, it is likely to be confused with flank wear. Crater wear, on the other hand, is detected well, since it only occurs at one cutting position and at a specific cutting shape and is therefore easier to identify. Finally, the Mask R-CNN trained for all images (six classes and three cutting positions, see Fig. 11) performs with 55.42% less than the three specific networks.

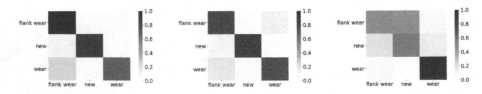

Fig. 9. Accuracy results of the three-class problem on the guide chamfer (82.03%, left), cutting face (77.29%, middle) and flank face (64.64%, right)

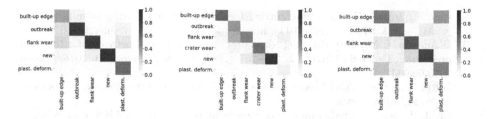

Fig. 10. Accuracy results of the six-class problem on the guide chamfer (70.59%, left), cutting face (51.21%, middle) and flank face (59.42%, right)

Due to the different cutting positions, wear can reflect differently and thus, allowing more mistakes to occur. The results with three specific networks are more precise than with one general network, i.e. a differentiation of the cutting positions would result in an increase of accuracy of up to 5% for six classes. For better results, however, more images of the less frequently occurring classes are needed - and for the general network - also an equal distribution of cutting positions. This could further optimize the results of the network, thereby reducing effort and improving performance.

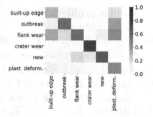

Fig. 11. Accuracy result of the final network on the six-class problem

6 Conclusion

In conclusion, we have successfully implemented a Mask R-CNN model to classify wear on images of tool cutting edges of solid carbide mills and drills. In total 35 neural networks were trained on 5,577 labeled images to achieve an accuracy of 55.42% up to 82.03%. The detection of overlapping wear types is a significant improvement to existing approaches. For accurate results, first each tool cutting position was trained with a separate network, which was specially configured for the properties of this region. Second, a single model containing all positions was trained to build one comprehensive industrial solution.

For further improvements, the output of the Mask R-CNN can be optimized using machine learning methods such as Support Vector Machines (SVMs) or Decision Trees. By the use of the output masks, it is also possible to calculate the size of the detected wear. In conclusion, our solution can be adapted to various applications in industry. Generally speaking, it can be considered as a foundation for further research and product development.

References

1. Abdulla, W.: Mask R-CNN for Object Detection and Segmentation (2017). https://github.com/matterport/Mask_RCNN

2. Cuš, F., Župerl, U.: Real-time cutting tool condition monitoring in milling. Strojniski Vestnik **57**, 142–150 (2011). https://doi.org/10.5545/sv-jme.2010.079
3. D'Addona, D.M., Teti, R.: Image data processing via neural networks for tool wear prediction. Procedia CIRP **12**, 252–257 (2013). https://doi.org/10.1016/j.procir.2013.09.044
4. Dutta, A., et al.: VGG Image Annotator (VIA) (2019). http://www.robots.ox.ac.uk/~vgg/software/via/
5. Erdik, T., şen, Z.: Prediction of tool wear using regression and ANN models in end-milling operation a critical review. Int. J. Adv. Manuf. Tech. **37**, 765–766 (2008). https://doi.org/10.1007/s00170-008-1758-0
6. e.V., D.: DIN 4000–82. Beuth Verlag GmbH (2011)
7. e.V., D.: DIN 4000–81. Beuth Verlag GmbH (2012)
8. Fernández-Robles, L., et al.: Machine-vision-based identification of broken inserts in edge profile milling heads. RCIM **44**, 276–283 (2017). https://doi.org/10.1016/j.rcim.2016.10.004
9. García-Ordás, M.T., et al.: Tool wear monitoring using an online, automatic and low cost system based on local texture. MSSP **112**, 98–112 (2018). https://doi.org/10.1016/j.ymssp.2018.04.035
10. Girshick, R.: Fast R-CNN. In: Proceedings of the IEEE ICCV, pp. 1440–1448 (2015). arXiv: 1504.08083
11. Girshick, R., et al.: Rich feature hierarchies for accurate object detection and semantic segmentation. CoRR, pp. 580–587 (2013). arXiv: 1311.2524
12. He, K., et al.: Mask R-CNN. CoRR, pp. 2980–2988 (2017). arXiv: 1703.06870
13. IDS Imaging Development Systems GmbH: iDS UI-1460SE (2020). https://de.ids-imaging.com/store/ui-1460se.html
14. International Organization for Standardization: ISO 8688–1. ISO, 1st edn. (1989)
15. International Organization for Standardization: ISO 8688–2. ISO, 1st edn. (1989)
16. Kong, D., et al.: Gaussian process regr. for tool wear pred. MSSP **104**, 556–574 (2018). https://doi.org/10.1016/j.ymssp.2017.11.021
17. Lin, T.Y., et al.: Microsoft COCO. In: ECCV, pp. 740–755 (2014). arXiv: 1405.0312
18. M-Service & Geräte Peter Müller e.K.: Bedienungsanleitung Modell CV-KLQ-LED-9 (2016). https://www.m-service.de/seiten/d/downloads_d_only/d_Bedienungsanleitung_CV-KLQ-LED.pdf
19. MAPAL Fabrik für Präzisionswerkzeuge Dr. Kress KG: Kompetenz Vollbohren mit Vollhartmetall, PKD und Wechselkopf-Systemen (2020). https://www.mapal.com/de/produkte/vollbohren-aufbohren-senken/vollbohren/
20. Mikołajczyk, T., et al.: Neural network approach for automatic image analysis of cutting edge wear. MSSP **88**, 100–110 (2017). https://doi.org/10.1016/j.ymssp.2016.11.026
21. Pfeifer, T., Wiegers, L.: Reliable tool wear monitoring by optimized image and illumination control in machine vision. Measurement **28**(3), 209–218 (2000). https://doi.org/10.1016/S0263-2241(00)00014-2
22. Shi, X., Wang, X., Jiao, L., Wang, Z., Yan, P., Gao, S.: A real-time tool failure monitoring system based on cutting force analysis. Int. J. Adv. Manuf. Technol., 2567–2583 (2017). https://doi.org/10.1007/s00170-017-1244-7
23. Simonyan, K., Zisserman, A.: Very deep convolutional networks for large-scale image recognition. In: ICLR (2015). arXiv: 1409.1556
24. Sukeri, M., et al.: Wear detection of drill bit by image-based technique. IOP Conf. Ser. Mat. Sci. Eng. **328** (2018). https://doi.org/10.1088/1757-899X/328/1/012011
25. Thakre, A.A., et al.: Measurements of tool wear parameters using machine vision system. Model. Simul. Eng. **2019**, 1–9 (2019). https://doi.org/10.1155/2019/1876489

A Hybrid Deep Learning Approach for Forecasting Air Temperature

Gregory Gygax and Martin Schüle$^{(\boxtimes)}$

Institute for Applied Simulation, Zurich University of Applied Sciences ZHAW,
Wädenswil, Switzerland
{gregory.gygax,martin.schuele}@zhaw.ch

Abstract. Forecasting the weather is a great scientific challenge. Physics-based, numerical weather prediction (NWP) models have been developed for decades by large research teams and the accuracy of forecasts has been steadily increased. Yet, recently, more and more data-driven machine learning approaches to weather forecasting are being developed. In this contribution we aim to develop an approach that combines the advantages of both methodologies, that is, we develop a deep learning model to predict air temperature that is trained both on NWP models and local weather data. We evaluate the approach for 249 weather station sites in Switzerland and find that the model outperfoms the NWP models on short time-scales and in some geographically distinct regions of Switzerland.

Keywords: Weather forecasting · Deep learning · Hybrid modeling

1 Introduction

To forecast the weather is a long-standing scientific challenge. Also, accurate weather forecasts have great economic impact and mitigate costs to lives and assets in the case of high-impact weather.

Our weather is produced by a physical atmospheric system with complex dynamics. The usual metereological models used in numerical weather prediction (NWP) are based on modeling the atmospheric dynamics and atmosphere-land-sea couplings. The simulation of these models, initialized by a wealth of measured and inferred data, then allows to forecast various parameters such as air temperature or precipitation. In recent years and decades, the models and simulation techniques have been developed to the point where they allow a fairly accurate weather forecast for up to 10 days.

Despite these successes of the common meteorological models, more and more work is recently being undertaken that aims to produce weather forecasts using a machine learning (ML) approach. It is hoped that this further improves weather forecasting, especially in terms of fast and accurate spatio-temporal resolution. Such forecasting on a relatively short time-scale is also called nowcasting.

Instead of aiming to replace the elaborate physical models completely by ML approaches, we propose in this contribution a hybrid approach, combining the

© Springer Nature Switzerland AG 2020
F.-P. Schilling and T. Stadelmann (Eds.): ANNPR 2020, LNAI 12294, pp. 235–246, 2020.
https://doi.org/10.1007/978-3-030-58309-5_19

NWP models with a deep learning (DL) approach. We believe this combines the advantages of NWP models such as an accurate representation of atmospheric physics and a global approach to weather forecasting with the advantages of data-driven ML approaches such as fast and comprehensive local data analysis.

Our approach develops a DL model which is trained on local measurement data of weather stations and the corresponding forecasts of the NWP models plus local past weather data. The main contributions of this work can be summarised as follows: 1. We analyze the performance of the main NWP models for Switzerland in regard to forecast accuracy of air temperature at 249 weather station sites, 2. We design a DL model that learns to make local air temperature forecasts based on the performance of the NWP models and additional local data. As we will discuss in more detail, the results indicate that our approach allows to generate improved forecasts on short time-scales and for some geographically distinct regions of Switzerland.

The rest of the work is structured as follows: Next, we discuss the background and related work. In Sect. 3 we describe the data situation and the NWP models used. In Sect. 4 we describe the technical details of our approach. In Sect. 5 we present the results of experiments with the model on Swiss weather data. In Sect. 6 we conclude with a discussion of our approach and results.

2   Background and Related Work

Weather prediction has a long and successful history. In numerical weather prediction (NWP) the methodology usually centers on modeling the physics of the atmosphere and taking samples from numerical simulations of these models to generate forecasts [1,2]. The approach essentially consists in modeling the physics of the atmosphere and it couplings e.g. to sea and land, in model initialization schemes, and running large simulations on super-computing facilities. Ensemble modeling allows for an estimate of the uncertainty of forecasts. Advanced computational techniques are used in order to run simulations of models. The current state of the art in numerical weather prediction is reviewed by Bauer et al. [3]. The NWP models are developed at large research centers such as the European Centre for Medium-Range Weather Forecasts (ECMWF). We will also use ECMWF models in this work (see Table 1), but refrain from explaining these models in more detail as this does not constitute the focus of this contribution.

In recent years there have been more and more approaches to weather forecasting with ML models. These models sometimes try to predict a number of parameters [4,5], as NWP models, but often the focus is on certain parameters, e.g. on precipitation forecasts [7,8]. Also, some authors devise a hybrid approach, by combining different models or modeling strategies, where others rely on a straight ML or DL modeling approach. Our approach is in the spirit of Reichstein et al. [6] where the authors argue for hybrid modeling strategies for the earth sciences, combining physical models with ML approaches.

Some work on weather forecasting with DL methods we would like to mention specifically: Grover et al. [4] develop a hybrid approach where they combine machine learning algorithms locally trained on key weather variables with a deep learning model that models the joint statistics of the variables and a statistical method for spatial interpolation. The model predicts wind, temperature, pressure and dew point for weather station locations in the US and in some cases outperforms the NOAA (National Oceanic and Atmospheric Administration) models. Weyn et al. develop a convolutional neural network (CNN) approach to model the atmospheric state [5]. Xingjian et al. develop a convolutional LSTM model for precipitation nowcasting and outperfom an operational precipitation forecasting algorithm using radar map data [7]. Hernández et al. use an autoencoder and FNN (feedforward neural network) to forecast accumulated daily precipitation for a meteorological station in Colombia [8]. The cited work shows that the main DL models such as FNN, CNN, conv-LSTM, etc. are currently being explored for weather forecasting tasks.

3   Description of Data and Weather Prediction Models

In order to build and evaluate our approach we use weather data collected by weather stations and historic weather forecast data generated by some of the main NWP models for Europe.

In regard to the measurement data we selected a number of key weather parameters and collected these for 249 weather stations locations in Switzerland for the time period 1990–2020. In this contribution we however only consider the mean temperature 2 m above ground in 1 h frequency. The data was collected by MeteoSwiss (Swiss Federal Institute of Meteorology and Climatology) weather stations and provided by Meteomatics, a private weather data provider.[1]

The forecast data was collected for the NWP models listed in Table 1 for the time period 2019-09-17 to 2020-03-24. These models constitute some of the main NWP models for Europe. Some regional models such as COSMO are however missing.

We analysed the performance of the NWP models by the mean squared error (MSE) of their forecasts per forecast horizon for the 249 weather station sites for the air temperature 2 m above ground. Figure 1 shows an example for the location Wädenswil, Switzerland. We can see that, as expected, the predictions become worse with growing forecast horizon.

We aim to beat these models or the best of these models in accuracy. However, what is the best model for a given time and location? At some point in time t it is not a priori clear which model will perform best for the next hours and days. We therefore constructed a benchmark model in the following way: given a location and some point in time we determine the model that has performed best in the past 60 h. The forecasts of that model will be picked as comparison to the forecasts made by our own model for the prediction made at time t. This procedure is repeated for every location and point in time. In Fig. 1 the thereby generated benchmark predictions, averaged over the entire time-period, are shown, yielding the lowest prediction errors compared to the original models.

---

[1] Meteomatics, https://www.meteomatics.com. Last accessed June 2020.

Table 1. NWP models used in the work.

Short name	Description
cmc-gem	Global Environmental Multiscale model operated by the Canadian Meteorological Center
ecmwf-ifs	The European Center for Medium-Range Weather Forecasts' (ECMWF) Integrated Forecasting System (IFS). Atmospheric global circulation model used for medium-range forecasts
ecmwf-ens	Ensemble Prediction System (EPS) by ECMWF
ecmwf-mms	Long-range seasonal forecast by ECMWF
ecmwf-vareps	Long-range ensemble forecast by ECMWF
knmi-hirlam	High Resolution Limited Area Model from the Royal Netherlands Meteorological Institute.
mf-arome	Regional model by Meteo France
mix	Mixture model combining different models designed by Meteomatics
mm-swiss1k	High-resolution model for Switzerland designed by Meteomatics
ncep-gfs	Global Forecasting System by the National Centers for Environmental Prediction (NCEP)
ncep-gfs-ens	Ensemble model of Global Forecasting System by NCEP
ukmo-euro4	European model by the UK MetOffice

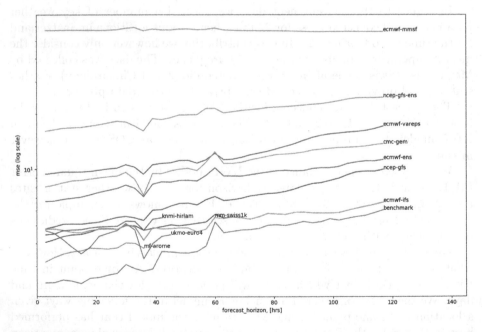

Fig. 1. MSE of the NWP model predictions of the air temperature 2 m above ground for the location Wädenswil, Switzerland for different forecast horizons. The lowest line indicates a benchmark model. Data considered for the time-period 2019-09-17 to 2020-03-24.

4   Method

The main idea is to train locally, i.e., at each selected weather station location, a DL model that takes as input time-series forecast data generated by the NWP models and the measurement data at that site. There are thus several time series, one for the measurements and several for the NWP models, which we collect in one feature time series vector. The target is the forecast h steps into the future which is then compared to the actual, measured values. That is, for each location where data is available (the site of a weather station), the model aims to close the gap between forecasted values and actually recorded values by training the model accordingly on the historic data.

Formally, let $M(t)$ denote the measured value at time t and let $F_h^{(i)}(t)$ denote the prediction made by model i at time t for time $t + h$ In other words, $F_h^{(i)}(t)$ is an estimate of $M(t + h)$. In this contribution, we only consider the air temperature 2 m above ground as value. Because M is available on an hourly basis, but forecasts are available on a 3 h basis only, we split M into 3 time series with a 1 h lag relative to each other and add these new time series to the feature vector. Given a forecast horizon h, we then construct for every time t the target $Y_h(t) = (M(t + h))$ and the input or feature vector

$$X(t) = \begin{pmatrix} M(t) \\ M(t-1) \\ M(t-2) \\ F_h^{(0)}(t) \\ \vdots \\ F_h^{(n)}(t) \end{pmatrix}$$

We further transform, as the time series is not stationary, $X(t)$ and $Y(t)$ by subtracting $M(t)$ from each value, yielding $Y_h(t) = (M(t + h) - M(t))$ and

$$X(t) = \begin{pmatrix} 0 \\ M(t-1) - M(t) \\ M(t-2) - M(t) \\ F_h^{(0)}(t) - M(t) \\ \vdots \\ F_h^{(n)}(t) - M(t) \end{pmatrix}$$

Finally, we are using $X(t), \cdots, X(t - l)$, with look back period $l = 20$, for predicting $Y(t)$. The value for l seems reasonable to us, amounting to a 60 h look back window, but we did not investigate that parameter value further.

The described data preparation was then carried out for each location.

4.1 Model

With relatively small amount of data (ca. 1000 time steps), a small model with one GRU (Gated Recurrent Unit) layer with 2 nodes and a subsequent dense layer seemed appropriate. Larger models quickly overfitted. However we did not yet look systematically into this matter.

4.2 Benchmark

We defined the benchmark as the prediction of the NWP model which performed best during the past 20 time steps (the same time window with a look back period $l = 20$) corresponding to 60 h. Formally, we have

$$bench(t) = F_h^{(i)}(t), \text{ where } i = \arg\min_i \sum_{j=0}^{20} (F_h^{(i)}(t - j) - M(t - j))^2$$

4.3 Training

Data was split for cross validation using the last 10% of the data for testing and the rest for training.

A model was trained using each possible combination of the following parameters:

– Forecast Horizon: 3 h, 6 h, 12 h, 18 h, 24 h
– Station: 1 of 249
– Look back: 20
– Weather parameter: hourly mean temperature 2 m above ground

This results in $5 \cdot 249 = 1245$ models. Each model was trained for 1000 epochs.

4.4 Implementation

All computation was done using python3.8 on linux. Models were built, trained and evaluated on an NVIDIA RTX 2060 GPU using Tensorflow 2.1. We used Tensorflow's standard implementation of the 'Adam' optimizer and the mean squared error (MSE) loss function.

4.5 Evaluation

For each time t, the predicted values were evaluated against the benchmark model predictions. The error (MSE) was computed for the testing and training sets for both the newly trained models and the benchmark model for every station $s \in S$, resulting in $mse_{train}(s), mse_{test}(s), bench_{train}(s), bench_{test}(s)$.

For each station we can then build the differences $mse_{train}(s) - bench_{train}(s)$ and $mse_{test}(s) - bench_{test}(s)$. If the difference is smaller than 0, the new model is better than the benchmark for the given station.

5 Results

The performance of the new model was assessed by looking at the difference of the prediction error to the benchmark error. The error used was the mean squared error (MSE) either over the training period or the testing period. Here, we focus on the results for the testing set.

5.1 Nowcasting and Geographic Differences

We evaluated the model for 249 weather station sites in Switzerland. We note two main points: 1. The model performs consistently better for a forecast horizon of 3 h, i.e., in the "nowcasting" range; 2. For larger forecast horizons the model performs better for some locations or some regions of Switzerland but not for whole country.

Figure 2 shows two examples for two distinct locations: in a) the model does not perform better while in b) the model performs better over all forecast horizons. As expected, the model performs better than the benchmark model on the training set in both cases.

Figure 3 shows an overview of the errors per station for each forecast horizon on a national scale. The model performs well for the 3 h forecast for most stations. Forecasts quality seems to deteriorate for increasing forecast horizon and seem worst for 12 h forecast horizon. Interestingly, there seem to be clusters of locations, e.g. in the canton of Valais, where the model seems to be consistently better than the benchmark model. We therefore decided to look at the Valais example in more detail. Figure 4 shows a zoomed in view on the stations in the canton of Valais.

5.2 Evaluation Metrics and Error Distribution

We analyzed the differences between our model's performance to the benchmark model performance by looking at the mean and median differences of the MSE of the predictions by our model and the benchmark model, further on referred to as MBP and MEBP, respectively. Also, we assessed the ratio of stations that performed better under the model than with the benchmark model. A station is assumed to perform better than the benchmark, if the difference of MSE(model) - MSE(benchmark) < 0. We performed this analysis on the level of the forecast horizons once for all available stations and once for the Valais stations. Results are summarized in Table 2. Both tables show that the majority of stations (81% for Switzerland and 89% for Valais) have better forecasts with the new model than with benchmark for the 3 h forecast. While the MPB seems to indicate

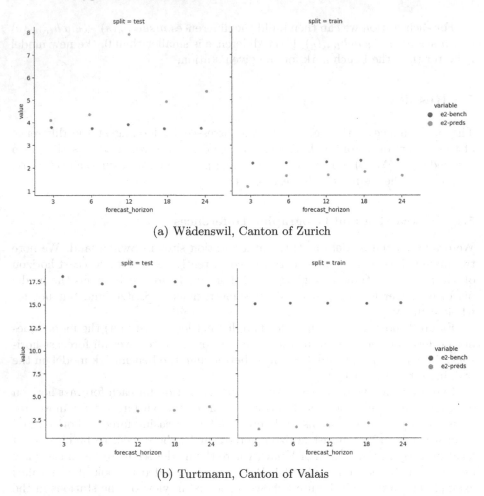

(a) Wädenswil, Canton of Zurich

(b) Turtmann, Canton of Valais

Fig. 2. Prediction errors of our model and benchmark model over forecast horizons for
a) Wädenswil and b) Turtmann.

better performance for all forecast horizons except the 12 h, the ratio of improved
models and the MEBP indicate that these values are probably caused by outliers.
In the case of the stations in Valais, there seems to be improvement for all forecast
horizons, although the improvement for 24 h forecast is very small.

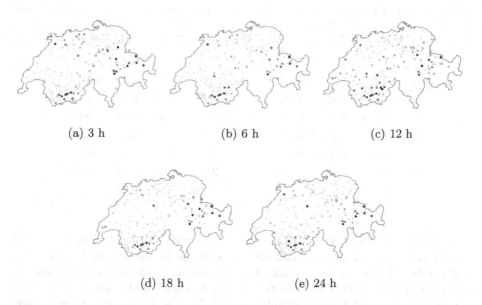

(a) 3 h   (b) 6 h   (c) 12 h

(d) 18 h   (e) 24 h

Fig. 3. Geographic error distribution: Each dot corresponds to one station. Red indicates station where model performed worse than the benchmark model while blue dots indicate stations where the model was better. Darker shades indicate larger absolute differences. (Color figure online)

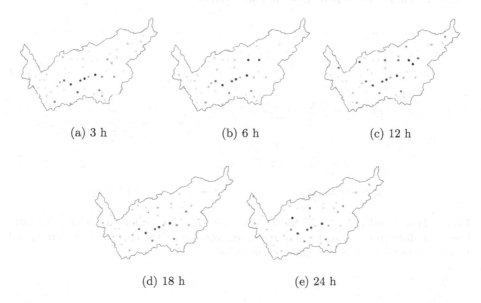

(a) 3 h   (b) 6 h   (c) 12 h

(d) 18 h   (e) 24 h

Fig. 4. Geographic error distribution in the canton of Valais. Coloring analogous to Fig. 3.

Table 2. FH: forecast horizon (h), R: Ratio of stations where the model is better than benchmark model, MPB: Mean difference of MSE of the predictions model vs. benchmark model, MEPB: median difference of MSE of the predictions model vs. benchmark model.

FH	MPB	R	MEPB		FH	MPB	R	MEPB
3	-2.01	0.81	-1.06		3	-2.37	0.89	-1.54
6	-0.54	0.45	0.20		6	-1.09	0.61	-0.64
12	0.73	0.36	0.70		12	0.46	0.56	-0.22
18	-0.30	0.47	0.10		18	-0.99	0.54	-0.34
24	-0.54	0.51	-0.02		24	-0.43	0.51	-0.02
(a) Switzerland					(b) Valais			

Figure 5 shows the distribution of the difference MSE(model)−MSE (benchmark) for all stations and for the stations in the Valais. Furthermore, the colors indicate the mean error of the station over all forecast horizons. We can easily recognize that stations either perform consistently bad or well over all forecast horizons on both geographic scales. That is, if a station benefits from the model forecasts for any forecast horizon, it is likely to benefit for the other forecast horizons as well. This seems to support the thesis that ML boosted models can improve forecast quality in difficult to model locations while other locations might not benefit from our approach.

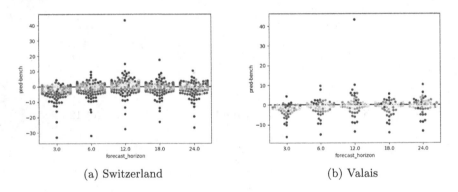

(a) Switzerland   (b) Valais

Fig. 5. MSE (model) and MSE (benchmark) distribution per forecast. Colors indicate the mean difference over all forecast horizons. Note that these colors do not correspond to the colorscale used on the map visualizations.

6   Discussion

We have developed in this contribution an approach that combines numerical weather prediction (NWP) models with a machine learning (ML) approach.

Specifically, we developed a deep learning (DL) model to predict air temperature 2 m above ground that is trained both on NWM models and local weather data and evaluated the approach for 249 weather station sites in Switzerland. Our preliminary results show that the approach has potential: in the nowcasting domain, i.e., for short time-scales, the model performs better almost everywhere, for longer forecast horizons it seems that the approach could bring improvements for some but not all regions. A new task may therefore be to identify the locations that could benefit from our approach, e.g., a classifier based on geographic features might come into play.

We currently interpret the results as shown on the map (Fig. 3) as follows: in mountainous regions such as the Valais, the potential for improvement seems to be highest, because there you might find yourself in a special micro weather situation, possibly created by the mountains that shield the region from the coarse-meshed macro weather situation simulated by the NWP models. However, this hypothesis should be examined more closely. Unfortunately, we have not collected forecast data for all mountain valley regions in Switzerland such as the Engadin.

We have not yet systematically analyzed the DL model in terms of architecture and parameter tuning. Therefore we think that with further experiments and analyses of the model substantial improvements are still possible.

In future work we will work further on the following approaches: 1. To forecast further parameters, for example to predict precipitation, 2. To use data from neighboring stations to forecast the weather at a particular station and 3. Collect more data, e.g. provided by small weather stations at local farmers, and develop the model further.

We believe that the blend of NWP models and ML models has great potential and will continue to find its way into the science of weather forecasting.

Acknowledgment. This work was supported by Innosuisse grant 26301.1 IP-ICT and Hydrolina Sarl.

References

1. Richardson, L.F.: Weather Prediction by Numerical Process. Cambridge University Press, New York (2007)
2. Kalnay, E.: Atmospheric Modeling, Data Assimilation and Predictability. Cambridge University Press, New York (2003)
3. Bauer, P., Thorpe, A., Brunet, G.: The quiet revolution of numerical weather prediction. Nature **525**(7567), 47–55 (2015)
4. Grover, A., Kapoor, A., Horvitz, E.: A deep hybrid model for weather forecasting, In: Proceedings of the 21th ACM SIGKDD International Conference on Knowledge Discovery and Data Mining, pp. 379–386 (2015)
5. Weyn, J.A., Durran, D.R., Caruana, R.: Can machines learn to predict weather? Using deep learning to predict gridded 500-hPa geopotential height from historical weather data. J. Adv. Model. Earth Syst. **11**(8), 2680–2693 (2019)

6. Reichstein, M., Camps-Valls, G., Stevens, B., Jung, M., Denzler, J., Carvalhais, N.: Deep learning and process understanding for data-driven Earth system science. Nature **566**(7743), 195–204 (2019)
7. Xingjian, S.H.I., Chen, Z., Wang, H., Yeung, D.Y., Wong, W.K., Woo, W.C.: Convolutional LSTM network: a machine learning approach for precipitation nowcasting. In: Advances in Neural Information Processing Systems, pp. 802–810 (2015)
8. Hernández, E., Sanchez-Anguix, V., Julian, V., Palanca, J., Duque, N.: Rainfall prediction: a deep learning approach. In: Martínez-Álvarez, F., Troncoso, A., Quintián, H., Corchado, E. (eds.) HAIS 2016. LNCS (LNAI), vol. 9648, pp. 151–162. Springer, Cham (2016). https://doi.org/10.1007/978-3-319-32034-2_13

Using CNNs to Optimize Numerical Simulations in Geotechnical Engineering

Beat Wolf[1]([✉]) [ID], Jonathan Donzallaz[1] [ID], Colette Jost[2] [ID], Amanda Hayoz[1] [ID], Stéphane Commend[2] [ID], Jean Hennebert[1] [ID], and Pierre Kuonen[1]

[1] iCoSys, University of Applied Sciences Western Switzerland, Fribourg, Switzerland
beat.wolf@hefr.ch
[2] iTEC, University of Applied Sciences Western Switzerland, Fribourg, Switzerland

Abstract. Deep excavations are today mainly designed by manually optimising the wall's geometry, stiffness and strut or anchor layout. In order to better automate this process for sustained excavations, we are exploring the possibility of approximating key values using a machine learning (ML) model instead of calculating them with time-consuming numerical simulations. After demonstrating in our previous work that this approach works for simple use cases, we show in this paper that this method can be enhanced to adapt to complex real-world supported excavations. We have improved our ML model compared to our previous work by using a convolutional neural network (CNN) model, coding the excavation configuration as a set of layers of fixed height containing the soil parameters as well as the geometry of the walls and struts. The system is trained and evaluated on a set of synthetically generated situations using numerical simulation software. To validate this approach, we also compare our results to a set of 15 real-world situations in a t-SNE. Using our improved CNN model we could show that applying machine learning to predict the output of numerical simulation in the domain of geotechnical engineering not only works in simple cases but also in more complex, real-world situations.

Keywords: Numerical simulation · Geotechnical engineering

1 Introduction

A common problem in geotechnical engineering is to evaluate the stability of a given excavation as well as its influence in terms of displacements on the neighbouring structures. In our case, we want to limit the settlement, the wall deviation as well as the wall deflection while ensuring that the excavation will not collapse. Today this task is often realized with the help of finite element numerical simulations, which can be time-consuming. As a result, only a few manually designed excavation configurations can be validated by numerical simulations. However, if we want to further automate the process of defining the

© Springer Nature Switzerland AG 2020
F.-P. Schilling and T. Stadelmann (Eds.): ANNPR 2020, LNAI 12294, pp. 247–256, 2020.
https://doi.org/10.1007/978-3-030-58309-5_20

most cost-effective excavation configuration, we need to evaluate hundreds - if not thousands - of different configurations and ensure that they respect the desired safety and displacement standards.

In this work, we explore, based on the previous work [4], the possibility to predict the key results of a numerical simulation for supported excavations using a machine learning approach. Our main goal is to show that the results previously achieved with simple excavation configurations can be reproduced using more complex, real life-like excavation configurations. Instead of predicting the outcome of the complete numerical simulation, we focus on 5 key values that indicate the viability of a solution in terms of stability and displacements.

Our work is based on the numerical simulation software ZSwalls [1]. The latter performs a 2D analysis of retaining walls for deep excavations using the nonlinear finite element kernel of the ZSOIL software [2], which has proven to be effective in predicting the behaviour of large excavations in urban areas [3,8].

Figure 1 shows an example of the input parameters of the software and Fig. 2 shows the results of a simulation.

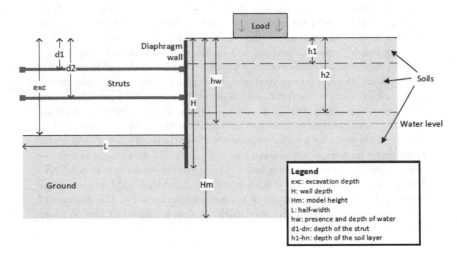

Fig. 1. Problem description

As our goal is not to reproduce the full results of the numerical simulation, but rather to estimate its behavior, we can greatly reduce the complexity of the prediction problem. To determine if a configuration is viable or not we predict only the following five key values:

- The convergence is a boolean value that determines if the retaining system is stable or not, or basically if the finite element numerical simulation can find an equilibrium.
- The maximum settlement (m) is the maximal vertical displacement at the terrain level, behind the wall, at the end of the excavation.

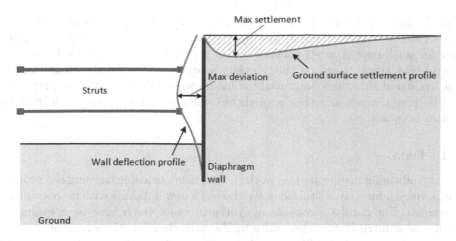

Fig. 2. Output of the simulation

- The maximum deviation (m) is the maximal horizontal displacement along the wall, at the end of the excavation.
- The maximum and minimum moments (kNm) represent extreme values of the bending moment along the wall. These internal forces allow structural engineers to design the wall, i.e. determine the wall's thickness and/or type.

Based on our previous work [4], we present a new and improved machine learning model for this type of data which can handle complex excavation configurations. This new machine learning model also includes a new way to encode the complex physical simulation data used as input for the neural network.

2   Methods

In recent years various machine learning techniques have been used in the domain of numerical simulations. While some go as far as completely replacing the numerical simulations with a machine learning model [5], others use machine learning to speed up parts of the numerical calculations [11]. In some domains, such as industrial systems or fluid dynamics systems, similar ideas to ours of using machine learning approximations as the objective functions of optimizations methods have also been explored [7,9]. In the same vein as those other works, we described in [4] the possibility of using machine learning approaches to predict the outcome of a numerical simulation for unsupported excavations. The previous work showed the potential of this approach when used on simple excavations. We obtained good results for the binary convergence value, as well as errors between 1 to 5% for the 3 quantities of interest (deviation, settlement and absolute bending moment). The used metric was based on the mean relative error, truncated for very small values. A simple 3 layer multilayer perceptron (MLP) model with 30/20/10 neurons, replicated for all 4 values, has been used.

The dataset was limited to very simple situations, which have limited practical use in the real world. Indeed, unsupported excavations are only used in practice for small depths, while anchors and/or struts are usually needed to support deep excavations. Based on this previous work we increased the complexity of the evaluated situations, improved the model to handle the more complex cases in the new dataset and split a predicted value (bending moment) in two for better accuracy.

2.1 Data

After confirming in our previous work [4] that using machine learning is a promising approach for simple situations, we created a new database with more complex scenarios. The database consists of synthetic cases which have been evaluated using the numerical simulation software ZSwalls. Based on the advice of geotechnical engineers and real-life scenarios, we made the following changes to the previous database:

- The support structure can have up to six struts
- The stratigraphy has up to five different soil types
- Addition of a potential groundwater table
- Deeper excavations (up to 25 m deep)
- Diaphragm walls (in addition to sheet-pile walls)

For each situation, an excavation height, a retaining structure and a soil model are generated. Excavations with a height of 5 to 25 m are considered. Each bearing structure consisted of either a sheet pile or a diaphragm wall with a maximum height of twice the excavation height. The wall's stiffness is chosen out of a set of standardized sheet-pile or diaphragm walls. The struts are distributed with an equal distance between 3 to 5 m along with the excavation height. The distance between the top strut and the surface ranges between 1 to 3 m depending on the excavation height. The bottom 3.0 m are free of struts. With a maximum excavation height of 25 m, a maximum amount of 7 struts can be present. In 25% of the situations, a buttress with an unloading factor of 75 to 100% is included. Three otherwise similar situations with different prestress forces in struts are generated.

Five soil types were defined (deposit, granular, cohesive, moraine and rock) with individual ranges for each soil parameter: Young's modulus, cohesion and friction. The thickness of each layer was chosen following a normal distribution. To generate realistic soil configurations, the soil type of each layer depends on the type of the overlying soil layer. To build our dataset, we use ZSwalls as numerical simulation software. ZSwalls does 2D deep excavation retaining wall analysis based on the finite element method through the ZSOIL software, successfully used to predict the behaviour of large excavations in urban environments [3,8]. We used almost the same output values of the simulation as our prediction targets, with one notable change to our previous work. While the convergence, settlement and deviation were kept, the moment was split into two distinct values. We now predict the minimum and maximum moment.

We simulated 20'000 excavations using ZSwalls, of which 14'548 (72.7%) converge (the construction holds) and 5'450 (27.2%) do not (the constructions breaks down). In addition to the synthetic cases, we defined a set of 15 real cases, which were also simulated with ZSwalls, to have a better understanding of the behaviour of the system in real-life scenarios. One of these real cases is illustrated in Fig. 3, with all varying parameters depicted in yellow.

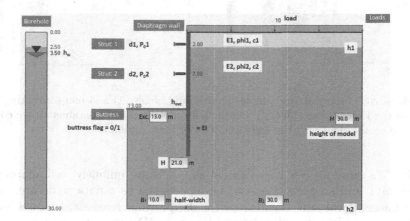

Fig. 3. Illustration of one real case, with varying parameters depicted in yellow

To better understand the high dimensional data of the different excavation configurations, we decided to visualize the training and real-life data using t-SNE [6], a machine learning algorithm which allows the visualization of high-dimensional data in lower dimensions (in our case 2). Figure 4 shows a t-SNE analysis (perplexity 30, learning rate 200, iterations 250) of the 15 real-life cases compared to our simulated database. The results of the t-SNE analysis have been supported by an additional PCA analysis not shown here.

We can observe that the real cases fall within the clusters of the training database, but with some at the edges of those clusters which might indicate a need for a more diverse database.

2.2   ML Model

In our previous work [4] a single layer of one soil type was supported. The move to support multiple soil layers with different properties required a new way to encode the information for the model as well as a new model. Moving away from the previously used MLP model, an LSTM approach was first explored. This implied to adapt the encoding of input data, especially for different soil layers. The recurrence steps in the LSTM cells were expected to match with the sequence of soil layers. Unfortunately, this approach yielded disappointing results. We then opted for a CNN based model that ultimately achieved better

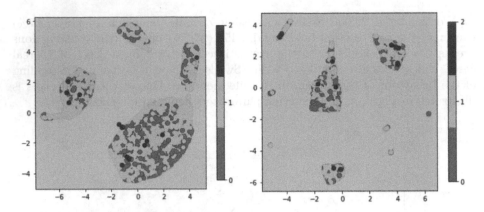

Fig. 4. t-SNE of the training samples (converging (green, 1) and non converging (red, 0)) and the 15 real cases (blue, 2). On the left only the general parameters are considered and on the right only the soil layers (Color figure online)

results. We chose a CNN based model as CNNs are uniquely well adapted for image data, and an excavation can be represented as a data structure similar to an image. Because of this, instead of the previous fixed-sized inputs for the model, we opted to describe the problem as a 1D array representing a cross-section of the excavation.

We model the excavations to a depth of 75 m, with the maximum depth of the excavation itself going to a maximum of 25 m but the wall being able to go below that depth. The height of the modelled case is split into 90 layers. The minimum resolution (or precision) of this model, which corresponds to the maximum *physical* height of a layer, is about 0.83 m. Each layer contains the data about the excavation at a specific depth, divided into 7 channels and consisting of the following information:

- Young's modulus of the soil layer
- Friction angle of the soil layer
- Cohesion of the soil layer
- Presence/absence of a strut with its pre-stress
- Presence/absence of water
- Presence/absence of the wall with its stiffness value
- Lack of ground (defining the depth of the excavation)

Additional information, such as the half-width, load, presence/absence of buttress and the total height of excavation are input directly into the dense layer of the network. We can see that the complexity and diversity of the situations are greatly improved. There is no more limit in the number of soil layers or struts applicable to a given excavation. Currently, we train one model for every value we predict separately. Figure 5 shows the architecture of the CNN model used.

The model starts with 2 blocks of 2 1D Convolutions with dropout (0.2) and average pooling. After extracting the features of the 1D input array we

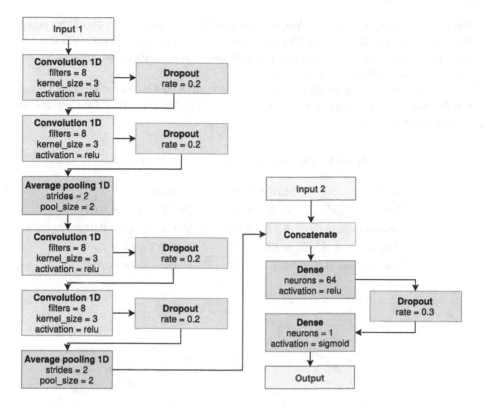

Fig. 5. Architecture of the CNN model used for the prediction

concatenate them with the additional information mentioned previously (half-width, load, etc.) to then go through a dense layer with dropout (0.3) and do the final prediction of the network with a single neuron dense layer. We use the Adam optimizer with a learning rate of 0.0007.

3   Results

In this section, we present the results of our method in terms of accuracy. For the binary convergence value, we measure the Area Under Receiver Operating Characteristics (AUROC) which allows us to measure how well the model can distinguish the two cases. For the four linear values, we use the following metric (1):

$$Error = \frac{|y - y_{pred}|}{k + |y|} \tag{1}$$

Here y is the output of the simulation, y_{pred} is the predicted value of the model and k is a fixed value detailed below. The goal of this metric is to measure the relative error, while at the same time not giving too much weight to errors on small values. For example, a deviation error of 1 cm on a deviation of 1 cm, while

100% is not very important. Yet an error of 30 cm on a 30 cm total deviation, also 100%, is a real problem. For both the settlement and deviation metric we used $k = 0.04$ and for the maximum and minimum moment we used $k = 200$, based on the feedback of domain experts.

For the binary convergence value, we get an AUROC value of 0.992 which can be considered a very good result. The result for the other four values are summarized in Table 1.

Table 1. Accuracy measures of our CNN model

Value	Error	Std. dev.	Measure
Settlement [m]	5.7%	6.4%	Metric (1) ($k = 0.04$)
Deviation [m]	8.3%	9.6%	Metric (1) ($k = 0.04$)
Max moment [kNm]	6.8%	6.8%	Metric (1) ($k = 200$)
Min moment [kNm]	6%	6.4%	Metric (1) ($k = 200$)

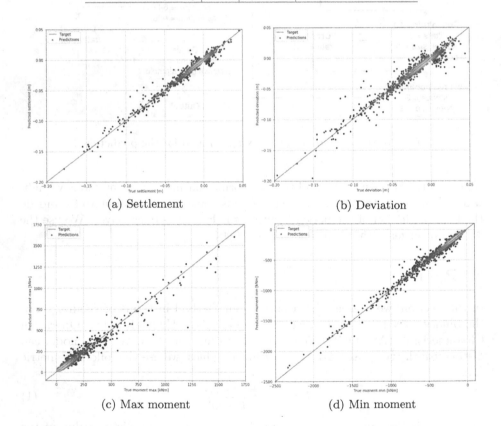

(a) Settlement   (b) Deviation

(c) Max moment   (d) Min moment

Fig. 6. Scatter-plots prediction vs ground-truth on the 4 linear metrics

In Table 1, we can observe that all four other values are predicted in a similar accuracy range (5.7 to 8.3%). To better understand these results, we look at the scatterplots for the four linear metrics. We can see the settlement (Fig. 6a) and deviation (Fig. 6b) at the top and min (Fig. 6c) and max (Fig. 6c) at the bottom of Fig. 6.

The dots in the scatterplot are coloured based on their density, which shows that the ranges with most cases are also the ones that are best predicted. For the binary convergence value, which is a classification problem, we can see the confusion matrix and the AUROC in Fig. 7.

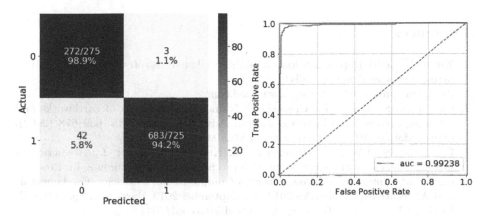

Fig. 7. Confusion matrix and AUROC of the binary classification value

Given the complexity of the new situations compared to the ones in [4], the accuracy is overall very good for all 5 values we are predicting.

4 Discussion

Through this work, we were able to demonstrate that the results described in [4] could be replicated in more complex, real life-like scenarios. We were able to show that by encoding the input of the physics simulation and by using a CNN machine learning model, the more complex scenarios could be predicted with an error of below 10%, which is an acceptable error rate according to geotechnical engineering domain experts, given the uncertainties the numerical simulations work with.

In our work, we did not compare our approach with other AI techniques such as KNN or random forest. However, in [7] it was shown that the neural network approach is the most suitable for this type of problem.

In future work, we would like to increase the complexity of the predicted situations even further, while at the same time exploring better models as well as improve the quality and diversity of the training database, in particular in regards to approaches like active learning [10]. A major part of our future work is also the use of the trained model as an objective function for optimisations, for example using a genetic algorithm. A first version of the genetic algorithm has already been used to create an internal prototype which allows optimising real constructions in seconds instead of days.

In terms of network architectures, we want to explore the possibility and its impact on the accuracy of having a single model that predicts all five values, which would simplify the training and deployment of the model.

Acknowledgment. This publication is written in the framework of the Optisoil project co-funded by the swiss Hasler Foundation.

References

1. Zswalls™, a 2D deep excavation - retaining wall analysis software program (2018). https://www.zsoil.com/zswalls/
2. Zsoil, elmepress (2019). http://www.zsoil.com
3. Commend, S., Geiser, F., Crisinel, J.: Numerical simulation of earthworks and retaining system for a large excavation. Adv. Eng. Softw. **35**, 669–678 (2002). https://doi.org/10.1016/j.advengsoft.2003.10.011
4. Commend, S., Wattel, S., Hennebert, J., Kuonen, P., Vulliet, L.: Prediction of unsupported excavations behaviour with machine learning techniques. In: Proceedings of XIV International Conference on Computational Plasticity, Fundamentals and Applications, COMPLAS 2019, 3–5 September 2019, Barcelona, Spain (CONFERENCE), 7 p. (2019). http://hesso.tind.io/record/4094
5. Goncharova, Y.A., Indrupskiy, I.M.: Replacement of numerical simulations with machine learning in the inverse problem of two-phase flow in porous medium. J. Phys. Conf. Ser. **1391**, 012146 (2019). https://doi.org/10.1088/1742-6596/1391/1/012146
6. van der Maaten, L., Hinton, G.: Visualizing data using t-SNE. J. Mach. Learn. Res. **9**, 2579–2605 (2008)
7. Moiz, A., Pal, P., Probst, D., Pei, Y., Zhang, Y., Som, S., Kodavasal, J.: A machine learning-genetic algorithm (ML-GA) approach for rapid optimization using high-performance computing. SAE Int. J. Commercial Veh. (2018). https://doi.org/10.4271/2018-01-0190
8. Obrzud, R., Preisig, M.: Large scale 3D numerical simulations of deep excavations in urban areas-constitutive aspects and optimization. Mitt. der Geotechnik Schweiz **167**, 57–68 (2013)
9. Paternina-Arboleda, C., Montoya-Torres, J., Fabregas, A.: Simulation-optimization using a reinforcement learning approach, pp. 1376–1383 (2008). https://doi.org/10.1109/WSC.2008.4736213
10. Settles, B.: Active learning literature survey. Computer Sciences Technical report 1648, University of Wisconsin-Madison (2009)
11. Tompson, J., Schlachter, K., Sprechmann, P., Perlin, K.: Accelerating Eulerian fluid simulation with convolutional networks. In: Precup, D., Teh, Y.W. (eds.) Proceedings of the 34th International Conference on Machine Learning. Proceedings of Machine Learning Research, vol. 70, pp. 3424–3433. PMLR, International Convention Centre, Sydney, Australia (2017). http://proceedings.mlr.press/v70/tompson17a.html

Going for 2D or 3D? Investigating Various Machine Learning Approaches for Peach Variety Identification

Anna Wróbel[1]($\boxtimes$), Gregory Gygax[1], Andi Schmid[2], and Thomas Ott[1]

[1] School of Life Sciences and Facility Management, Zurich University of Applied Sciences ZHAW, Wädenswil, Switzerland
{wrob,gygg,ottt}@zhaw.ch
[2] Realisation Schmid, Scharans, Switzerland
andi@realisation-schmid.ch

Abstract. Machine learning-based pattern recognition methods are about to revolutionize the farming sector. For breeding and cultivation purposes, the identification of plant varieties is a particularly important problem that involves specific challenges for the different crop species. In this contribution, we consider the problem of peach variety identification for which alternatives to DNA-based analysis are being sought. While a traditional procedure would suggest using manually designed shape descriptors as the basis for classification, the technical developments of the last decade have opened up possibilities for fully automated approaches, either based on 3D scanning technology or by employing deep learning methods for 2D image classification. In our feasibility study, we investigate the potential of various machine learning approaches with a focus on the comparison of methods based on 2D images and 3D scans. We provide and discuss first results, paving the way for future use of the methods in the field.

Keywords: Peach variety identification · ML classification · 3D scans

1 Introduction

The identification of plant species and varieties has always been an important skill in human culture and a driving factor for agricultural success. The diversity of crops is important for resilient cultivation and ecosystems as well as healthy nutrition, but has come under pressure, e.g. due to monoculture farming methods. Thus, various initiatives aim to preserve the diversity of crops. For specific crops such as peaches or apples the task of identifying varieties requires very specific expertise, and even experts can get ambiguous results. Therefore, genetic identification has become a major tool for variety identification [1]. However, DNA analysis takes time, is relatively expensive and is not entirely unambiguous [2].

Obviously, machine learning methods offer an alternative for the identification of crop varieties. It does not come as a surprise that the identification and classification of different plant species has a long tradition in the literature of neural network applications

© Springer Nature Switzerland AG 2020
F.-P. Schilling and T. Stadelmann (Eds.): ANNPR 2020, LNAI 12294, pp. 257–265, 2020.
https://doi.org/10.1007/978-3-030-58309-5_21

and machine vision (e.g., [3, 4]). However, the identification of varieties within a species is a difficult problem as subtle differences can point to relevant discriminating features [5]. Given its spectacular success and development over the past few years, it is tempting to use deep neural network (DNN) technology based on 2D images for this task [6]. Traditionally, however, for many crops manually calculated 3D shape descriptors have been used by experts since relevant differentiating features of varieties seem to be mainly encoded in the 3D structure of the plants or their seeds or stones respectively [7]. As the advances in technology render an automated approach using 3D scanning feasible, some authors have suggested using 3D scans in combination with targeted feature engineering as a basis for crop identification, e.g. [8].

In this contribution, we present the first, to the best of our knowledge, study to investigate and compare the potential of 2D image-based and 3D scanning-based machine learning approaches for identifying varieties of peaches. As with all crops, the identification problem is of great importance for peach breeding. Peach varieties are expressed in differences in the structure of the peach stones. The goal of the study is to pre-evaluate the potential of different methods that can offer breeders a cheap and fast tool which can complement, if not replace, DNA analysis. Based on the established literature in the field of plant identification, we decided to focus on two different methodological lines. We evaluated the performance of convolutional neural networks (CNN) based on 2D images and we assessed several classification methods (support vector classifier SVC, random forest classifier, linear discriminant analysis LDA, k-nearest neighbor classifier KNN) in combination with 3D descriptors gained from a Fourier analysis of the 3D scans.

Finally, in regard to future practical usage in the field, we focus on "cheap" equipment. Thus, spectral imaging technology was not considered.

2   Materials and Methods

2.1   Peach Sample Preparation, Data Acquisition and Preprocessing

A selection of eight representative varieties of peaches with a total of 190 different fruits were used as the data basis for this study. After carefully cleaning and drying the stones of the peaches, 3D scans and 2D images were taken, according to the protocols described below.

2D Images and Imaging Protocol. For 2D images a commercially available camera (Sony DSC-HX300) was used. Objects were placed in a light tent, always in the same place. Pictures of a resolution of 5184×2920 pixels were taken with the same camera and light settings. Every object was flipped in a repeatable manner resulting in 6 samples per object. The images were further preprocessed in order to obtain centered 256×256 images. Each picture was cropped based on the weighted center of mass and then resized to size 256×256 padded with edge pixels values in height. It was necessary as some of the objects were much longer than wide. The aspect ratio of the images remained unchanged.

3D Scanner and Scanning Protocol. For 3D scanning a commercially available scanner (PT-M scanner from Isra Vision) was used. For scanning, objects were placed in the middle of a turning table. The scanner was set to turn the turntable automatically 16 times. After a full turn, the object was flipped around its longest axis and scanned again using 16 turntable steps. The scans were automatically aligned by the software that comes with the scanner. In case of gross errors (misalignments), the data were discarded and the object was scanned again. The resulting scans were in a form of a collection of meshed points in 3D space. They were exported in the STL format.

As the orientation in space of this representation is not identical for every scanned object, each object was first shifted such that its center of gravity lied in the origin of the 3D coordinate system. Then, it was rotated such that its principal rotation axes were aligned with the axes of the coordinate system. For the first part, the center of gravity was defined as the component-wise mean of every vector. In the next step, the components of the inertia tensor were calculated (assuming the same weight for every point). The inertia tensor is described by the symmetric matrix

$$I = \begin{pmatrix} I_{11} & I_{12} & I_{13} \\ I_{21} & I_{22} & I_{23} \\ I_{31} & I_{32} & I_{33} \end{pmatrix} \tag{1}$$

with $I_{l \neq k} = -\sum x_l x_k$ and $I_{11} = \sum (x_2^2 + x_3^2)$, $I_{22} = \sum (x_1^2 + x_3^2)$, $I_{33} = \sum (x_2^2 + x_1^2)$.

The data were rotated such that the principal axes of the inertia tensor, i.e. the eigenvectors, align with the axes of the coordinate system (see Fig. 1). As the points are not equally spaced, the alignments are not perfect and there can be some variation between scans. Furthermore, since only the axes are aligned and the stones are not symmetric, not all of the stones have the same orientation. Therefore, every object was flipped once around every axis, such that 4 samples per scan were created (original and rotated around the three axes). The data were then transformed from vectors to a spatial grid with binary encoding cells, representing the surface of the stone. The resolution of the grid was 0.1 mm.

The data basis that was used for the classification tasks is summarized in Table 1. For the classes, i.e. peach varieties, an encoding scheme used by the breeder A. Schmid was applied. Additionally, 3 varieties have specific names.

2.2 CNN Approach for 2D Images

Considering the small data set size, transfer learning based on a pretrained convolutional net was used for the classification of the images [9]. We evaluated several image classification models, including InceptionV3, Resnet50, MobileNetV2, VGG16 and VGG19 as the convolutional base, all pretrained on the ImageNet dataset [10, 11]. Classification with VGG16 model was the only successful approach. We experimented with adding 2 to 3 dense layers consisting of 128 up to 512 neurons with dropout ranging from 0.2 to 0.5 on top of the nontrainable convolutional base. Adam and SGD optimizers with categorical cross-entropy loss function and accuracy metrics were investigated. In the final architecture, the VGG16 convolutional base was followed by global max pooling, 30% dropout and two dense layers. The first dense layer consisted of 256 neurons with ReLU activation function while the classifier on top of it was an 8-neuron dense layer with a

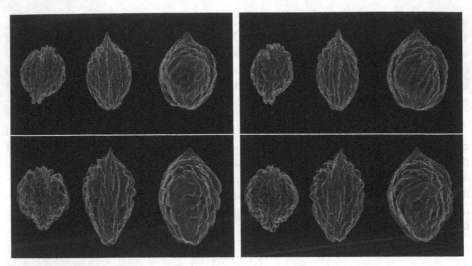

Fig. 1. Projections of 3D scans onto each principal axis for two different varieties. In each row, two different stones of the same variety are shown.

Table 1. Overview data basis.

Peach variety	Number of stones	Number of images	Number of scans
146	25	150	100
303	25	150	100
929	22	132	84[a]
930	25	150	100
349	18	108	72
101 zephir	25	150	100
102 nectaross	25	150	100
103 sweet dream	25	150	100
Total	**190**	**1140**	**756**

([a]for class 929 one stone had to be excluded from the scans due to damaged data).

softmax activation function. The final model was compiled with an SGD optimizer with a learning rate of 0.001 and a momentum of 0.9. The target vector was one-hot encoded. Training data were augmented with ImageDataGenerator.

For each training, 10% of the images were used as a test set, another 10% of the remaining samples were used as a validation set and the rest was used as a training set. The model was trained for 100 epochs with an early stopping condition. The performance of the model was assessed using 10-fold cross-validation with a stratified fold split.

The model was built using Python with TensorFlow2 and trained on a Tesla P100 GPU.

2.3   Classification Methods for 3D Scans

We evaluated several different machine learning algorithms in combination with a feature selection procedure based on a 3D Fourier analysis. The evaluated approaches are Linear Discriminant Analysis (LDA), Support Vector Classifier (SVC), Random Forest Classifier (RF) and k-Nearest Neighbors Classifier (KNN). All these approaches were applied after a preceding feature engineering step based on 3D Fourier coefficients. To this end, the spatial grid of the scans was transformed using a fast Fourier transform. From the obtained Fourier domain, the (50, 50, 50) 'corners' of the Fourier spectrum were considered. Since the Fourier spectrum exhibits point symmetry and thus redundancy, only the 4 lower corners were taken into account, the imaginary parts of the coefficients were discarded and only the real parts were used. Important features were then selected using ANOVA by keeping only the frequencies with a p-value below the 0.9999 quantile of all p-values. In this way, 100 frequencies were selected that were then used as a feature vector. Before training the classifiers, each component was scaled by the z-transform (centered around 0, with std $= 1$). Only the training data were used for fitting the scaler.

Hyperparameters of the models were tuned with GridSearchCV. For final classification we used LDA with singular value decomposition solver, RF with Gini impurity criterion and KNN with 5 nearest neighbors and Euclidean distance. The SVC model achieved the highest classification accuracy with radial basis function kernel type, regularization parameter increased from 1 to 40 and kernel coefficient gamma set to 0.01.

The models were built using Python with scikit-learn and trained on an Intel Xeon-based cluster computing node. They were evaluated using 10-fold cross-validation with a stratified fold split.

3   Results

The results for all the methods based on 10-fold cross-validation are summarized in Table 2. Generally, the accuracy of the best methods is around 90% with the best 3D-based methods slightly above (92.2% for LDA and 91.9% for SVM) and the 2D CNN method slightly below this value (89.2%). In comparison, the accuracy of the RF and KNN models is significantly lower (84.1% and 83.1% respectively).

Table 2. Accuracy of different methods based on 10-fold cross-validation

Method	Accuracy (mean ± stdv)
2D CNN	0.892 ± 0.036
3D SVC	0.919 ± 0.052
3D LDA	0.922 ± 0.054
3D Random Forest	0.841 ± 0.095
3D KNN	0.831 ± 0.084

To further investigate and understand these results we take a look at the normalized confusion matrices, averaged over the 10-fold cross-validation (Fig. 2: 2D-based CNN; Fig. 3: 3D-based methods)

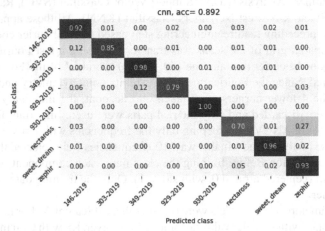

Fig. 2. CNN normalized confusion matrix averaged over all 10 folds of the k-fold cross-validation with a stratified fold split.

For the 2D-based CNN approach, the main difficulty seems to occur for the discrimination of the variety *nectaross* from *zephyr* as in average 27% of images of the *nectaross* class are classified as *zephyr* (Fig. 2). This problem seems to be much less pronounced for the discrimination based on 3D scanning (Fig. 3). In particular, in the case of SVC, small misclassification errors seem to occur for various classes in a rather arbitrary fashion, not hinting at a specific problem of two classes. A possible explanation for the problem of discriminating *nectaross* from *zephyr* in the 2D case is revealed when looking at the actual stones and their respective images (Fig. 4). The stones of the two varieties exhibit a similar structure. They mainly differ in size. The size information is, however, lost for the 2D images as they are automatically rescaled. In turn, the analysis of the Fourier-based feature vectors in the 3D approach shows that the low frequencies describing the coarse-grained structure of the stones play an important role for the classification. Hence, size is a feature that is definitely exploited in the 3D-based approach.

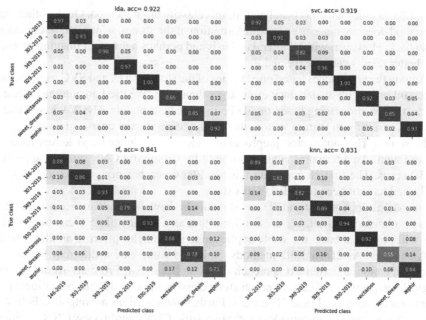

Fig. 3. Normalized confusion matrices for the 3D-based machine learning methods (from top left to bottom right: LDA, SVC, RF, KNN). Each matrix is averaged over all 10 folds of the k-fold cross-validation with a stratified fold split.

Fig. 4. Example of the 2 sorts that are misclassified most frequently. On the left *nectaross*, on the right *zephyr*.

4   Conclusions and Future Work

Species and variety identification is an important problem in crop breeding, which can potentially benefit greatly from machine learning-based pattern recognition methods. We investigated several machine learning approaches for peach variety identification based on 2D images and 3D scans of peach stones. The goal of the study was to learn more about the potential, strengths and weaknesses of different approaches for this particular problem. Our findings can be summarized as follows.

Despite a relatively small data basis with 190 peach stones, a 2D-based CNN approach looks promising. In fact, when looking at the images, an accuracy of nearly 90% seems surprising and speaks for the potential of the method. On the one hand, many stone sorts differ in color which is beneficial for this classification method. On the other hand, the method does not make use of size information of the stones, which can be considered a drawback. However, with a larger data basis and perhaps a different imaging protocol, even better results can be expected.

Among the approaches based on 3D scans of the stones, LDA and SVC were the most successful ones with an accuracy even larger than 90%. We used an automated feature engineering preprocessing based on 3D Fourier analysis and an ANOVA-based feature selection, resulting in feature vectors that mainly describe the coarse-grained structure of the stones. This automated preprocessing can be challenged and leaves room for improvement. An alternative could be offered by volumetric CNN directly applied on the 3D scans. The first attempt with this idea, however, has not yet shown conclusive results and has not been included in this study.

In conclusion, both 2D and 3D based methods showed promising accuracies on the basis of a limited data set. We are confident that an accuracy of 95% can be achieved. This will provide a basis for stable applications in the field, offering an alternative to DNA analysis. As 2D and 3D methods exploit to some extent different features for the classification, a combined approach could be beneficial. From an overall method-ological perspective, the case of peach stone classification could thus be an ideal play-ground for combining 2D and 3D images. Further investigations are ongoing. Following approaches are taken into consideration: volumetric CNNs, multi-view CNNs aggregat-ing 2D projections of the 3D scans, combination of the two [12] or a combination with 2D images.

Acknowledgements. We thank Doris Berchtold and Matteo Delucchi for useful hints and their support. This work was supported by the Innosuisse Innocheck Nr 33954.1 INNO-LS.

References

1. Arús, P., Verde, I., Sosinski, B., Zhebentyayeva, T., Abott, A.G.: The peach genome. Tree Genet. Genomes **8**, 531–547 (2012). https://doi.org/10.1007/s11295-012-0493-8
2. Singh, B.D., Singh, A.K.: Marker-Assisted Plant Breeding: Principles and Practices. Springer, New Delhi (2015). https://doi.org/10.1007/978-81-322-2316-0
3. Wäldchen, J., Mäder, P.: Plant species identification using computer vision techniques: a systematic literature review. Arch. Comput. Methods Eng. **25**(2), 507–543 (2017). https://doi.org/10.1007/s11831-016-9206-z
4. Gan, Y.Y., Hou, C.S., Zhou, T., Xu, S.F.: Plant identification based on artificial intelligence. Adv. Mater. Res. **255–266**, 2286–2290 (2011)
5. Wäldchen, J., Rzanny, M., Seeland, M., Mäder, P.: Automated plant species identification – trends and future directions. PLoS Comput. Biol. **14**(4), e1005993 (2018)
6. Carranza-Rojas, J., Goeau, H., Bonnet, P., Mata-Montero, E., Joly, A.: Going deeper in the automated identification of Herbarium specimens. BMC Evol. Biol. **17**, 181 (2017)
7. Rivera, A., Roselló, S., Casanas, F.: Seed curvature as a useful marker to transfer morphologic, agronomic, chemical and sensory traits from Ganxet common bean (Phaseolus vulgaris L.). Sci. Hortic. **197**, 476–482 (2015)
8. Karasik, A., Rahimi, O., David, M., Weiss, E., Drori, E.: Developement of a 3D seed mor-phological tool for grapevine variety identification, and its comparison with SSR analysis. Sci. Rep. **8**, 6545 (2018)

9. Soekhoe, D., van der Putten, P., Plaat, A.: On the impact of data set size in transfer learning using deep neural networks. In: Boström, H., Knobbe, A., Soares, C., Papapetrou, P. (eds.) IDA 2016. LNCS, vol. 9897, pp. 50–60. Springer, Cham (2016). https://doi.org/10.1007/978-3-319-46349-0_5

10. Krizhevsky, A., Sutskever, I., Hinton, G.E.: ImageNet classification with deep convolutional neural networks. In: Advances in Neural Information Processing Systems, pp. 1097–1105 (2012)

11. He, K., Zhang, X., Ren, S., Sun, J.: Deep residual learning for image recognition. arXiv:1512.03385v1 (2015)

12. Hedge, V., Zadeh, R.: FusionNet: 3D object classification using multiple data representations. arXiv:1607.05695v3 (2016)

A Transfer Learning End-to-End Arabic Text-To-Speech (TTS) Deep Architecture

Fady K. Fahmy$^{(\boxtimes)}$, Mahmoud I. Khalil , and Hazem M. Abbas

Department of Computer and Systems Engineering,
Ain Shams University, Cairo, Egypt
fadykhalaf01@gmail.com, {mahmoud.khalil,hazem.abbas}@eng.asu.edu.eg

Abstract. Speech synthesis is the artificial production of human speech. A typical text-to-speech system converts a language text into a waveform. There exist many English TTS systems that produce mature, natural, and human-like speech synthesizers. In contrast, other languages, including Arabic, have not been considered until recently. Existing Arabic speech synthesis solutions are slow, of low quality, and the naturalness of synthesized speech is inferior to the English synthesizers. They also lack essential speech key factors such as intonation, stress, and rhythm. Different works were proposed to solve those issues, including the use of concatenative methods such as unit selection or parametric methods. However, they required a lot of laborious work and domain expertise. Another reason for such poor performance of Arabic speech synthesizers is the lack of speech corpora, unlike English that has many publicly available corpora (LjSpeech, https://keithito.com/LJ-Speech-Dataset/., Blizzard 2012, http://www.cstr.ed.ac.uk/projects/blizzard/2012/phase_one/.) and audiobooks. This work describes how to generate high quality, natural, and human-like Arabic speech using an end-to-end neural deep network architecture. This work uses just ⟨text, audio⟩ pairs with a relatively small amount of recorded audio samples with a total of 2.41 h. It illustrates how to use English character embedding despite using diacritic Arabic characters as input and how to preprocess these audio samples to achieve the best results.

Keywords: Tacotron 2 · WaveGlow · Arabic text-to-speech · Speech synthesis · Deep learning · Neural networks

1 Introduction

Speech synthesis has been a challenging task for decades. Conventional text-to-speech (TTS) systems are usually made up of several components connected through a pipeline that includes text analysis frontends, acoustic models, and audio synthesis models. Building each component in a conventional TTS system often requires comprehensive domain expertise and a lot of laborious work like feature engineering and annotation. Besides, errors generated by each component

© Springer Nature Switzerland AG 2020
F.-P. Schilling and T. Stadelmann (Eds.): ANNPR 2020, LNAI 12294, pp. 266–277, 2020.
https://doi.org/10.1007/978-3-030-58309-5_22

propagate to later stages, making it hard to identify the source of the final perceived error.

Researchers have adopted the use of concatenative speech synthesis [1,2] for years. The idea is based on selecting and concatenating units (phonemes) from a large database to generate intelligible speech. Such units could be any of the following: phones[1] diphones[2] half-phones, syllables, morphemes, words, phrases, or sentences. Generally, the longer the unit, the larger the size of the database that must cover the unit with different prosodies. The drawbacks of concatenative methods for speech synthesis are (a) they need massive databases for large unit size, (b) noise captured while recording units may degrade the quality of synthesized speech since units recorded are represented as it is while synthesizing, and finally (c) the massive amount of labeling and recording.

Statistical parametric speech synthesis based on Hidden Markov Model (HMM) [3,4] showed an increase in adoption rate and popularity over time. It solved a lot of problems of concatenative methods such as (a) modeling prosodic variation by modifying HMM parameters, thus solving the problem of large databases, (b) it has proved to have fewer word error rates which lead to better understandably, and (c) it is more robust because the pre-recorded units in unit selection synthesis could be recorded in different environment adhering different noise profiles. The drawbacks of HMM-based synthesis may include (a) requiring a lot of feature engineering and domain expertise, and (b) generated speech sounds more robotic than speech generated by unit selection speech synthesis.

Deep neural network architectures have proved extraordinary efficient at learning the inherent features of data. WaveNet [5] is a generative model for generating waveforms based on PixelCNN [6]. It has outperformed production level parametric methods in terms of naturalness. Still, it has two significant drawbacks: (a) it requires conditioning on linguistic features from an existing TTS system, so it is not a fully end-to-end system, and (b) it synthesizes speech very slowly due to the auto-regressive nature of the architecture. Deep voice [7] is another example of deep neural architectures. It has proven high performance, production-level quality, and real-time synthesis. It consists of five stages, namely a segmentation model for locating phoneme boundaries, a grapheme-tophoneme conversion model, a phoneme duration, a fundamental frequency prediction model, and an audio synthesis model. Deep Voice is a step towards a genuinely end-to-end neural network architecture.

With the introduction of end-to-end architectures such as Tacotron [8], much laborious work to synthesize speech is alleviated. Such examples for laborious work include feature engineering, and human annotation (although a slight human annotation is needed to prepare the ⟨text, audio⟩ pairs for training). Tacotron is a generative text-to-speech model based on a seq-to-seq model with attention mechanism [9] taking characters as input and producing

---

[1] Distinct speech sound or gesture, regardless of whether the exact sound is critical to the meanings of words.

[2] Consists of two connected half phones that start in the middle of the first phone and end in the middle of the second phone.

audio waveforms. Tacotron uses content-based attention [10], where it concatenates context vector with attention RNN cell output to provide an input to decoder RNNs. Tacotron 2 [11] is a natural evolution of Tacotron. It offers a unified purely neural network approach and eliminates the non-neural network parts used previously by Tacotron, such as the Griffen-Lim reconstruction algorithm to synthesize speech. Tacotron 2 consists of two main components, (a) recurrent seq-to-seq generative model with attention, and (b) a modified Wavenet acting as a vocoder to synthesize speech signal. Tacotron 2 uses hybrid attention [12] (both location-based and content-based attention).

This paper describes how to use a modified deep architecture from Tacotron 2 [11] to generate mel-spectrograms from Arabic diacritic text as an intermediate feature representation followed by a WaveGlow architecture acting as a vocoder to produce a high-quality Arabic speech. The proposed model is trained using a published pre-trained Tacoron 2 English model using a dataset with a total of 2.41 h of recorded speech.

The rest of this paper is organized as follows: Sect. 2 presents a review of related works in the Arabic TTS domain. Sect. 3 describes the proposed model architecture, including the two main components, feature prediction network and WaveGlow, while Sect. 4 introduces the training setup and procedures, issues faced in training, and quantitative and qualitative analysis evaluation of the results. Finally, the paper is concluded in Sect. 5.

2   Arabic TTS Works

Many works are covering Arabic text-to-speech synthesis to generate a good and human-like speech. In [13], Y. A. El-Imam uses a set of sub-phonetic elements as the synthesis units to allow synthesis of unlimited-vocabulary speech of good quality. The input to the system is an Arabic diacritic spelling or simple numeric expressions.

Abdel-Hamid Ossam et al. in [14], managed to improve the synthesized Arabic speech using an HMM-based approach. They used a statistical model to generate Arabic speech parameters such as spectrum, fundamental frequency (F0), and phonemes duration. Then, the authors applied a multi-band excitation model and used samples extracted from spectral envelop as spectral parameters.

Speech synthesis using diacritic text such as [15] has gained a lot of momentum because there is a lack of Arabic diacritic database for speech synthesis. The work discusses two methods to recognize appropriate diacritic marks for Arabic text: a machine learning approach and a dictionary method. This work uses a statistical parameter approach using non-uniform unit size for speech synthesis. It employs variable-sized units, as it has proven to be more effective than using fixed-size units such as phonemes and diphonemes. It partially solves some problems of classical statistical parameter methods. Such issues are speech quality, articulatory effect, and discontinuity effect. This work aimed to build an Arabic TTS system with the integration of diacritization system.

Studying Arabic phonetics [16] for speech synthesis and corpus design is vital to provide a corpus that has excellent coverage of phonetics and phonology. We have used the corpus generated from [16] in the training phase of the spectrogram prediction network model. We have also used another technique in this work to phonetize diacritic Arabic characters as part of training the spectrogram prediction network.

The work [17], by Imene Zangar and Zied Mnasri, uses Deep neural networks (DNN) for duration modeling for Arabic speech synthesis. In this work, the authors compare duration modeling using Hidden Markov Model (HMM) and duration modeling based on deep neural network of different architectures to minimize the root mean square prediction error (RMSE). They concluded that using DNN for modeling duration outperformed HMM-based modeling from the HTS toolkit and the DNN-based modeling from the MERLIN toolkit.

3 Model Architecture

Unlike conventional methods for speech synthesis, end-to-end neural network architectures not only alleviate the need for extensive domain expertise and laborious feature engineering, but they also require minimal human annotation. They can be conditioned for any language, gender, or sentiment. Conventional TTS synthesizers consist of many stages, each trained separately. This can give rise to making each component's error cascade to later stages. End-to-end architectures are structured as a single component and thus can become more robust.

In this work, a slightly modified model that is described in [11] is adopted where the *Wavenet* part is replaced with a flow-based implementation of *Waveglow* [18]. Hence, the proposed model shown in Fig. 1 consists of two components:

1. A sequence-to-sequence architecture spectrogram prediction network, with attention which takes a diacritic Arabic text as input and predicts the corresponding mel-spectrogram as output.
2. A flow-based implementation of WaveGlow which takes the mel-spectrograms as input and generates a time-domain waveform of the input text.

There are many advantages of using mel-frequency spectrograms[3] as an intermediate feature representation between spectrogram prediction network and WaveGlow. They include

(a) mel-spectrograms can be computed easily from time domain waveforms, making it easy to train each of the two components separately.
(b) they are easier to train compared to waveforms as they are phase invariant and thus training can be done using simple loss functions such as squared loss.

---

[3] A spectrogram is a visual representation of the spectrum of frequencies of a signal as it varies with time.

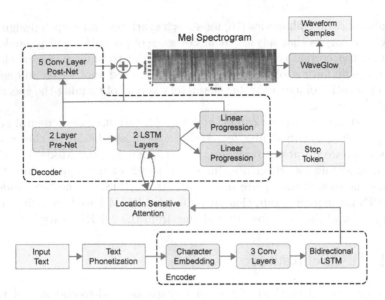

Fig. 1. Block diagram of the spectrogram prediction network with WaveGlow, it takes diacritic Arabic characters as input and produces audio waveform as output [11].

(c) mel-frequency spectrograms are related to linear-frequency spectrograms. One can obtain a mel-frequency spectrogram from a linear-frequency spectrogram by converting the frequency axis to log scale and the "colour" axis, the amplitude, to decibels.

(d) Mel-frequency spectrograms use mel-frequency scale, they can emphasize details on lower frequencies, which is essential for speech naturalness. It also gives less attention to higher frequencies which are not critical for human perception.

(e) It is straightforward for WaveGlow to be conditioned on mel-frequency spectrogram to generate a good quality speech.

3.1   Spectrogram Prediction Network

As shown in Fig. 1, the spectrogram prediction network is a sequence-to-sequence architecture. It consists of an encoder that creates an internal representation of the input signal, which is fed to the decoder to generate the predicted mel-spectrogram. The encoder is made of three parts: character embedding, three convolution layers, and bidirectional LSTM. It takes character sequence as input and produces a hidden feature vector representation. The decoder is made of a two-layer LSTM network, two-layer pre-net, five Conv-layer post-net, and linear progression. It consumes the hidden feature vector representation produced by the encoder and generates the mel-spectrograms of given input characters. Since the diacritic Arabic text is used as an input, a text phonitization block is employed to transform the Arabic characters to another Unicode character set.

The following block in the architecture is an embedding layer (512-dimensional vector) which represents each character symbol numerically. The output of the embedding layer is fed to three convolutional layers, each of 512 filters of dimension 5×1 to span five characters and model long-term contexts (N-gram). Each convolutional layer is followed by batch normalization [19] and a ReLU activation [20]. Tensors produced by the convolutional blocks are fed to bi-directional LSTM of 512 units (256 in each direction). Forward and backward results are concatenated to generate encoded features to be supplied to the decoder.

Spectrogram prediction network uses a hybrid attention model described in [12]. The reason why an attention mechanism is necessary for the spectrogram prediction network is solving long sequence problems (long character sequence), as it is hard for encoder-decoder architecture without attention to memorize a very long input sequence. Accordingly, the performance of the architecture without attention mechanism will eventually deteriorate with long sequences. Attention mechanism solves the problem of long sequences by attending on a part of the sequence (using attention weights) just like what human does when trying to figure out a long sequence. As shown in Fig. 2. At each decoder step, to form the context vector and update the attention weights attention uses the following: (a) the projection of the previous hidden state of decoder RNN's network onto a fully connected layer, (b) the projection of the output of the encoder data on a fully connected layer, and (c) the additive attention weights. The context vector C_i is computed by multiplying the encoder outputs, h_j, and the attention weights, α_{ij}, as in Eq. 1

$$c_i = \sum_{j=1}^{T_x} \alpha_{ij} h_j \tag{1}$$

$$\alpha_{ij} = \frac{\exp(e_{ij})}{\sum_{k=1}^{T_x} \exp(e_{ik})} \tag{2}$$

$$e_{ij} = \mathbf{w}^T \tanh\left(W_{s_{i-1}} + V_{h_i} + \mathbf{b}\right) \tag{3}$$

where α_{ij} is attention weight, and e_{ij} is an energy function. W and V are matrices, while $\mathbf{w}$ and $\mathbf{b}$ are vectors and they are all trainable parameters.

The output of the decoder layer is then fed to pre-net, which consists of two fully connected layers of 256 hidden ReLU units, then passed through 2 unidirectional LSTM of 1204 units. The concatenation of LSTM output and context vector is projected to a linear transformation to predict mel-spectrogram, which is passed to a five-layer post-net. A scaler (stop token) is calculated in parallel by projecting concatenation of context vector with the decoder LSTM output and passing them through a sigmoid activation to predict when to stop generating speech at inference time. Mel-spectrograms are computed using 50 ms frame hub, and a "han" window function.

All convolutional layers are regulated using dropout [21], while LSTM layers are regulated using zoneout [22].

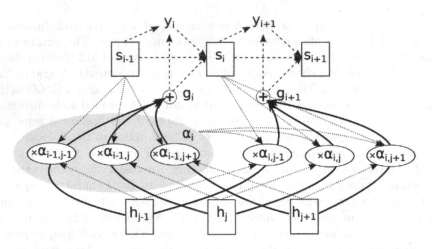

Fig. 2. Hybrid attention mechanism used in spectrogram prediction network [12]

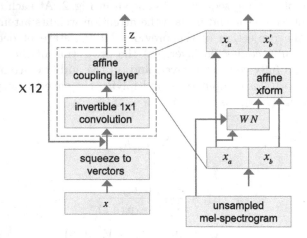

Fig. 3. Block diagram of WaveGlow Vocoder [18], it takes a spectrogram as input and produces an audio waveform.

3.2 WaveGlow Vocoder

WaveGlow is a flow-based generative network that combines insights from glow [23] and Wavenet. According to the authors of [18], it generates speech with quality as good as the best open-source implementations of WaveNet. However, it is much faster as it is not auto-regressive and could fully utilize GPUs. It is trained alongside with the spectrogram prediction network using the original mel-spectrograms as an input and the audio clips as the output. WaveGlow can be easily conditioned on mel-spectrograms to generate high-quality waveforms.

The forward path, as shown in Fig. 3, takes a group of eight samples as a vector as in [23], then passes the output into twelve steps of flow, each step consisting of 1×1 convolution followed by an affine coupling layer. The affine coupling layer acts as an invertible neural network [24]. Half of the channels are used as inputs, while the block, WN, can be any transformation. The coupling layer preserves invertibility for the overall network. Invertible 1×1 convolution is added before the affine layer to mix information between channels. The weights W of the invertible convolution are initialized to be orthogonal[4] and thus they are also invertible.

كلمةٌ	{k l m t u1 n}
عَلَى حَدِّ قَوْلِ الْبَاحِثَةِ	{E a l a} {H a dd i0} {q A w l i0} {< a l b aa H i0 ∧ a t i0}
لَا تَثِقْ فِي كُلِّ مَا تَرَاهُ	{l a} {t a ∧ l1 q} {f ii0} {k u0 ll i0} {m a} {t a r aa h u0}

Fig. 4. Phonitization examples. The left side of the graph represents diacritic Arabic words, while the right side represents the corresponding Unicode character symbols.

4 Experimental Results and Analysis

4.1 Training Setup

We have trained the Spectrogram prediction network on Nawar Halabi's Arabic Dataset[5] [16], which contains about 2.41 h of Arabic speech, a total of 906 utterances, and 694556 frames. The dataset consists of ⟨text, audio⟩ pairs. The input text is diacritic Arabic characters, while the output is a 16-bit 48 kHz PCM audio clip is with a bit-rate of 768 kbps. Since the dataset is relatively small, it is split into a 95% training set and a 5% validation set. The training was executed on a supercomputing environment.[6]

The spectrogram prediction network was trained separately using diacritic Arabic characters as input and original mel-spectrograms at the decoder side as the target. Because of the small dataset size, we were not able to learn character embedding, nor the attention between encoder and decoder perfectly. Also, the quality of the generated speech was poor. As a result, we utilized transfer learning from English by (a) transforming diacritic Arabic words into English characters using an open-source phonitization algorithm,[7] refer to Text Phonitization in Fig. 1, phonitization examples at Fig. 4, (b) using a pre-trained English model[8]

---

[4] Orthogonal matrix is a square matrix whose columns and rows are orthogonal unit vectors.

[5] http://en.arabicspeechcorpus.com/.

[6] https://hpc.bibalex.org/.

[7] https://github.com/nawarhalabi/Arabic-Phonetiser.

[8] https://drive.google.com/file/d/1c5ZTuT7J08wLUoVZ2KkUs_VdZuJ86ZqA/view.

with the learned English character embedding to be able to fully train the attention mechanism. The audio training clips have been down-sampled to 22050 Hz in to employ the same audio parameters as those in the open-source implementation[9] (trained on LJSpeech dataset) such as the hop length and the filter length. Silence moments (below 60 dB) of each training sample were removed using a frame size of 1024 and a hop size of 256, which greatly helped to align the attention graph shown in Fig. 5.

Other training parameters are: a batch size of 8 on 2 GPUs, Adam optimizer [25] with $\beta_1 = 0.9$, $\beta_2 = 0.999$, and $\epsilon = 10^{-6}$, a constant learning rate of 10^{-3}, and L_2 regularization with weight 10^{-6}. A training epoch took, on average, about 15 min while only about 2 s were needed to generate a waveform.

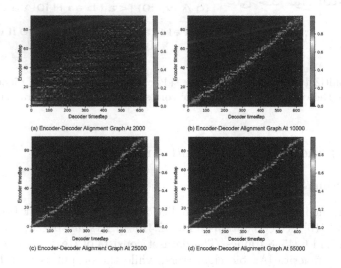

(a) Encoder-Decoder Alignment Graph At 2000

(b) Encoder-Decoder Alignment Graph At 10000

(c) Encoder-Decoder Alignment Graph At 25000

(d) Encoder-Decoder Alignment Graph At 55000

Fig. 5. Alignment graphs at different steps of training

4.2 Analysis

For quantitative analysis, both training and validation losses were assessed as metrics. Simple mean square error loss (MSE) between predicted and target mel-spectrograms was calculated.

For qualitative analysis, the attention alignment graph was used as a metric. The Attention alignment graph is an indication of how the decoder is attending correctly to encoder input. Encoder reads input step-by-step and produces status vectors. Decoder reads all status vectors and produces audio frames step-by-step. A good alignment simply means: An "A" sound generated by the decoder should be the result of focusing on the vector generated by the encoder from reading "A" character. The diagonal line is the result when audio frames are generated by focusing (paying attention) on the correct input characters.

---

[9] https://github.com/NVIDIA/tacotron2.

Figure 5 shows that the spectrogram prediction network was continually improving in learning attention throughout the training process. It helped in eliminating some pronunciation errors as well as removing some pauses in the generated speech. Our model started to pick up alignment after about 40 epochs of training.

Further qualitative analysis was carried out by using human ratings similar to Amazon's Mechanical Turk.[10] We used a pre-trained model of WaveGlow[11] to infer ten randomly selected samples of spoken sentences. Each sample is rated by 26 raters on a scale from 1 to 5 with a step of 0.5 to calculate a subjective mean opinion score (MOS) for audio naturalness. Each evaluation is conducted independently from each other. Table 1 compares the proposed architecture with other architectures samples from [26] such as concatenative methods with HMMs and Tacotron with the Griffin-Lim algorithm as a synthesizer. Figure 6 shows the detailed raters' review for each of the test samples where each entry is the sum of all 26 rates divided by the number of raters (26).

Table 1. MOS evaluation for different system architectures.

System Architecture	MOS
Concatenative methods with HMMs	3.89
Tacotron 1 with Griffin-Lim algorithm	4.02
Tacotron 2 with WaveGlow (proposed)	4.21

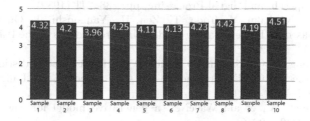

Fig. 6. Human judgement over ten randomly selected test samples.

5   Conclusions and Future Work

This paper describes how to use the Tacotron 2 architecture to generate intermediate feature representation from Arabic diacritic text using a pre-trained English model and a total of 2.41 h of recorded speech, followed by WaveGlow as a vocoder to synthesize high-quality Arabic speech. It also shows the viability of how to apply transfer learning from English text-to-speech to Arabic

---

[10] https://www.mturk.com/.
[11] https://drive.google.com/file/d/1rpK8CzAAirq9sWZhe9nlfvxMF1dRgFbF/view.

text-to-speech successfully in spite of the fact that the two languages are quite different in terms of character level embedding and language phonemes. It also describes how to preprocess audio speech training data to gain a plausible generated speech.

There are many possible future enhancements for this work. They may include integrating Arabic diacrtizer, which will reduce the amount of manual work needed to diacrtise a given Arabic text. Another possible enhancement is to model speech prosody (intonation, stress, and rhythm) for expressive and more human-like speech. Modeling prosody could be done using an architecture similar to Tacotron but with additional neural networks to embed prosody into the encoded text before encoding the information using the same sequence-to-sequence architecture. And last but not least, using a much larger dataset to train the model which will generally produce more plausible speech quality.

Acknowledgment. The authors would like to thank The Bibliotheca Alexandrina for providing the computing resources through their Supercomputing Facility https://hpc.bibalex.org.

References

1. Hunt, A.J., Black, A.W.: Unit selection in a concatenative speech synthesis system using a large speech database. In: 1996 IEEE International Conference on Acoustics, Speech, and Signal Processing Conference Proceedings, pp. 373–376 (1996)
2. Hamon, C., Mouline, E., Charpentier, F.: A diphone synthesis system based on time-domain prosodic modifications of speech. In: International Conference on Acoustics, Speech, and Signal Processing, pp. 238–241 (1989)
3. Tokuda, K., Nankaku, Y., Toda, T., Zen, H., Yamagishi, J., Oura, K.: Speech synthesis based on hidden Markov models. In: Proceedings of the IEEE, pp. 1234–1252 (2013)
4. Yu, K., Young, S.: Continuous F0 modeling for HMM based statistical parametric speech synthesis. In: IEEE Transactions on Audio, Speech, and Language Processing, pp. 1071–1079 (2011)
5. van den Oord, A., et al.: WaveNet: a generative model for raw audio. CoRR arXiv:1609.03499 (2016)
6. van den Oord, A., Kalchbrenner, N., Kavukcuoglu, K.: Pixel recurrent neural networks. CoRR arXiv:1601.06759 (2016)
7. Arik, S.O., et al.: Deep voice: real-time neural text-to-speech. CoRR arXiv:1702.07825 (2017)
8. Wang, Y., et al.: Tacotron: a fully end-to-end text-to-speech synthesis model. CoRR arXiv:1703.10135 (2017)
9. Sutskever, I., Vinyals, O., Le, Q.V.: Sequence to sequence learning with neural networks. In: Advances in Neural Information Processing Systems, pp. 3104–3112 (2014)
10. Bengio, S., Vinyals, O., Jaitly, N., Shazeer, N.: Scheduled sampling for sequence prediction with recurrent neural networks. In: Advances in Neural Information Processing Systems, pp. 1171–1179 (2015)
11. Shen, J., et al.: Natural TTS synthesis by conditioning WaveNet on Mel Spectrogram predictions. CoRR arXiv:1712.05884 (2017)

12. Chorowski, J.K., Bahdanau, D., Serdyuk, D., Cho, K., Bengio, Y.: Attention-based models for speech recognition. In: Advances in Neural Information Processing Systems, vol. 28, pp. 577–585 (2015)
13. El-Imam, Y.A.: An unrestricted vocabulary Arabic speech synthesis system. In: IEEE Transactions on Acoustics, Speech, and Signal Processing, pp. 1829–1845 (1989)
14. Abdel-Hamid, O., Abdou, S., Rashwan, M.: Improving Arabic HMM based speech synthesis quality. In: INTERSPEECH 2006 and 9th International Conference on Spoken Language Processing, INTER- SPEECH 2006 - ICSLP (2006)
15. Rebai, I., BenAyed, Y.: Arabic speech synthesis and diacritic recognition. Int. J. Speech Technol. **19**, 485–494 (2016)
16. Halabi, N., Wald, M.: Phonetic inventory for an Arabic speech corpus. In: Proceedings of the Tenth International Conference on Language Resources and Evaluation, LREC'16, pp. 734–738 (2016)
17. Zangar, I., Mnasri, Z., Colotte, V., Jouvet, D., Houidhek, A.: Duration modeling using DNN for Arabic speech synthesis. In: 9th International Conference on Speech Prosody (2018)
18. Prenger, R., Valle, R., Catanzaro, B.: WaveGlow: a flowbased generative network for speech synthesis. CoRR arXiv:1811.00002 (2018)
19. Ioffe, S., Szegedy, C.: Batch normalization: accelerating deep network training by reducing internal covariate shift. In: International Conference on Machine Learning (ICML) (2015)
20. Li, Y., Yuan, Y.: Convergence analysis of two-layer neural networks with ReLU activation. In: Advances in Neural Information Processing Systems, vol. 30, pp. 597–607 (2017)
21. Srivastava, N., Hinton, G., Krizhevsky, A., Sutskever, I., Salakhutdinov, R.: Dropout: a simple way to prevent neural networks from overfitting. J. Mach. Learn. Res. **15**, 1929–1958 (2014)
22. Krueger, D., et al.: Zoneout: regularizing RNNs by randomly preserving hidden activations. In: International Conference on Learning Representations (ICLR) (2016)
23. Kingma, D.P., Dhariwal, P.: Glow: generative flow with invertible 1 × 1 convolutions. CoRR arXiv:1807.03039 (2018)
24. Dinh, L., Sohl-Dickstein, J., Bengio, S.: Density estimation using Real NVP. CoRR arXiv:1605.08803 (2016)
25. Kingma, D.P., Ba, J.: Adam: a method for stochastic optimization. In: Bengio, Y., LeCun, Y. (eds.) 3rd International Conference on Learning Representations, ICLR 2015, San Diego, CA, USA, 7–9 May 2015, Conference Track Proceedings (2015)
26. Ali, A.H., Magdy, M., Alfawzy, M., Ghaly, M., Abbas, H.: Arabic speech synthesis using deep neural networks. In: Proceedings of the International Conference on Communications, Signal Processing and their Applications, ICCSPA' 20 (2020, to appear)

ML-Based Trading Models: An Investigation During COVID-19 Pandemic Crisis

Dalila Hattab[✉]

Financial Services Lab, Worldline (Seclin), Noyelles-lès-Seclin, France
dalila.hattab@worldline.com

Abstract. The aim of this paper is to investigate the effect of volatility surges during the COVID-19 pandemic crisis on long-term investment trading rules. These trading rules are derived from stock return forecasting based on a Multiple Step Ahead Direct Strategy, and built on the combination of machine learning models and the Autoregressive Fractionally Integrated Moving Average (ARFIMA) model. ARFIMA has the feature to account for the long memory and structural change in conditional variance process. The machine learning models considered are a particular Neural Network model (MLP), K-Nearest Neighbors (KNN) and Support Vector Regression (SVR). The trading performances of the produced models are evaluated in terms of economical metrics reflecting profitability and risk like: Annualized Return, Sharpe Ratio and Profit Ratio. The hybrid model performances are compared to the simple machine learning models and to the classical ARMA-GARCH model using a Volatility Proxy as external regressor. When applying these long-term investment trading rules to the CAC40 index, from May 2016 to May 2020, the finding is that both MLP-based and hybrid ARFIMA-MLP-based trading models show higher performances with a Sharpe Ratio close to 2 and a Profit Ratio around 40% despite the COVID-19 crisis.

Keywords: Long-memory process · Stock return forecasting · Multi-step Ahead Direct Strategy · Long-term investment trading strategies

1 Introduction and Problem Statement

A standard approach to daily volatility and stock return forecasting, once a given proxy has been selected, is to apply a statistical Generalized AutoRegressive Conditional Heteroskedasticity (GARCH)-like model [1,12]. There are many machine learning techniques available that have been identified to be successful in modelling volatility process [5,7]. In order to improve the performance of forecasting volatility and stock return models, the Autoregressive Fractionally Integrated Moving Average (ARFIMA) was introduced by Granger and

© Springer Nature Switzerland AG 2020
F.-P. Schilling and T. Stadelmann (Eds.): ANNPR 2020, LNAI 12294, pp. 278–290, 2020.
https://doi.org/10.1007/978-3-030-58309-5_23

Joyeux in [11]. Numerous authors published a lot of papers using ARFIMA model [13,14,16] or combining ARFIMA with other models on improving the accuracy of stock return forecasting [4,19]. In all the aforementioned cases, the modelling techniques are optimized using mathematical criteria. The forecast error may have been minimized during model estimation, but the evaluation of the true merit should be based on the performance of a trading strategy. In the literature, various forecasting models have been applied to find the best trading signals [6,15,17]. The aim in this work, is to investigate and assess, during high-risk market conditions due to the COVID-19 pandemic crisis, the trading strategies for a long-term investment of more than 3 years. These trading strategies are derived from Stock Return Forecasting using Multi-step Ahead Direct Strategy and based on ARFIMA and machine learning models. The performances are compared based on economical metrics reflecting their profitability and risk. The rest of the paper will be structured as follows: Sect. 2 will introduce the notation and definition of the return and the volatility. Section 3 will present the long memory characteristics of stock returns and their formulation. Section 4 describes the Multi-step Ahead Forecasting strategies and the formulation of our proposed hybrid modelling. Section 5 describes three simple trading strategies to be assessed from their performance point of view. Section 6 describes the experiments with a discussion of the performance results. Section 7 concludes the paper and provides some future research directions.

2   Return and Volatility: Definition and Notation

Modelling financial time series is a complex problem mainly due to the existence of statistical regularities (stylized facts). They can be observed more or less clearly depending on the nature of the series and its frequency. The properties are mainly concerned with daily stock prices and returns.

Return. Let us consider the following quantities of interest, each of them on a daily time scale: $P_t^{(o)}, P_t^{(c)}, P_t^{(h)}, P_t^{(l)}$, respectively the stock prices at the opening, the closing, the maximum and the minimum value for each trading day. Let r_t be the continuously compounded or log return (also simply called the return). In contrast to the prices, the returns or relative prices do not depend on monetary units which facilitates comparisons between assets. The return formulation in (1) represents the daily continuously compounded return for day t computed from the closing prices $P_t^{(c)}$ and $P_{t-1}^{(c)}$.

$$r_t = \ln\left(\frac{P_t^{(c)}}{P_{t-1}^{(c)}}\right) \tag{1}$$

Volatility as a Proxy. The studies in [18] introduce a family of estimators based on the normalization of the maximum, minimum and closing values by the opening price of the considered day. We can then define:

$$u = \ln\left(\frac{P_t^{(h)}}{P_t^{(o)}}\right) \qquad d = \ln\left(\frac{P_t^{(l)}}{P_t^{(o)}}\right) \qquad c = \ln\left(\frac{P_t^{(c)}}{P_t^{(o)}}\right) \qquad (2)$$

where u is the normalized high price, d is the normalized low price and c is the normalized closing price. We consider the estimator cited in [9] defined by the Eq. (3):

$$\sigma_t^1 = 0.511(u - d)^2 - (2\ln 2 - 1)c^2 \qquad (3)$$

3   Long Memory Characteristics

To see how much of an impact past returns for a security have on its future returns, autocorrelation is used to represent the degree of similarity between a given time series and a lagged version of itself over successive time intervals. The examination of the autocorrelation functions (ACF) in Fig. 1 of the historical CAC40 absolute returns, including the COVID-19 period, can serve as a prelude to study long memory characteristics. In this paper, we cope with long term memory processes that are non-stationary, and whose ACF decay more slowly than short-memory processes. We propose ARMA-GARCH and ARFIMA models to estimate the long memory return process with potential structural breaks [14].

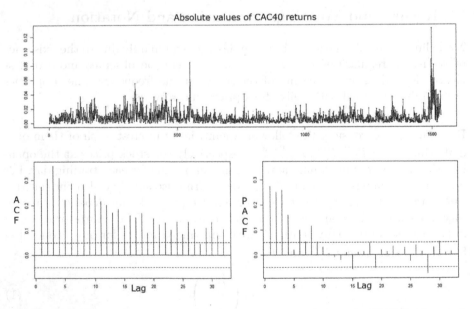

Fig. 1. The ACF of historical CAC40 absolute returns, including the COVID-19 period, decays slowly as the lag increases

3.1   ARMA-GARCH-Based Return

The GARCH(p,q) (Generalized AutoRegressive Conditional Heteroskedasticity) family of models [1,8,12] is generally employed for volatility forecasting. All GARCH models assume that the daily return time series r_t defined in Eq. (1) can be expressed as the sum of two components ARMA-GARCH, noted by r_t^{ag}:

$$r_t^{ag} = E(r_t|\Omega_{t-1}) + \varepsilon_t \tag{4}$$

where $E(.|.)$ denotes the conditional expectation operator, Ω_{t-1} the information set at time t–1, and ε_t the innovations of the time series. In the first phase, the best of the ARMA models is used to model the linear data of time series. In the second phase, the GARCH is used to model the nonlinear pattern of the residuals ε_t. This hybrid model which combines an ARMA model with GARCH error components is applied to predict an approximation of the future return series.

ARMA Mean Equation. The ARMA(P,Q) process of autoregressive order P and moving average order Q can be described as

$$r_t^{ag} = r_t = \mu + \sum_{i=1}^{P} a_i r_{t-i} + \sum_{j=1}^{Q} b_j \varepsilon_{t-j} + \varepsilon_t \tag{5}$$

with mean μ, autoregressive coefficients a_i and moving average coefficients b_i. In the case of this study, the estimation is made by a ARMA equation model with $(P = 1, Q = 1)$.

GARCH Variance Equation. The mean equation cannot take into account the heteroskedastic effects of the time series process as clustering of volatilities. The ε_t terms in the ARMA mean equation are the innovations of the form

$$\varepsilon_t = \sigma_t z_t, z_t \sim \mathbb{N}(0,1) \tag{6}$$

The stochastic component ε_t is expressed as the product between a variable z_t with null mean and unit variance and a time varying scaling factor σ_t.

The core of the model used, is the variance equation using the volatility proxy σ_t^1 defined in (3) as an external regressor, describing how the residuals ε_t and the σ_t past volatility affects the future volatility

$$\sigma_t = \sqrt{(\omega + \zeta\sigma_t^1) + \sum_{j=1}^{p} \beta_j (\sigma_{t-j})^2 + \sum_{i=1}^{q} \alpha_i \varepsilon_{t-i}^2} \tag{7}$$

The coefficients $\omega, \alpha_i, \beta_j, \zeta$ are fitted according to the maximum loglikelihood estimated procedure proposed in [1]. The GARCH order is defined by (q, p) (ARCH, GARCH) with the external regressor σ_t^1 of Eq. (3). In the case of this study, the estimation of the volatility made by a GARCH ($p = 1, q = 1$) model is considered. The skew-generalized error distribution is used when estimating and forecasting GARCH models. To estimate the parameters, the "hybrid" strategy solver is used as a model optimization of the loglikelihood function available in rugarch package under R [10].

3.2 ARFIMA-Based Return

An autoregressive fractionally integrated moving average (ARFIMA) model introduced in [11], provides a parsimonious parameterization of long-memory processes. The ARFIMA model is a generalization of the conventional autoregressive moving average (ARMA) model to fractional differences investigating structural breaks that have been useful in fields as diverse as hydrology, energy and economics [13,14]. The general form of an ARFIMA(p,d,q) to model the daily returns r_t can be given as:

$$\Phi(B)(1 - B)^d r_t = \theta(B)\varepsilon_t \tag{8}$$

where the parameter d is a non-integer value between [–0.5,0.5], r_t is the daily return at time t, ε_t is distributed normally with mean 0 and variance σ^2, and $\Phi(B)$ and $\theta(B)$ represent AR and MA components with lag operator B, respectively. In a fractional model, the power is allowed to be fractional, with the meaning of the term identified using the following formal binomial series expansion:

$$(1 - B)^d = \sum_{k=0}^{\infty} \binom{d}{k}(-B)^k \tag{9}$$

$$= \sum_{k=0}^{\infty} \frac{\prod_{a=0}^{k-1}(d - a)(-B)^k}{k!} \tag{10}$$

$$= 1 - dB + \frac{d(d - 1)}{2!}B^2 - \ldots \tag{11}$$

In addition, B^k is the backshift or lag operator, i.e., given r_t, $B^k r_t = r_{t-k}$. The formulation of the daily return r_t by ARFIMA model noted by r_t^{fi} can be rewritten by:

$$r_t^{fi} = r_t = \Phi(B)^{-1}(1 - B)^{-d}\theta(B)\varepsilon_t \tag{12}$$

The ARFIMA process is one of the best-known classes of long-memory models, having the parameter d for the fractional order, which could capture both the long-run dependency and the short-run dependency. In general, the estimators of d can be achieved by using Maximum Likelihood.

4   Multiple Step Ahead Forecasting

A multi-step ahead (also called long-term) time series forecasting task consists of predicting the next H values $[r_{m+1}, \cdots, r_{m+H}]$ given a univariate time series

$[r_1, \cdots, r_m]$ comprising m historical observations. When facing a multi-step-ahead forecasting problem, several choices are possible among strategies introduced in [2, 20] among them the recursive and the direct forecasting strategy.

With the recursive strategy, forecasts are generated using a one-step-ahead model, applied iteratively for the desired number of steps. With the direct strategy, an horizon-specific model is estimated and forecasts are computed directly by the estimated model for each forecast horizon. In the next sections, a description is given of these two strategies used to forecast the stock returns.

4.1 Recursive Strategy

In this strategy, a single model f is trained to iterate H times, a one-step ahead forecast. This model assumes an autoregressive dependence of the future value of the time series on the past m (lag or embedding order) values and additional null-mean noise term ω as follows:

$$r_{t+1} = f(r_t, \cdots, r_{t-m+1}) + \omega \tag{13}$$

In this work, the one step ahead forecasting approach based on the ARMA-GARCH autoregressive model, presented in the previous Sect. 3.1, is applied to estimate the next value $\hat{r}_{t+1}$. Let the trained one-step ahead model be $\hat{f}$ under maximum loglikelihood criteria. We formulate the recursive strategy, based on the definition r^{ag} given in Eq. (5), applied to the past 250 returns r_t and volatilities σ_t by:

$$\hat{r}_{t+1} = r_{t+1}^{ag} = \hat{f}(r_t, \cdots, r_{t-m+1}, \sigma_t, \cdots, \sigma_{t-m+1}) \tag{14}$$

4.2 Direct Strategy

The Direct strategy (DirStr) estimates a set of H forecasting models, each returning a forecast for the horizon $h \in [1, \cdots, H]$. In other terms, H models f_h are learned (one for each horizon) from the time series $[r_1, \cdots, r_m]$ where

$$r_{t+h} = f_h(r_t, \cdots, r_{t-m+1}) + w \tag{15}$$

with $t \in [1, \cdots, n - H]$ and $h \in [1, \cdots, H]$. The estimated forecasts $\hat{r}_{t+h}$ are obtained by using the H learned models $\hat{f}^h$ as follows:

$$\hat{r}_{t+h} = \hat{f}_h(r_t, \cdots, r_{t-m+1}) \tag{16}$$

We extend the direct strategy formulation with external input (DirStrX). We incorporate to the model, the ARFIMA expression of the stock returns, r^{fi} defined in Eq. (12), as follows:

$$\hat{r}_{t+h} = \hat{f}_h^x(r_t, \cdots, r_{t-m+1}, r_t^{fi}, \cdots, r_{t-m+1}^{fi}) \tag{17}$$

For this investigation, the two direct strategy approaches DirStr and DirStrX are compared for embedding orders $m \in \{2, 5\}$, forecasting horizons $h \in \{2\}$

and for different estimators $\hat{f}$ and $\hat{f}^x$ based on the implementation of the three machine learning approaches: Artificial Neural Networks (MLP), k-Nearest Neighbors (KNN) and Support Vector Machine based regression (SVR). In this experiment, both the R package nnet and gbcode package [3] have been used. In the next paragraph, the formulation of $\hat{f}$ and $\hat{f}^x$ using the MLP approach is given.

MLP-Based Return with DirStr and DirStrX. To illustrate the two direct strategy approaches derived from the different ML estimators $\hat{f}$ and $\hat{f}^x$, the reformulation of direct strategy equation is given in (16) and its extention to external input in Eq. (17) whose return forecasting model is based on the multi-layer perceptron (MLP) with a single hidden layer. Equations (18) and (19) describe the structure of the model for a single forecasting horizon $t + h$ in the context of the Direct Strategy, respectively for a DirStr and a DirStrX model with external regressor. As shown in Eq. (18), the model can be decomposed into a linear autoregressive component of order m and a nonlinear component whose structure depends on the number of hidden nodes H (selected through k-fold cross-validation). When the external regressor r^{fi} is added, to build the hybrid ARFIMA-MLP model, it will affect both the linear and nonlinear component, as shown in (19). In both cases the activity functions $f_o^h(\cdot)$ and $f_h(\cdot)$ are logistic functions. The preference on MLP first before exploring recurrent methods is justified by the fact that MLP shows already successful results in modeling volatility process [7]. The ambition in this work, is to check the performance of MLP in modeling stock returns and enabling successful long-term trading strategies including the COVID-19 crisis.

$$
\hat{r}_{t+h} = \hat{f}_h = f_0^h \left(\underbrace{b_o + \sum_{i=1}^{m} w_{io} r_{t-i}}_{\text{MLP linear component}} + \underbrace{\sum_{j=1}^{H} w_{jo} \cdot f_h \left(\sum_{i=1}^{m} w_{ij} r_{t-i} + b_j \right)}_{\text{MLP non-linear component}} \right)
\tag{18}
$$

$$
\hat{r}_{t+h} = \hat{f}_h^x = f_0^h \left(b_o \begin{array}{c} + \underbrace{\sum_{i=1}^{m} w_{io} r_{t-i} + w_{(i+m)o} r_{t-i}^{fi}}_{\text{Hybrid ARFIMA-MLP linear component}} \\ + \underbrace{\sum_{j=1}^{H} w_{jo} \cdot f_h \left(\sum_{i=1}^{m} w_{ij} r_{t-i} + w_{(i+m)j} r_{t-i}^{fi} + b_j \right)}_{\text{Hybrid ARFIMA-MLP non-linear component}} \end{array} \right)
\tag{19}
$$

5   Trading Strategies

For many traders and analysts, market direction is more important than the value of the forecast itself, as investments can be made simply by knowing the direction the returns will move. We propose a trading model based on the expected next day's market direction. Three different trading strategies are defined: the Naive Strategy described in Sect. 5.1, the Buy and Hold strategy described in Sect. 5.2 and the ML-Based Trading Strategy described in Sect. 5.3. To compare these different trading strategies, several trading performance measures presented in Sect. 5.4 are used and the performance results are discussed in Sect. 6.

We note τ_t, the buy/sell signal generated for the next day t+1. The expected return at day t+1 is defined by $\tilde{r}_{t+1} = \tau_{t+1} \times r_{t+1}$, where r_{t+1} is the true return formulation in Eq. (1).

5.1   Naive Strategy

We call it Naive Strategy because it is simplistic. We buy the stock at next opening market, if the current day t is bullish, i.e., uptrend where the closing price $P_t^{(c)}$ of the current day is greater than its opening price $P_t^{(o)}$. In trading, investors buy a stock or go long if they believe its value will increase. This way, they can sell it for a higher value than they paid and reap a profit. By the same token, the stock is sold at next opening, if the current day t is burrish, i.e., downtrend where the closing price $P_t^{(o)}$ of the current day is lower than its opening price $P_t^{(o)}$. In this case, investors sell first in the hope the price of the stock will decline in value and of being able to buy the asset back at a lower price later.

$$\tau_{t+1} = \begin{cases} +1, & P_t^{(c)} \geq P_t^{(o)} \\ -1, & P_t^{(c)} < P_t^{(o)} \end{cases} \qquad \tilde{r}_{t+1} = \tau_{t+1} \times r_{t+1} \qquad (20)$$

5.2   Buy and Hold Strategy

Buy and hold is a long term investment strategy where an investor buys stocks and holds them for a long time, with the goal that stocks will gradually increase in value over a long period of time. This is based on the view that in the long run, financial markets give a good rate of return even while taking into account a degree of volatility. Buy and hold says that investors will never see such returns if they bail out after a decline, so it is better for them to simply buy and hold, ie: $\tilde{r}_{t+1} = r_{t+1}$

5.3   ML-Based Trading Strategy

The only difference in the ML-Based Trading Strategy compared to the naive strategy, is in the fact that the forward trend of the predicted return is considered. The next day market direction and position to take is determined as follows: if the predicted return $\hat{r}_{t+1}$ for the next day t+1 is positive, go long at opening market on the next day t+1; else if it is negative, go short at opening market on the next day t+1.

$$\tau_{t+1} = \begin{cases} +1, & \hat{r}_{t+1} \geq 0 \\ -1, & \hat{r}_{t+1} < 0 \end{cases} \qquad \tilde{r}_{t+1} = \tau_{t+1} \times r_{t+1} \qquad (21)$$

Note that for the sake of assessment, the hypothesis is made that all positions are liquidated at the closing market. The predicted returns $\hat{r}_{t+1}$ are the outcomes based on the ML training models using forecasting strategies described in Sect. 4, with a time-series cross validation approach. This approach involves training the model on a window of data and predicting the outcome of the next period, then shifting the training window forward in time by one period. This process is repeated along the length of the time series. The cross-validated performance of the forecasting strategy is simply the performance of the next-days predictions using accuracy measures as Root Mean Squared Error (RMSE) presented in Sect. 5.4.

5.4 Forecasting Accuracy and Trading Performance Measures

To compare the accuracy and the trading performance, all models are maintained with an identical out-of-sample period. Among the widely known statistical measures, the Root Mean Squared Error (RMSE) is used to assess the forecasting accuracy. It is observed that these statistical accuracy measures are not enough to analyze the results on financial criteria [15]. The forecast error may have been minimized during the model estimation, but the evaluation of the true merit should be based on economical metrics reflecting profitability and risk.

Some of the more important economical performance measures used are summarized in Table 1. The Sharpe Ratio is a risk-adjusted measure of the return. Usually, any Sharpe Ratio greater than 1.0 is considered acceptable to good by investors. A ratio higher than 2.0 is rated as very good. The Profit Ratio is measured as the ratio between the sum of expected return $\tilde{r}_t$ being positive and the sum of expected return $\tilde{r}_t$ being negative over the predicted period $\{1, \cdots, n\}$. The expected return $\tilde{r}_t$ at time t is the formulation given in (20) and (21), n is the number of predictions and $\bar{r}_n$ represents the expected returns' average over $\{1, \cdots, n\}$.

6 Experimental Results

6.1 Dataset Description

The study is based on the daily CAC40 index prices respectively at the opening, the maximum, the minimum and the closing $P_t^{(o)}, P_t^{(h)}, P_t^{(l)}, P_t^{(c)}$. The series spans from the period of 1st May 2014 to 6th May 2020 covering the COVID-19 period and totalling 6 years history with 1536 trading days. This series has been transformed to returns r_t as defined in Eq. (1). To compare the performance

Table 1. Trading performance measures

Cumulative return	$R_t^c = (1 + \tilde{r}_t) \times R_{t-1}^c, t \in \{1, \cdots, n\}$
Annualized return	$R^A = 255 \frac{1}{n} \sum_{t=1}^{n} \tilde{r}_t$
Annualized volatility	$\sigma^A = \sqrt{255} \frac{1}{n-1} \sum_{t=1}^{n} (\tilde{r}_t - \bar{r})^2$
Sharpe ratio	$\frac{R^A}{\sigma^A}$
Profit ratio	$\frac{\sum_{t=1}^{n} \tilde{r}_t > 0}{|\sum_{t=1}^{n} \tilde{r}_t < 0|}$
RMSE	$\sqrt{\frac{1}{n} \sum_{t=1}^{n} (r_t - \hat{r}_t)^2}$

of the proposed ML-Based forecasting models, the recursive one-day-ahead or direct two-days-ahead return predictions are performed, based on a rolling origin with a moving window fixed to 250 trading days as a test sample. The average performance is considered out of 30 independent runs starting with a 250 trading days window taken randomly in the first two years of the data samples. The different forecasting models ARMA-GARCH, MLP, KNN, SVR and hybrid ARFIMA-KNN, ARFIMA-SVR, ARFIMA-MLP are considered to assess the impact of COVID-19 crisis on the trading performances compared to the Buy and Hold and naive benchmarks.

6.2 Forecast Evaluation and Trading Performance Results

As described in Table 2, most of the proposed models show higher performances compared to the buy and hold strategy. The best models are the two MLP-based models, MLP-DirStr and hybrid ARFIMA-MLP-DirStrX with a Sharpe Ratio close to 2, while the buy and hold strategy gives a Sharpe Ratio of -0.10. MLP-based models capture well the non-linearity nature of the returns. Another result shown in Fig. 2 is that the hybrid model ARFIMA-MLP-DirStrX (in blue color) clearly outperforms the simple MLP-DirStr (in grey color) with a higher log cummulative returns $log R_t^c$ along the period of time 2016 to 2020. The simple MLP-DirStr suffers on capturing well the long-term memory process since it is improved with ARFIMA in its hybrid version. Nevertheless, both models succeed to capture local irregularities due to the volatility surge during the COVID-19 period and the level of log cumulative returns is maintained and increased. The other models as KNN-DirStr and SVR-DirStr show lower performances. They fail in capturing long memory characteristics since by introducing the ARFIMA component in the hybrid ARFIMA-KNN-DirStrX and ARFIMA-SVR-DirStrX, the performances are increased with a Profit Ratio of 8% respectively 7%. Concerning the ARMA-GARCH model in Fig. 2 (in red color), based on recursive forecasting strategy, it suffers from low performance and shows limitation in capturing well the surge of volatility during the COVID-19 period with a decrease of the log cummulative returns. Concerning the naive strategy, despite the poor forecast accuracy with a largest RMSE, the profit ratio is 11%. As shown in Fig. 2 (in yellow color), this profit ratio is generated mostly during the pandemic crisis as prices are expected to go in the same direction with a decrease. This sudden gain leads to a Sharpe Ratio less than 1 showing a risky strategy.

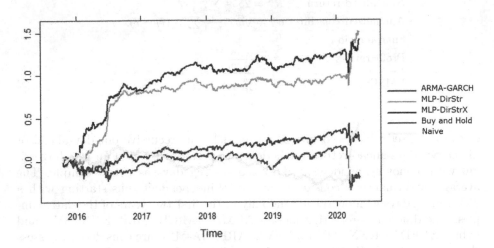

Fig. 2. log R_t^c, cummulative returns over the period t from October 2016 till May 2020 (Color figure online)

Table 2. Trading performance results

Model strategy	RMSE	R^A	σ^A	Sharpe ratio	Profit ratio
ARMA-GARCH	0.01241	7%	19.49%	0.36	7.4%
MLP-DirStr	**0.01127**	**35%**	19.37%	**1.81**	**43%**
MLP-DirStrX	**0.01196**	**33%**	19.38%	**1.70**	**40%**
KNN-DirStr	0.01269	−12%	19.47%	−0.65	−13%
KNN-DirStrX	0.01228	8%	19.48%	0.40	8%
SVR-DirStr	0.01247	3%	19.49%	0.14	3%
SVR-DirStrX	0.01247	7.3%	19.49%	0.36	7%
Naive	0.01899	10.6%	19.48%	0.54	11%
Buy and Hold	-	−2%	19.78%	−0.10	−2%

7   Conclusion and Future Work

This study reports an empirical work which investigates how forecasting and trading performances could be improved by using machine learning models coupled with multi-step ahead direct strategy forecasting. This work has resulted mainly in the creation of sustainable long-term trading algorithms based on the advantages of the simple MLP model and the ARFIMA models despite the high-risk market conditions. These trading models show higher performances with a Sharpe Ratio close to 2 which is potentially profitable and attractive for investments from an economic point of view. For the future, Deep Neural Networks will be investigated to see how they can contribute to increase the performance for long-term investment strategies.

References

1. Andersen T.G., Bollerslev T: Arch and Garch models. In: Encyclopedia of Statistical Sciences (1998)
2. Bontempi, G., Ben Taieb, S., Le Borgne, Y.-A.: Machine learning strategies for time series forecasting. In: Aufaure, M.-A., Zimányi, E. (eds.) eBISS 2012. LNBIP, vol. 138, pp. 62–77. Springer, Heidelberg (2013). https://doi.org/10.1007/978-3-642-36318-4_3
3. Gianluca, B.: Code from the handbook "statistical foundations of machine learning". https://github.com/gbonte/gbcode
4. Chaâbane, N.C.: A hybrid ARFIMA and neuronal network model for electricity price prediction. Int. J. Electr. Power Energy Syst. **55**, 187–194 (2014)
5. Dash, R., Dash, P.K.: An evolutionary hybrid fuzzy computationally efficient EGARCH model for volatility prediction. Appl. Soft Comput. **45**, 40–60 (2016)
6. Dash, R., Dash, P.K.: A hybrid stock trading framework integrating technical analysis with machine learning techniques. J. Finan. Data Sci. **2**, 42–57 (2016)
7. De Stefani, J., Caelen, O., Hattab, D., Bontempi, G.: Machine learning for multi-step ahead forecasting of volatility proxies. In: The 2nd Workshop on Mining Data for Financial Applications (2017)
8. Engle, R.F.: Autoregressive conditional heteroscedasticity with estimates of the variance of United Kingdom inflation. Econometrica **50**, 987–1007 (1982)
9. Garman, M.B., Klass, M.J.: On the estimation of security price volatilities from historical data. J. Bus. **53**, 67–78 (1980)
10. Ghalanos, A.: Introduction to the rugarch package. https://cran.r-project.org/web/packages/rugarch/vignettes/Introduction_to_the_rugarch_package.pdf
11. Granger, C.W., Joyeux, R.: An introduction to long-memory time series models and fractional differencing. J. Time Ser. Anal. **1**, 15–29 (1980)
12. Hentschel, L.: All in the family nesting symmetric and asymmetric GARCH models. J. Financ. Econ. **39**, 71–104 (1995)
13. Hosking, J.R.M.: Fractional differencing. Biometrika **68**, 165–176 (1981)
14. Jibrin, S.A., Musa, Y., Zubair, U.A., Saidu, A.S.: ARFIMA modelling and investigation of structural break(s) in West Texas Intermediate and Bret series. CBN J. Appl. Stat. **6**, 59–79 (2015)
15. Kumar, M., Thenmozhi, M.: Stock index return forecasting and trading strategy using hybrid ARIMA-neuronal network model. Int. J. Financ. Manage. **2**, 1 (2012)
16. Liu, K., Chen, Y., Zhang, X.: An evaluation of ARFIMA programs. In: International Design Engineering Technical Conferences and Computers and Information in Engineering Conference, Cleveland, USA (2017)
17. Papaioannou, P., Dionysopoulos, T., Russo, L., Giannino, F., Jantzko, D., Siettos, C.: S&P500 forecasting and trading using convolution analysis of major asset classes. In: The 2nd International Workshop on Data Mining in IoT System (2017)

18. Poon, S.H., Granger, C.W.: Forecasting volatility in financial markets: a review. J. Econ. Lit. **41**, 478–539 (2003)
19. Sadaei, H.J., Enayatifar, R., Guimarães, F.G., Mahmud, M., Alzamil, Z.A.: Combining ARFIMA models and fuzzy times series for the forecast of long memory time series. Neurocomputing **175**, 782–796 (2015)
20. Taieb, S.B., Bontempi, G., Atiya, A.F., Sorjamaa, A.: A review and comparison of strategies for multi-step ahead time series forecasting based on the NN5 forecasting competition. Exp. Syst. Appl. **39**, 7067–7083 (2012)

iNNvestigate-GUI - Explaining Neural Networks Through an Interactive Visualization Tool

Fabio Garcea[✉][ID], Sina Famouri[ID], Davide Valentino, Lia Morra[ID],
and Fabrizio Lamberti[ID]

Dipartimento di Automatica e Informatica, Politecnico di Torino, Turin, Italy
{fabio.garcea,sina.famouri,lia.morra,fabrizio.lamberti}@polito.it,
davide.valentino@live.it

Abstract. In recent years, deep neural networks have reached state of the art performance across many different domains. Computer vision in particular has benefited immensely from deep learning. Despite their high performance, deep neural networks often lack interpretability and are mostly regarded as a black box. Therefore, the availability of tools capable to provide insights into the models and identify potential errors is crucial. Such tools need to seamlessly integrate within the workflow of data scientists and ML researchers. In this paper we propose iNNvestigate-GUI, an open-source graphical toolbox which offers an extensive set of functionalities for users to compare different networks behavior and give an explanation to their outputs.

Keywords: Deep learning · Convolutional neural networks · Visualization · eXplainable Artificial Intelligence

1 Introduction

Deep learning has become a fundamental tool for a variety of applications. Thanks to easy-to-use libraries like Keras [5], a deep neural network (DNN) can be implemented in few lines of code, and many practitioners from a variety of fields can use DNNs despite limited knowledge and background in machine learning. However, practitioners still need crucial insights into the trained models. Typical examples that occur within the development life cycle include debugging models that do not converge or perform poorly on the target labels, or finding samples which the model cannot handle correctly for the specified task.

Given that DNNs are large models with millions of parameters trained on thousands of data points, visual analytics is emerging as a powerful tool to aid the inspection of DNNs and tackle the amount of data generated during their training [6,13]. For instance, tools as Tensorboard allow to visualize gradients, activations, losses, etc. focusing on the training and optimization process.

Another important aspect of training DNNs is having insights into their properties, e.g., determining the most important features for classification.

© Springer Nature Switzerland AG 2020
F.-P. Schilling and T. Stadelmann (Eds.): ANNPR 2020, LNAI 12294, pp. 291–303, 2020.
https://doi.org/10.1007/978-3-030-58309-5_24

To compensate for the black-box nature of DNNs, many eXplainable Artificial Intelligence (XAI) techniques have been designed to provide post hoc explanations of predictions. Techniques and algorithms that can provide visual interpretations are particularly effective, especially for DNNs targeting image interpretation [13]. Yet, there is a lack of tools to easily integrate such techniques in the development cycle. Activis [8] is one of the most comprehensive GUIs (Graphical User Interfaces) available for this purpose, but unfortunately, it is not publicly available. We argue that similar open source and publicly available interfaces are crucial to support the integration of XAI techniques in the development of deep learning models.

In this paper, we present iNNvestigate-GUI[1], an open-source, user-friendly, visual analytics tool for DNNs, especially tailored to computer vision. It is built upon the open-source iNNvestigate library [2], which provides a reference implementation of several visualization algorithms. Our aim is to provide a GUI which simplifies model interpretation by providing easy, code-free access to a comprehensive pool of visualization methods.

Designing such software has its challenges. First of all, visual comparison of several DNNs will be computationally expensive. The target users are diverse, with varying levels of machine learning knowledge and needs, so visual interpretability is a key factor. The tool should be easily integrated in the research and development workflow. Last but not least, as a crucial step towards XAI, when comparing DNNs one should not only take into account the final performance but also the reason behind the outputs.

We will explain how iNNvestigate-GUI fits into the literature in Sect. 2. In Sect. 3, the design challenges are explained in detail. Section 4 gives an overview of implementation details and design goals. The tool was tested with a group of non-expert users (computer science students). As described in Sect. 5, it reached a usability score of 73.34 according to the SUS scale.

2   Related Work

In this section we will first describe the currently available GUI-based tools and libraries to perform visual analysis in deep learning applications. Previous works have been categorized according to their license, availability and target audience, as detailed in Table 1.

2.1   Visual Analytic Tools for Deep Learning

Open-Source Tools for Experienced Users. This group covers the majority of visual analytic tools in literature and collects most of the most popular applications used for deep learning applications. The well known TensorFlow Graph Visualizer [21] for instance, which is part of the widely adopted TensorFlow framework, allows visualizing a neural network as a directed graph embedding other crucial information like layers hyperparameters in a single scalable

---

[1] Code is available at https://gitlab.com/grains2/innvestigate-gui/.

view. Embedding Projector [17] is another recent visualization tool developed by Google and included in the TensorFlow framework that allows plotting tensors in space through different dimensionality reduction techniques. Chung et al. proposed a dynamic real-time visual system to monitor the 2D representation of the filters learned by different layers and an interactive approach to steer the model configuration during the training process [4]. The Deep Visualization Toolbox [22] provides a matrix-like grid-view representation of the activations of the neurons in a given layer for a specific input image or video. A similar approach has been recently adopted in Summit [7], a tool developed to let practitioners and experts visualize neuron activations, thus enhancing the interpretability of the models. Several visualization tools targeting neural networks focus on the training and optimization phase. For instance, DeepEyes [11] supports the interpretation of the features learned by a CNN model during the training phase. As evident from Table 1, most of these tools allow visualization of gradients and activations and are more suited to effectively monitoring the training process than towards XAI.

Other recent visualization tools like LSTMvis [18] and GANviz [19] are instead focused on specific types of networks (such as LSTM or GANs), whereas in this work we are aiming at a more general purpose tool.

Proprietary Tools for Experienced Users. ActiVis [8] is an example of proprietary Web application developed by Facebook that represents a comprehensive alternative to TensorFlow Graph Visualizer. It allows visualizing a neural network through a node-graph representation and performing behavioral analysis at different levels, from a subset of samples to a single instance and down to the activation of a single neuron. However, it is deployed on FBLearner Flow, the machine learning platform of Facebook, and is available only for internal researchers and practitioners. In [9], Shixia Liu et al. proposed a tool named CNNVis, which allows to visualize a clustered representation of the features learned by neurons and the connections between neurons at different layers with a minimal representation aimed at reducing the visual clutter caused by a high number of links between nodes. This tool has however no public implementation and only an online demo is currently available.

Educational Tools. Some visualization tools are designed to help students better understand how neural networks work and, more in general, to be used for educational purposes [16,20]. An example, TensorFlow Playground is a web application developed by Google researchers, that allows manipulating interactively a simple model, including the structure, hyper-parameters and data points, to appreciate directly their effect on the decision boundaries learnt by the network. However, such tools are not adequate to visualize complex networks and datasets typical of real-life projects, even for inexperienced researchers.

2.2   Libraries of Visualization Algorithms

DNNs are generally regarded as black boxes due to their lack of explicit interpretability. To tackle this issue, several visualization algorithms have been

Table 1. Summary of the main deep learning visualization tools and comparison with the proposed tool.

	TF graph visualizer	Embedding projector	ActiVis	DeepVis	CNNVis	ReVACNN	DeepEyes	Summit	iNNvestigate-GUI
Visualization									
Node-link graph	✓		✓						
Embeddings		✓	✓			✓	✓	✓	
Activations			✓	✓	✓	✓	✓	✓	✓
Gradients			✓	✓	✓	✓			
Hyperparameters	✓					✓		✓	
Attributions								✓	✓
Training history							✓		
Framework									
TensorFlow	✓	✓							✓
Keras	✓					✓			✓
FBlearner Flow			✓						
ConvNetJS						✓			
Caffe				✓			✓		
User interface									
GUI (web-app)	✓	✓	✓			✓		✓	✓
GUI				✓			✓		
Command line									
Availability									
Open source	✓	✓		✓		✓	✓	✓	✓
Proprietary			✓						

proposed to help understanding why a model is producing a certain output for a given input [13]. These techniques visualize aspects such as the filters learned by a specific layer of the network, the activation of a certain neuron, or the gradients flowing through the layers. Perturbation-based methods stimulate and visualize changing network behavior by perturbing the input of the model. These methods rely on different visual paradigms varying from heatmaps to pixel display grids. Since the seminal work by Zeigler and Fergus [10], the number of available visualization techniques has been increasing steadily given the growing interest in XAI. A complete review is outside of the scope of this paper, and the reader is referred to many excellent surveys available [6,13].

In practice, the applicability of visualization techniques is often hindered by a lack of publicly available reference implementations [2]. A few libraries have been recently proposed to gather and unify different visualization techniques in a common framework, including Keras Explain [1], DeepExplain [3] and iNNvestigate [2]. As detailed in Table 2 all include a variety of gradient-based (like DeepLIFT [15]), model-independent methods (like LIME [12]) and perturbation-based methods. Still, their integration in the model development cycle can be greatly simplified by providing a graphical interface, and presents several challenges which are discussed in Sect. 3.

3 Design Challenges

A series of joint design sessions were conducted by involving researchers and practitioners with different levels of expertise. Moving from the analysis of existing tools, illustrated in Sect. 2, we highlighted several critical gaps to be addressed, with an emphasis on open-source solutions. A seconded set of design challenges (C1–C5) was identified.

Table 2. Summary of the available libraries of visualization algorithms.

	Keras explain	DeepExplain	iNNvestigate
Visualization technique			
Gradient/saliency maps	✓	✓	✓
SmoothGrad			✓
DeconvNet			✓
Guided backpropagation	✓		✓
PatternNet			
GradCAM	✓		
Guided GradCAM	✓		✓
Input * Gradient		✓	✓
LRP	✓	✓	✓
Integrated gradients	✓	✓	✓
DeepTaylor			✓
DeepLIFT		✓	✓
Pattern attribution			
Prediction difference	✓		
Grey-box occlusion	✓	✓	
LIME	✓		
Shapley value sampling		✓	

C1. Resource demanding visualizations. Training and evaluating DNNs is computationally expensive, especially when working with images. The time required to produce the visualizations should be limited in order to enhance user acceptability, but many of the existing visualization techniques are computationally intensive. Likewise, large datasets are usually involved in running deep learning experiments. Targeting the open-source community, the proposed implementation should allow a variety of computing setups, to access the provided visualization techniques.

C2. Diversified set of users. Designing a tool that is easy to use and accommodates both expert and non-expert users is challenging. Many of the available tools either target very inexperienced users and are mostly intended as teaching aids, or are designed to work in an industrial R&D environment where users are likely to have similar experience levels. Open-source tools target users who may have different needs and preferences. The workflow should follow a clear and simple structure; the different views of the interface should be self-explanatory and the visualizations designed with clarity, yet being capable of producing useful insights in non trivial projects.

C3. Performing instance-based and dataset-based analysis. Several XAI techniques are designed to provide a post hoc explanation of the predictions on a specific instance. However, as mentioned before, DNNs are trained and tested on very large datasets, and it is impractical for the user to manually comb through the dataset to find critical samples. A suggestion system is needed to rapidly identify data instances that are worthy of inspection.

C4. Simplify integration in R&D. The practical adoption of XAI techniques is often hindered by i) the lack of a reference, readily-available implementation and ii) the need to design and implement specific code for their integration in the model development pipeline. A graphical tool should allow to produce the expected results significantly faster than writing code from scratch and, in general, generate the required visualization through a limited number of clicks.

C5. Model complexity and variety. Many visual analytics tools are designed to evaluate a single model, and often assume a relatively simple architecture. In practice, dozens of different models may need to be trained and compared, and we argue that this comparison should take into account not only performance but also the quality of the prediction and the presence of systematic biases [14].

4   Implementation

In this section, a detailed description of the functionalities offered by the iNNvestigate-GUI visualization tool is reported. In Sect. 4.1, the main design goals (G1–G4) are described and motivated. Then, we move on to illustrate how these goals were achieved by designing a workflow for easy visualization of DNNs (Sect. 4.2) and for navigating a large dataset for sample selection (Sect. 4.3).

4.1   Design Goals

G1. Offering to researchers and practitioners a fast code-free tool for interpreting their models. Available visualization libraries represent different attempts to create a common reference implementation to tackle models explainability and interpretability [2]. Their use, however, passes through an Application Programming Interfaces (API); hence, time is needed to read the documentations and write ad hoc code. A visual analytics tool offers a ready-to-use common GUI to multiple visualization methods. It has to support inspections at different levels of depth, from the entire model to single layers and units.

G2. Easy graphical comparison of multiple models. It is common during a deep learning project to train and evaluate multiple models with different architectures, hyper-parameters and configurations. Since there is no consensus as to which visualization methods have the most desirable properties [2], the visualization tool must provide an easy interface to compare the pool of XAI techniques on multiple models.

G3. Allowing navigation of large scale datasets and identification of poorly classified and borderline data instances. Deep learning models are in general trained and validated on large scale datasets, whereas most visualization methods (with the notable exception of embeddings) operate at the instance level. Selecting interesting input instances for analysis is not straightforward. Random sampling is time consuming and may be lead to missed errors. The proposed tool must provide an intuitive and effective way to select samples that are worthy of further analysis, e.g., instances that the DNNs cannot classify correctly. This approach could both save the time needed to perform an ad hoc analysis of the samples, and highlight crucial instances that may remain unnoticed, eventually increasing the capability of the tool to provide insights on the model behavior.

G4. Web-based implementation to tackle computationally demanding tasks. Although visualization is less computationally intensive than training, some visualization techniques still require an ad hoc training phase and indeed, a large number of samples may need to be processed. To empower users with different resources requirements and availability in terms of memory, disk space and computational power, we chose to develop the tool as a web application. The advantage of this approach is the capability to demand computationally demanding tasks to a back-end, possibly equipped with GPUs, while providing to the users a lightweight front-end accessible from anywhere through a web browser. This framework is already adopted by many popular tools (e.g., Tensorboard, Embedding Projector), and accommodates both users equipped with high-end workstations as well as those exploiting cloud computing services (e.g., Amazon Web Services).

4.2 Explaining Custom Models Through Visualization Methods

In this subsection, we describe the complete process to explain the behavior of DNNs through one or more visualization methods. Based on the analysis in Sect. 2, we selected the iNNvestigate [2] package as the reference implementation, and provided an ad hoc implementation for methods not included in this library, i.e., GradCAM and Guided GradCAM [14]. In addition, it is possible to visualize the output activations of a given neuron. These methods were included because they can produce particularly intuitive and easy-to-interpret visualizations especially suited to novice and non-expert users.

The iNNvestigate-GUI workflow starts by uploading the dataset and the pre-trained model(s). It is possible to analyze both models trained by the user or directly select the ImageNet-trained models available in the Keras library, which could be useful also for teaching purposes. For the visualization methods, all the options and all the required configurations parameters are provided as scroll-down lists to enhance the intuitiveness of the GUI. The tool also allows to specify a single layer or a single neuron to visualize the activation.

Once the setup is completed, the selected visualizations are generated and displayed to understand the DNNs behavior. The visualization panel is divided in multiple boxes, one for each of the models loaded in the configuration phase. The visualization method is applied to the output of the selected layer (the last convolutional layer by default) or neuron, for each data instance. Through an interactive panel it is thus possible to inspect the produced visualizations in a synchronized fashion, allowing fast and intuitive comparison between the behavior of multiple networks. For each data sample and DNN, the top predictions and their scores are shown next to the visualization output (see Fig. 1).

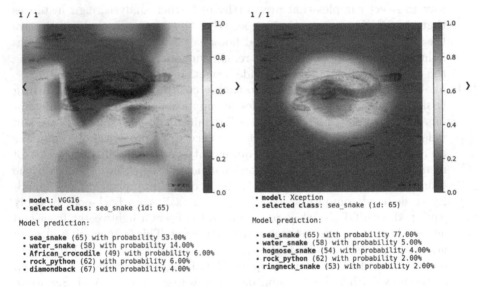

Fig. 1. Comparison of different model predictions for one of the images included in T1. The two models exploit different visual features to make the classification. The overlapping heatmaps have been produced using the GradCAM technique.

4.3   Suggesting Useful Data Samples for Analysis

iNNvestigate-GUI allows the user to easily identify useful samples to analyze thanks to the Suggestion panel (see Fig. 2). We assumed that the users should analyze the predictions for a mix of data instances with different properties: for instance, incorrectly classified samples allow the user to investigate the source of possible errors. Moving from these observations, the Suggestion panel categorizes available samples in order to allow the user to select a mix of samples with different properties for inspection. We identified two operating modalities based on i) whether a single or multiple models are compared and ii) whether the ground truth labels are available.

In the *multiple model* setting, the Suggestion panel shows a scatter plot of the input samples according to the confidence and agreement of the different models, as reported in Fig. 2. The mean prediction score across all models is reported on the x axis, and the number of predicted classes on the y axis. While hovering with the mouse over one of the data points in the scatter plot it is possible to have a visual preview of the examples. Based on their position in the chart different types of data instances can be distinguished.

Samples in the right-lower corner are images that are classified in the same way by all the evaluated models with a high prediction score. Samples in this are are classified in the same class with high confidence by all the models, and thus are likely to be correctly classified. Still, they could be interesting to inspect in order to exclude the presence of systematic biases in the dataset.

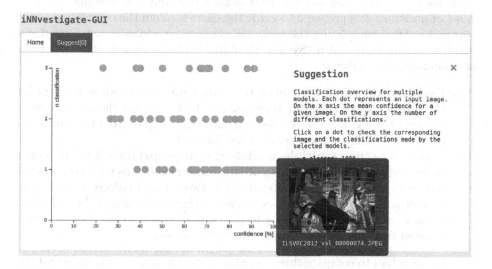

Fig. 2. *Suggestion* panel of iNNvestigate-GUI. A scatter plot represents the input dataset processed by multiple neural networks and guides the user towards selecting meaningful samples for further analysis. The average prediction score (x axis) is plotted against the number of different classes (y axis) predicted by the models. This visualization allows to identify data samples with high/low agreement among different models, as well as those predicted with high/low confidence.

Samples in the top-left corner (low agreement / low confidence) are probably borderline cases, or correctly classified by only a subset of the DNNs.

Samples in the top-right corner are predicted in different classes (low agreement) but with high prediction scores. They could include out-of- distribution samples on which DNNs are likely to misbehave, samples that may easily fool one or more of the models (including adversarial samples), or again, samples correctly classified only by a selection of models. Under this structure, users could focus their attention on the top-left and top-right quadrants.

In the *single model* setting, the Suggestion panel shows a simpler histogram plot. The user can choose to visualize the distribution of the activations for a pre-defined layer, e.g., to select the images that mostly excite a specific layer. Alternatively, for of labelled data, it is possible to plot the distribution of the difference between the prediction and the correct label (typically 1.0 for classification models) to identify samples that are correctly or incorrectly classified.

5   A Usability Test Case

To better evaluate the usability of the tool, two tasks have been prepared and submitted to a group of 9 computer engineering students (with a previous background in deep learning) at Politecnico di Torino. All students had at least a basic knowledge of CNNs and attended at least one course in machine learning. A set of questions was prepared to guide the users through the completion of the two tasks. After completing the tasks, all users were administered a questionnaire according to the SUS (System Usability Scale) approach. The two tasks are summarized as follows:

T1. Evaluate the visual features used by different models to predict the same subset of input images. This task emulates the comparison of different trained models. A subset of 100 data samples from the ImageNet ILSVRC2012 dataset was selected for this task.

T2. Assess whether multiple models use appropriate visual features to classify a subset of inputs of the same class. This task emulates the search for visual biases. For instance, a network may inadvertently learn to predict an object based on co-occurring background features. For this task a subset of 15 images belonging to the *golden_retriever* class was randomly selected from the ImageNet ILSVRC2012 dataset.

The users had to compare three popular models available in Keras (VGG16, ResNet50 and Xception) with varying level of complexity. In order to reduce the time needed to complete the task, available visualization algorithms where restricted to Gradient/Saliency, Guided Backpropagation, GradCAM e LRP-z.

During task T1, all users could successfully use the Suggestion panel to identify images where the models agreed/disagreed or had low/high prediction confidence. In particular, users focused their search in the right-lower quadrant (high agreement/high confidence, see Fig. 3) and left-top quadrant (low agreement/low confidence). In both cases, users found GradCAM to be the easiest method to interpret (66.7% and 55.6% of the users, i.e. 6 out of 9 and 5 out of 9 respectively).

In task T2, the participants were asked to identify samples that were classified correctly by all the models and samples with inconsistent behavior. Results showed that 77.8% (7/9) of the participants relied on the histogram chart to identify both, while 22.2% of the users (2/9) also relied on the scatter plot. In this case the most intuitive algorithm for the analysis of visual features of images classified with the correct labels by most of the models was Guided Backpropagation (55.6% of the votes, i.e. 5/9). On the other side, when analyzing images that were incorrectly classified, the most popular choice was again GradCAM (66.7% of users, i.e. 6/9).

In task T2, users were asked to rate the visual explanations on a scale from 1 (completely agreeing) to 5 (completely disagreeing). 55.6% (5/9) of the participants was completely satisfied by the visual explanations for the VGG16 and ResNet50 models (meaning they found the proposed attributions appropriate for the label), whereas only 44.4% (4/9) were satisfied with the Xception network.

The mean usability score was 73.34%, which according to the SUS usability scale is above average (any value above 68% is considered above the average).

Moreover, 77.8% (7/9) of the participants declared that they would have not been able to easily solve both tasks without iNNvestigate-GUI, while only 22.2% (2/9) stated they could solve the same tasks writing ad hoc code.

Fig. 3. Visual comparison of the predictions made by the three models in T1. The overlapping heatmap has been produced by the GradCAM algorithm. This is image was selected by the majority of the participants from the *Suggestion* view as an example of high confidence and high agreement classification. All the models are focusing on similar visual features to classify the frame. Best seen in RGB; consider brighter areas of the frames in case of gray scale visualization.

6   Conclusion

This paper proposed a new GUI-based Web application featuring a comprehensive pool of XAI visualization algorithms, and intended to compare and understand multiple DNN models. The tool leverages an existing open-source library, named iNNvestigate, for existing implementation of visualization algorithms.

In contrast to other existing tools, the proposed GUI allows a fast comparison between multiple models at different levels of depth and implements a suggestion strategy to highlight the most critical data samples. The target users for the tool span from the inexperienced learner to researchers and deep learning practitioners. As demonstrated by preliminary user experiments, it can be exploited by inexperienced users to improve and speed up DNNs development.

We plan to extend iNNvestigate-GUI by adding support for deep learning frameworks other than Keras and implementing additional visualizations, such as node-link diagrams, to simplify the selection and inspection of individual layers. More complex use cases are needed to demonstrate how visual analytics can prevent biases and errors to be introduced during model training.

References

1. Keras Explain (2018). https://github.com/primozgodec/keras-explain
2. Alber, M.: iNNvestigate neural networks!. J. Mach. Learn. Res. **20**(93), 1–8 (2019)
3. Ancona, M.: DeepExplain (2017). https://github.com/marcoancona/DeepExplain
4. Chung, S., Suh, S., Park, C., Kang, K., Choo, J., Kwon, B.C.: Revacnn: real-time visual analytics for convolutional neural network. In: KDD 16 Workshop on Interactive Data Exploration and Analytics, pp. 30–36 (2016)
5. Gulli, A., Pal, S.: Deep Learning with Keras. Packt Publishing Ltd. (2017)
6. Hohman, F., Kahng, M., Pienta, R., Chau, D.H.: Visual analytics in deep learning: an interrogative survey for the next frontiers. IEEE Trans. Vis. Comput. Graph. **25**(8), 2674–2693 (2018)
7. Hohman, F., Park, H., Robinson, C., Chau, D.H.P.: Summit: scaling deep learning interpretability by visualizing activation and attribution summarizations. IEEE Trans. Vis. Comput. Graph. **26**(1), 1096–1106 (2019)
8. Kahng, M., Andrews, P.Y., Kalro, A., Chau, D.H.P.: Activis: visual exploration of industry-scale deep neural network models. IEEE Trans. Vis. Comput. Graph. **24**(1), 88–97 (2017)
9. Liu, M., Shi, J., Li, Z., Li, C., Zhu, J., Liu, S.: Towards better analysis of deep convolutional neural networks. IEEE Trans. Vis. Comput. Graph. **23**(1), 91–100 (2016)
10. Matthew, D., Fergus, R.: Visualizing and understanding convolutional neural networks. In: Proceedings of the 13th European Conference Computer Vision and Pattern Recognition, Zurich, Switzerland, pp. 6–12 (2014)
11. Pezzotti, N., Höllt, T., Van Gemert, J., Lelieveldt, B.P., Eisemann, E., Vilanova, A.: Deepeyes: progressive visual analytics for designing deep neural networks. IEEE Trans. Vis. Comput. Graph. **24**(1), 98–108 (2017)
12. Ribeiro, M.T., Singh, S., Guestrin, C.: "why should i trust you?" explaining the predictions of any classifier. In: Proceedings of the 22nd ACM SIGKDD International Conference on Knowledge Discovery and Data Mining, pp. 1135–1144 (2016)
13. Seifert, C., et al.: Visualizations of deep neural networks in computer vision: a survey. In: Cerquitelli, T., Quercia, D., Pasquale, F. (eds.) Transparent Data Mining for Big and Small Data. SBD, vol. 11, pp. 123–144. Springer, Cham (2017). https://doi.org/10.1007/978-3-319-54024-5_6
14. Selvaraju, R.R., Cogswell, M., Das, A., et al.: Grad-cam: visual explanations from deep networks via gradient-based localization. In: Proceedings of the IEEE International Conference on Computer Vision, pp. 618–626 (2017)
15. Shrikumar, A., Greenside, P., Kundaje, A.: Learning important features through propagating activation differences. In: Proceedings of the 34th International Conference on Machine Learning, vol. 70, pp. 3145–3153. JMLR.org (2017)
16. Smilkov, D., Carter, S., Sculley, D., Viégas, F.B., Wattenberg, M.: Direct-manipulation visualization of deep networks. arXiv preprint arXiv:1708.03788 (2017)
17. Smilkov, D., Thorat, N., Nicholson, C., Reif, E., Viégas, F.B., Wattenberg, M.: Embedding projector: Interactive visualization and interpretation of embeddings (2016)

18. Strobelt, H., Gehrmann, S., Pfister, H., Rush, A.M.: LSTMVis: a tool for visual analysis of hidden state dynamics in recurrent neural networks (2016)
19. Wang, J., Gou, L., Yang, H., Shen, H.W.: Ganviz: a visual analytics approach to understand the adversarial game. IEEE Trans. Vis. Comput. Graph. **24**(6), 1905–1917 (2018)
20. Wang, Z.J., et al.: Cnn explainer: Learning convolutional neural networks with interactive visualization. arXiv preprint arXiv:2004.15004 (2020)
21. Wongsuphasawat, K.: Visualizing dataflow graphs of deep learning models in tensorflow. IEEE Trans. Vis. Comput. Graph. **24**(1), 1–12 (2017)
22. Yosinski, J., Clune, J., Nguyen, A., Fuchs, T., Lipson, H.: Understanding neural networks through deep visualization. arXiv preprint arXiv:1506.06579 (2015)

Author Index

Printed in the United States
by Baker & Taylor Publisher Services

Printed in the United States
by Baker & Taylor Publisher Services